MARITIME SATELLITE COMMUNICATIONS

MARITIME SATELLITE COMMUNICATIONS

Editor

Kim Berg

DPH

DISCOVERY PUBLISHING HOUSE

INDIA

Published by:

DISCOVERY PUBLISHING HOUSE
4383/4B, Ansari Road, Darya Ganj
New Delhi-110 002 (India)
Phone : +91-11-23279245; 23253475; 43596065
Mobile : +91 9811179893 / +91 9871656464
E-mail : discoverybooksindia@gmail.com
orderdphbooks@gmail.com
namitwasan9@gmail.com
web : www.discoverypublishinggroup.com

Maritime Satellite Communications
Edited by: **Kim Berg**

International Standard Book Number: **978-81-19523-68-9** (Hardback)

PREFACE

In the realm of terrestrial communication technologies, significant progress has been made in recent times, resulting in unprecedented data rates and opening up avenues for various new applications. The upcoming sixth generation (6G) era promises even greater improvements, with anticipated capacity growth of 10 to 100 times compared to current standards. However, while wireless communication on land has made substantial strides, providing reliable and high-speed data rates for maritime communication remains a formidable challenge.

Maritime communication plays a vital role due to the substantial increase in oceanic activities such as naval shipping and logistics, offshore oil exploration, wind farming, fishing, and tourism, among others. Moreover, maritime communication is crucial for Internet of Things (IoT) applications like environmental monitoring and climate change control. Unfortunately, the limitations of fourth generation (4G) and fifth generation (5G) technologies confine their use to onshore scenarios, as base stations cannot be installed far offshore or in oceans. Thus, satellites become a critical component in maritime communications, offering connectivity in unconnected oceanic regions. However, satellite-based solutions come with significant drawbacks such as high proprietary terminal and operation costs and limited available bandwidth.

In recent times, aerial wireless solutions, including unmanned aerial vehicles (UAVs) and helikites, have been introduced for maritime communication applications. These technologies aim to address the challenges faced in ship-to-ship and ship-to-shore communication. Currently, most communication technologies in this domain utilize medium frequency (MF), high frequency (HF), very high frequency (VHF), and ultra-high frequency (UHF) bands.

This book serves as a comprehensive guide to the diverse techniques employed in maritime communication using radio frequency (RF) and optical bands. It starts by introducing the different techniques and then focuses on essential topics such as channel models for RF and optical bands, modulation and coding schemes, coverage and capacity considerations, and radio resource management specific to maritime communications. The book also explores emerging use cases, including the Internet of Ships and the ship-to-underwater Internet of Things, shedding light on their potential and impact. Additionally, it highlights a number of exciting open challenges in the field and identifies future research directions, such as extending broadband connectivity to the deep sea, harnessing terahertz and visible light signals for on-board applications, and utilizing data-driven modeling for radio and optical marine propagation.

It caters to professionals seeking a comprehensive understanding of the techniques, challenges, and potential solutions in this domain. Additionally, graduate students and academics exploring advanced topics in wireless communication and IoT applications will find this book to be a valuable resource.

The editor has collected reliable and scientifically proven information from various credible sources to provide readers with the latest and most accurate information available. The book is a compendium of contributions from various experts in the field, and the editor expresses gratitude to these contributors for sharing their knowledge and expertise.

Editor

CONTENTS

CHAPTER 1

A Distributed Satellite System for Multibaseline AT-InSAR: Constellation of Formations for Maritime Domain Awareness Using Autonomous Orbit Control

Kathiravan Thangavel[1,2,5,6], Pablo Servidia[3], Roberto Sabatini[4,5,6], Pier Marzocca[1,3,6], Haytham Fayek[5,7] Santiago Husain Cerruti[3], Martin España[4] and Dario Spiller[1,2]

[1]*Sir Lawrence Wackett Defence & Aerospace Centre, RMIT University, Melbourne, VIC 3083, Australia*
[2]*School of Aerospace Engineering, Sapienza University of Rome, 00138 Rome, Italy*
[3]*Comisión Nacional de Actividades Espaciales (CONAE), Av. Paseo Colón 751, Buenos Aires* C1063ACH, *Argentina*
[4]*Department of Aerospace Engineering, Khalifa University of Science and Technology, Abu Dhabi P.O. Box 127788, United Arab Emirates*
[5]*SmartSat Cooperative Research Centre, North Terrace, Adelaide, SA 5000, Australia*
[6]*School of Engineering, RMIT University, Melbourne, VIC 3000, Australia*
[7]*School of Computing Technologies, RMIT University, Melbourne, VIC 3083, Australia*

ABSTRACT

Space-based Earth Observation (EO) systems have undergone a continuous evolution in the twenty-first century. With the help of space-based Maritime Domain Awareness (MDA), specially Automatic Identification Systems (AIS), their applicability across the world's waterways, among others, has grown substantially. This research work explores the potential applicability of Synthetic Aperture Radar (SAR) and Distributed Satellite Systems (DSS) for the MDA operation. A robust multi-baseline Along-Track Interferometric Synthetic Aperture Radar (AT-InSAR) Formation Flying concept is proposed to combine several along-track baseline observations effectively for single-pass interferometry. Simulation results are presented to support the feasibility of implementing this acquisition mode with autonomous orbit control, using low-thrust actuation suitable for electric propulsion. To improve repeatability, a constellation of this formation concept is also proposed to combine the benefits of the DSS. An MDA application is considered as a hypothetical mission to be solved by this combined approach.

Keywords: *astrionics; autonomous orbit control; autonomous systems; Constellation of formations; control systems; distributed satellite system; electric propulsion; Formation Flying; intelligence surveillance and reconnaissance (ISR); maritime domain awareness; multi-baseline AT-InSAR; Trusted Autonomous Satellite Operations (TASO)*

1. INTRODUCTION

Strong and robust maritime security arrangements are required to contribute to the Intelligence, Surveillance, and Reconnaissance (ISR) operations and Maritime Domain Awareness (MDA), which can be accomplished using satellite technologies. This approach based on space information becomes essential for countries in the southern hemisphere, with a sizeable maritime domain to protect in terms of sovereignty and sovereign rights, naval assets, infrastructure, resources, and people [1-3]. These capabilities can significantly help with resource and biodiversity preservation, economic and environmental sustainability, disaster mitigation, and security at marine, in addition to supporting safety and security at sea [4]. This is especially true for isolated regions such as Australasia [5,6] and resource-rich regions such as the Gulf of Guinea [7], the South China Sea [8], Micronesia [9], the Argentine Sea [10], the Mediterranean Sea [11] and the Indian Ocean [12], just to name but a few. According to the United Nations (UN), Illegal, Unreported and Unregulated (IUU) fishing is a major factor contributing to more than 90% of global fisheries stocks getting fully exploited, overexploited, or depleted, affecting regions most impacted by climate change. This practice also accounts for one-fifth of global fisheries catches, which can be worth up to USD 23.5 billion per year, making it the third most lucrative business natural resource crime after timber and mining [13]. For MDA, satellites can provide the data for tracking ship movements, i.e., for ISR operations and data for observing the marine environment, such as meteorological and oceanographic conditions. In 2014, the International Maritime Bureau (IMB) estimated that maritime piracy caused USD 16 billion in economic losses annually, mostly as a result of theft, transportation delays, insurance costs, anti-piracy measures, etc.

The Distributed Satellite Systems (DSS) involves a set of small satellites working together which can simultaneously cover larger areas and outperform a single large (i.e., monolithic) satellite, which is often more expensive and less effective. DSS has many advantages, including easier design, faster build time, lower replacement costs and increased redundancy [14-17]. One issue is to keep the formation geometry (required to accomplish the mission) while avoiding inadvertent collisions due to uncertainty in the state of the formation and/or failures. Recent

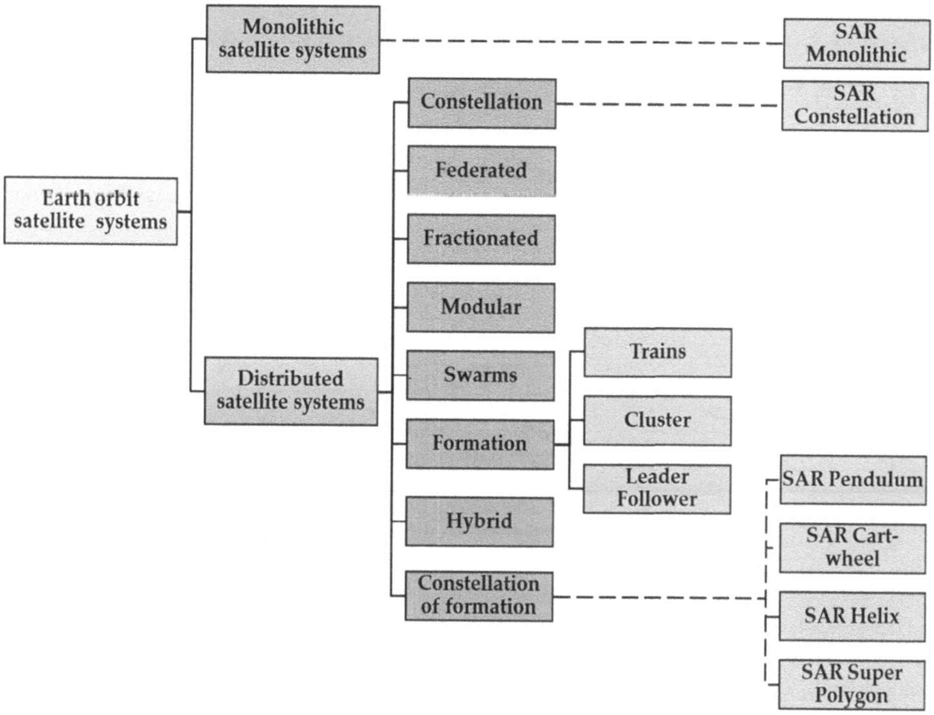

Figure 1. *A possible satellite system classification, with SAR satellite implementation examples.*

research focuses on various control strategies to address these changes, including the possible adoption of artificial intelligence (AI) techniques [18,19]. Figure 1 depicts a possible classification and example of Synthetic Aperture Radar (SAR) satellite mission types [20,21]. The satellite systems are classified into monolithic and distributed satellite systems, and the latter is divided into several possible implementation branches as the constellation (flying far from each other, without relative navigation/control), Formation Flying (close flight, requires relative control) and other options as swarms or hybrid approaches [18].

The implementation of a control law requires communication between the satellites in each formation, i.e., Inter Satellite Links (ISL), in such a way that all the absolute positions are known at least by one of the satellites, while all of the satellites receive all the relative errors and the center of mass acceleration command in order to implement the associated force. ISL allows for satellite-to-satellite communication on each DSS formation and there are many possible implementations, as shown by Liz Martinez et al. [22]. A direct solution is given by the Star topology, where the follower satellites of the formation communicate this navigation state to the chief, and hence the chief can broadcast this information to all the followers, also including its own navigation state and the relative navigation with respect to each of the followers. Figure 2 shows this and other feasible topologies, with the full-duplex ISL being represented

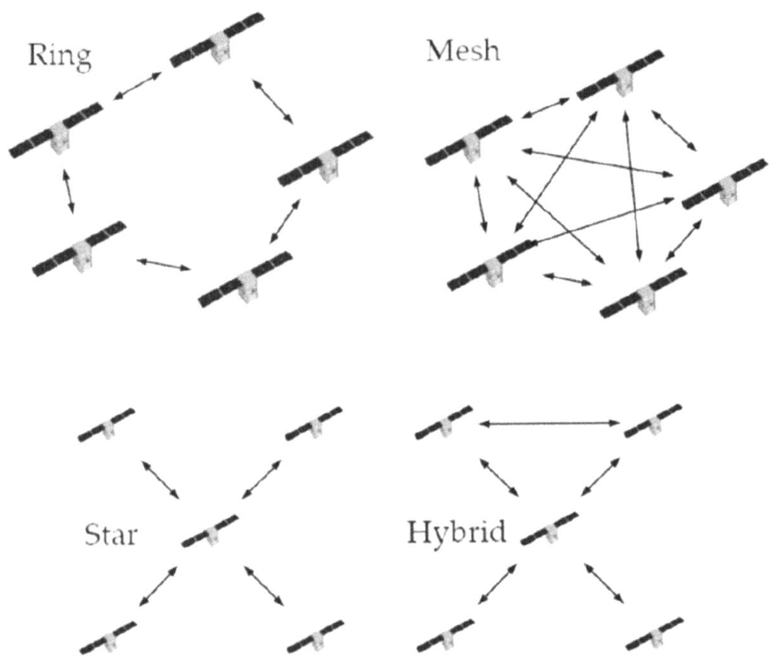

Figure 2. *Inter-Satellite Link (ISL) topologies within the DSS, adapted from [22].*

by double arrows. By including reactive components into the architecture, ISL allows the DSS operations to be enhanced and data to be processed on-board the satellite for timely operation, which makes intelligent DSS (iDSS). The ground station network and/or geostationary satellite service can be used to facilitate communications between satellites of different formations, which may be useful to perform constellation reconfigurations and process collision avoidance alarms from external objects [16]. As the communications become part of the control loop, a complete infrastructure to validate autonomous orbit control shall be able to emulate the inter-satellite links, as proposed in [23]. ISL is an essential component of the DSS astrionics architecture. They make it possible for rapid data sharing between satellites and AI-based on-board data processing [24-26], which relieves some of the tasks that were previously carried out by the ground segment. In this way, the information generated within each formation can be combined on-board (for instance, processed on one satellite and distributed within each formation using a Star ISL, as shown in the following figure) and delivered directly to the user, which is crucial on a surveillance application as MDA We propose a single pass and on-board generation of the interferogram created by the combination of the SAR acquisitions of all the satellites of each formation, which will allow to obtain a timely high-quality product. This improves both the efficiency of the downlink and the operational effectiveness of the system, as measured by a decreased amount of work for human operators and more streamlined mission management.

In this research work, a constellation of formations is proposed to combine the benefits of the repeat cycle given by the constellation with the single-pass products allowed only by the Formation Flying distribution. An example of a monolithic SAR satellite was the Envisat [27], which provided a repeat cycle of 35 days. To perform interferometry, the product constructs the interferogram from different acquisitions of the same scene, which in this case is separated for 35 days. The constellation solution reduces the revisit time to a few days as the Satellite System for Emergency Management (SIASGE) system (Satélite Argentino de Observación COn Microondas (SAOCOM-1) and Cosmo-Skymed SAR constellations). However, this could not be enough for applications needing real-time generation of the interferogram.

In this study, a real-time interferometry is required for the MDA application, which needs to be computed on every single pass over a given target zone or area of interest (AOI). This requirement may be derived from two main motivations: the need for a fast determination of the interferogram, and the need for high coherence in the interferometry, in order to avoid artifacts caused by differences in the background of the scene due to the atmospheric changes or other effects not related to this specific application. Furthermore, as a new DSS architecture type, a constellation of these formations is considered to keep this feature and reduce the revisit time. This work also investigates the possibility of allocating control accelerations among satellites on each formation as a function of the formation objective (relative geometry) and the constellation objective (ground track repetition cycle period). The following contributions were made:

- A safe multi-baseline shifted-Helix Formation Flying is proposed for Along-Track Interferometric Synthetic Aperture Radar (AT-InSAR) Distributed Satellite Systems (DSS), in the context of a Maritime Domain Awareness (MDA) mission over Australia.
- Autonomous orbital control is evaluated for reconfiguration and maintenance of this DSS formation.
- For an increased revisit of maritime surveillance, a novel DSS Archetype, "Constellation of Formations", is proposed, with an associated autonomous control law evaluated by simulations.

The objective of this work is not to define the constellation parameters but to propose a concept for its implementation adding autonomy by means of this two-level (constellation/formation) autonomous orbit control. Single pass interferometry is selected to maximize the coherence, which can be achieved by satellites flying very closely using a Formation Flying (FF) approach, as was implemented on TanDEM-X [28]. This pioneering mission generated new SAR products by defining the acquisition modes as a function of the relative orbits between both satellites, each of them having a complete SAR instrument. Examples of these relative orbits are given in Figure 3. Satellite Formation Flying (SFF) is the coordination of multiple neighboring satellites to accomplish an objective/goal stated in terms of the relative orbits between them. There are various configurations of Formation Flying missions in order to satisfy the user requirements. Each configuration can be

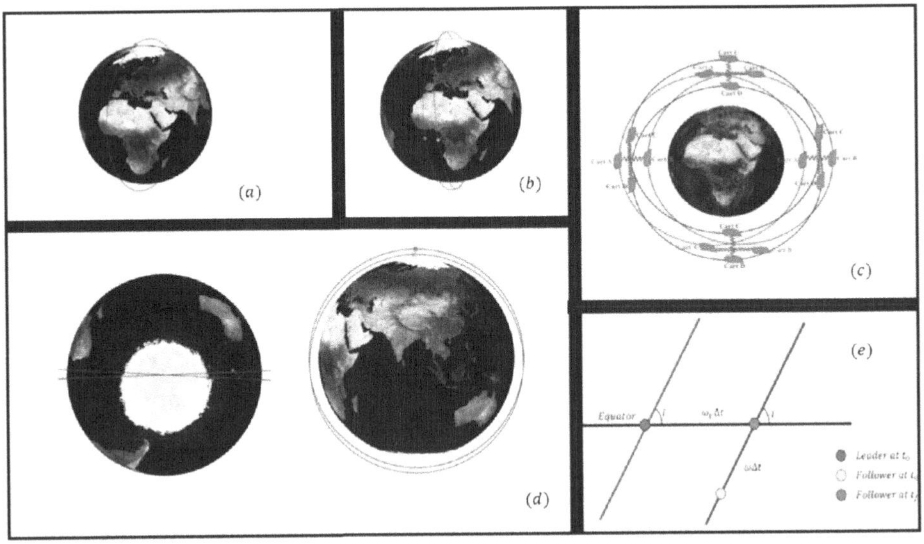

Figure 3. *Chief-deputy classification (a) leader-follower, (b) pendulum, (c) cartwheel, (d) helix configuration, (e) same-ground track.*

obtained by small changes in the orbital parameters of each deputy satellite with respect to the nominal parameters of the chief satellite. In order to meet the needs of the users, different configurations of Formation Flying missions have been proposed. SFF can be classified depending on the configuration, mode of operation, and other factors.

2. SYNTHETIC APERTURE RADAR

A satellite radar instrument produces and transmits its own energy using a known microwave signal pattern, then records the reception of that signal reflected back after interacting with the earth's surface. When this instrument moves with a known velocity relative to the earth's surface, the reflection also adds azimuthal information due to the Doppler frequency deviation and is referred to as SAR data collection. SAR data must be interpreted differently from optical images because the signal responds to surface characteristics such as structure and wetness rather than being a static image. Compared to optical technology, SAR technologies can "see" through the darkness and can operate at any time of the day. Moreover, for longer microwave bands such as the L-band, the SAR instrument can also see through clouds, fog, and rain. This robust operation allows for tracking the trends in habitat, water, and moisture levels, the consequences of natural or human disturbance and variations in the earth's surface as a result of quakes or sinkhole openings. These products are created by analyzing the reflections of signals off a target location and measuring the two-way transit time back to the satellite, its frequency deviation, and the polarization changes. The SAR interferometry technique "interferes" (differences) with two SAR images of the same area, producing maps called interferograms that reveal ground-surface displacement (range change) here

Table 1. *Some baseline types for interferometric SAR [28].*

Baseline	SAR Type	Measurement and Application
$\Delta\varphi$ (look angles difference)	Across-Track	Topography, digital elevation models
$\Delta t = ms, \ldots, s$	Along-Track	Ocean currents, moving object detection
$\Delta t = days$	Differential	Glacier/ice fields, lava flows, hydrology
$\Delta t = days, \ldots, years$	Differential	Subsidence, seismic events, volcanic activities, crustal displacements
$\Delta t = ms, \ldots, years$	Coherent Estimator	Sea-surface decorrelation times, land cover classification

between the two time periods. The phase differences are used to extract information about the captured objects (in comparison to a single image). As a result, at least one aspect ("Baseline") must differ between the images.

Future SAR missions will benefit from increased capability, reliability, and flexibility as a result of this spatial separation [28]. Applications for multistatic SAR systems include single-pass cross-track and along-track interferometry, spaceborne tomography, wideswath imaging, resolution augmentation, ground-moving target acquisition, interference suppression, and multistatic SAR imaging. Simultaneous data collection from numerous satellites reduces temporal and atmospheric disruptions, enhances performance, and allows the identification of rapid changes. AT-InSAR systems are employed to estimate the radial velocity of targets moving on the ground by combining the interferometric phases, which are acquired by combining the two intricate SAR images obtained by two antennas spatially separated along the platform moving direction [29]. The AT-InSAR can be used in various applications such as monitoring real-time traffic management, ocean currents, coastal surveillance, ice drift, etc. In AT-InSAR, the baseline difference is an along-track distance, with a magnitude depending on the mission type, and determines the time difference associated with the pass of the satellite over the particular target, hence measured in seconds for single pass interferometry (two consecutive satellites looking at the same target) or days/years for multiple-pass interferometry (i.e., to process a stack of images taken on different passes over the same scene by the same satellite, other different satellites in a SAR constellation). Figure 4 shows different types of SAR in a simplified classification. The two main branches, interferometric and polarimetric, can also be combined as in the polarimetric SAR interferometry (POLInSAR) techniques [30].

The interferometric SAR missions can be implemented by multi-static SAR, which is characterized by their relative position or equivalent time, known as the baseline. This subfield of SAR categorization will serve as the primary focus of the analysis. Some baseline types usually implemented on SAR interferometric missions are shown in Table 1, and the baseline position difference is shown in Figure 5 for Along-Track interferometry. These baselines can be implemented in multiple passes, on which the interferogram is constructed with data of points of view obtained after several days when there is a repetition cycle on the ground track, or along a single pass when the multi-static SAR is composed of neighboring satellites flying in formation. The latter may improve the SAR product in several ways, as the interferogram can be obtained in almost real-time.

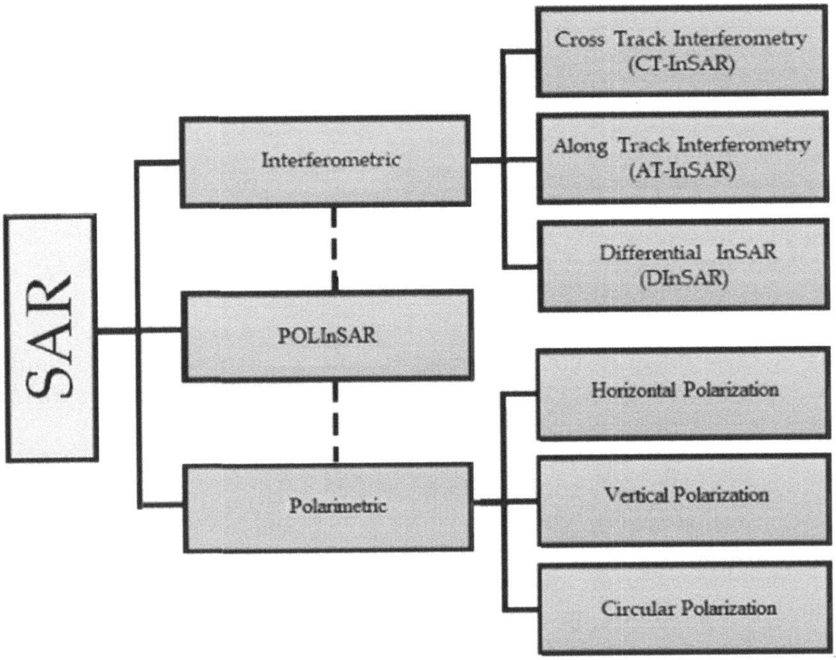

Figure 4. *A simplified SAR classification. Solid lines are definitive classification and dashed line are the possible combination.*

In SFF, there are several topologies to implement the relation between the satellites, for instance, the typical leader-follower approach. In this case, a Deputy satellite (also called follower or secondary) can choose from among the several formation geometries described previously to follow the chief, which can be reconfigured on-board [31]. The chief may also have an autonomous control objective to maintain the absolute orbit (for instance, dragfree), which is therefore followed by the Deputy. Beyond the topology, there are objectives to follow the absolute orbit (for example, to guarantee a constellation repeat cycle for a desired coverage) and other objectives to follow a specific relative orbit geometry within the formation. A more general approach will be proposed to deal with these two types of objectives, to be presented as a Constellation of Formations. The type of formation useful for the MDA application is first taken into account, followed by simulation results using low-thrust continuous control for formation maintenance and reconfiguration.

3. ROBUST MULTISCALE AT INSAR

As mentioned previously, the Along Track formation relies on precise control of the along-track separation in order to avoid collision between the leader (or chief) and follower. This entails a risk due to the typical drift between satellites in case the orbit control was not active for a period of time. On the other hand, the possible need for two different baseline scales simultaneously multiplies the risk, as there are now three possible collision events if only one more follower

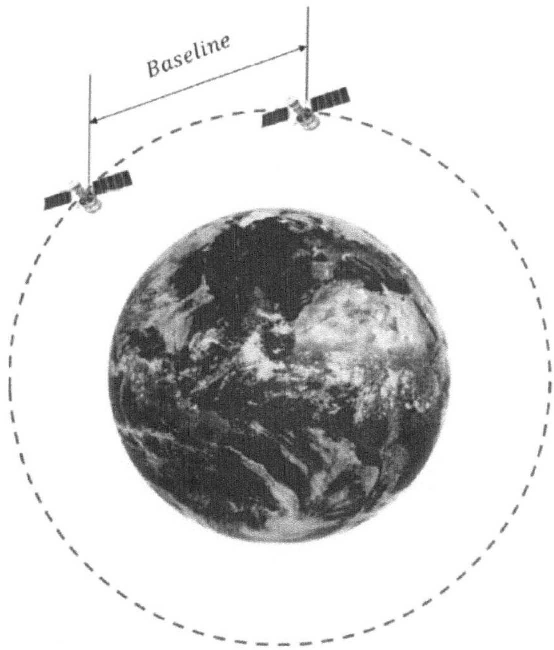

Figure 5. *AT-InSAR baseline difference.*

satellite were added to the configuration. D'Amico showed in [32] a description of a relative orbit in terms of Relative Orbital Elements $\delta\underline{\alpha} = (\delta a, \delta\lambda, \delta\underline{e}, \delta\underline{i})$, where $\delta\underline{e} = (\delta e_x, \delta e_y)$ is the relative eccentricity vector and $\delta\underline{i} = (\delta i_x, \delta i_y)$ is the relative inclination vector. The orbit phase difference is given by $\delta\lambda$, while the semi-major axes' relative difference is δa. A safe formation is guaranteed when $\delta\underline{e}$ and $\delta\underline{i}$ are parallel or anti-parallel, for $\delta a = 0$. A strict AT-InSAR formation only has an orbit phase difference between satellites, thus it is not possible to guarantee the safe condition (as $\delta\underline{e} = \delta\underline{i} = \underline{0}$, here parallelism cannot be evaluated). Here a variation of this along track formation is proposed and as follows:

1. To add a small helix component to each follower relative to the chief, where the across-track component is one order of magnitude smaller than the chief/follower along-track baseline.
2. To scale the chief/follower relative eccentricity and relative inclination vectors in order to generate low-risk "pipes" for each satellite.

Figure 6 shows a particular case of a parallel relative eccentricity and relative inclination vectors for two followers with respect to the chief.

Notice that the difference between the followers also preserves the relative eccentricity and relative inclinations vectors as collinear, therefore achieving a safe condition. This formation can include more followers by adding other scales on the same axis, preserving the collinearity between the relative eccentricity and inclination vectors for the given follower. On the other hand, different along-track

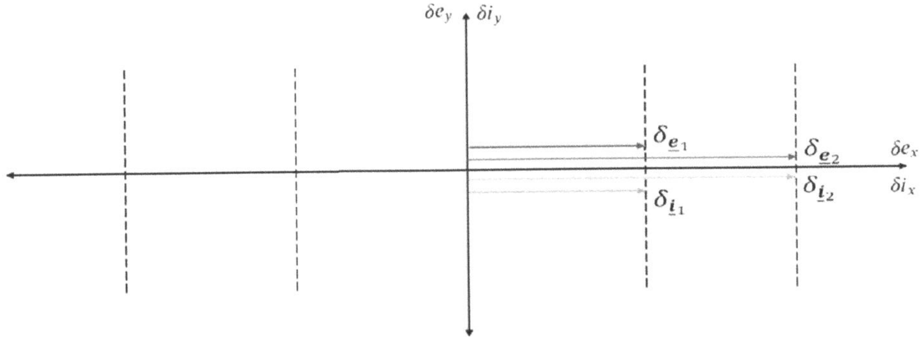

Figure 6. *Example of relative eccentricity $\delta_{\underline{e}_i}$ (in red) and relative inclination $\delta_{\underline{i}_i}$ (in green) vectors of two followers $i = 1, 2$.*

baselines could be chosen for each of these followers. As the chief orbit is sun-synchronous and frozen, the satellite altitude and relative orbit baselines are guaranteed to repeat for the same latitude; hence the interferogram products generated on each pass have geometric coherence between different passes, enabling to perform differential interferometry by taking a set of images generated by the SAR system for the same zone. Moreover, the multiple along-track channels make it possible to track different velocity ranges for the targets on the Earth's surface, see [29], which can also be compared along the same pass by adding followers with different alongtrack baselines with the safe configuration previously presented. The following equation, adapted from [32] (Equation (1)), defines a metric $\delta r_{rn}^{\min}$ to evaluate the minimum distance, on the radial/normal plane, between two satellites in a formation, by using the Relative Orbital Elements, as follows, for a chief orbit with semi-major axis (a_c) :

$$\delta r_{rn}^{\min} = \frac{\sqrt{2}a_c|\delta \underline{e} \cdot \delta \underline{i}|}{\sqrt{\delta e^2 + \delta i^2 + |\delta \underline{e} + \delta \underline{i}| \cdot |\delta \underline{e} - \delta \underline{i}|}} \tag{1}$$

where $\delta e = |\delta \underline{e}|$ and $\delta i = |\delta \underline{i}|$. This metric will be used in the following section to evaluate the stationary regime after a reconfiguration using an autonomous orbit control.

4. AUTONOMOUS ORBIT CONTROL

The relative orbital elements are used by an autonomous feedback orbit control law derived in [31]. This control law guarantees a bound control acceleration expressed in the Radial, Transverse, Normal (RTN) frame. Figure 4 shows the simulation results for two followers after a reconfiguration maneuver starting with a pure along track (unsafe) condition. Figure 7 a shows the eccentric vector $\delta \underline{e}$ and the inclination vector $\delta \underline{i}$ components, which achieves a final state close to the pattern defined in Figure 6 . Figure 7 b shows the evaluation of the radial/normal minimum distance metric $\delta r_{rn}^{\min}$. Figure 7c shows the along-track separation, which can be defined dynamically for each of the followers. As there is a small helix component added to the along-track formation, there will also be an oscillation on the along-track distance, whose amplitude doubles the amplitude of the radial/normal component. The proposed geometry only sketches the idea of the relative geometry, while the definition of the parameters should be given by the application and can be changed dynamically using autonomous orbit control. The reconfiguration between the unsafe along-track formation and the safe one proposed in the previous section is examined in this section.

Figure 8 shows the result for follower 1 with an orbit control period of 20 s, and assuming here ideal orbit navigation. For a satellite mass of 100 kg, the simulated thrust bound would be 1mN in all directions. Figure 9 shows the coordinates given by the difference between relative perigee angle φ and relative ascending node angle θ, with respect to the norms of the relative eccentricity $\delta \underline{e}$ and relative inclination $\delta \underline{i}$ vectors, scaled by the chief's semi-major axis a_c [32]. Notice that the beginning of the trajectory is at the origin of δe and δi, and thus it is not under a safe condition, while at the end, the difference $\phi - \theta$ is nearly zero (i.e., the relative eccentricity and inclination vectors are collinear) while the stationary relative eccentricity and inclination norms are approximately 40 m when scaled by the semi-major axis. Therefore, the reconfiguration achieves the desired baseline and the required safe condition.

5. CONSTELLATION OF FORMATIONS

The formation presented in the previous section has a ground-track repeatability described by the repeatability of any of the ground tracks of the satellites in the formation. These ground tracks limit the observation coverage for a given instrument's field of view. A constellation of satellite formations is proposed to increase the repeatability and coverage of the desired DSS while keeping the formation advantages. Figure 10 shows an example of two formations flying in a constellation. Notice that each formation preserves the single-pass SAR interferometry objective. On the other hand, the resulting constellation can be designed by using the center of mass of each formation as the constellation-equivalent satellite.

Let the constellation $C = \{F_1, F_2, \ldots, F_{Nc}\}$ be a set of N_c formations F_i, each of them composed by N_{F_i} satellites, i.e., $F_1 = \left\{s_{1,1}, s_{1,2}, \ldots, s_{1,N_{F_1}}\right\}, F_2 = \left\{s_{2,1}, s_{2,2}, \ldots, s_{2,N_{F_2}}\right\}$, etc. As each satellite $s_{i,j}$ belongs to the formation F_i and to the constellation C, there are at least two objectives for the orbit controller:

O_F-To keep the relative orbit of the satellite $s_{i,j}$ within the formation F_i.

O_C - To keep the formation F_i in the constellation C.

The formation objective O_F has been treated in the previous section by using the relative orbital elements to describe the feedback error. However, there is flexibility in the implementation of the control law that designers can now use, as for any given acceleration control determined by the follower relative orbit feedback law, designers can implement it on the follower alone, on the chief satellite alone, or on both. In this case, the freedom is used to implement the formation orbit control laws in such a way that it does not modify the dynamics of the formation's center of mass.

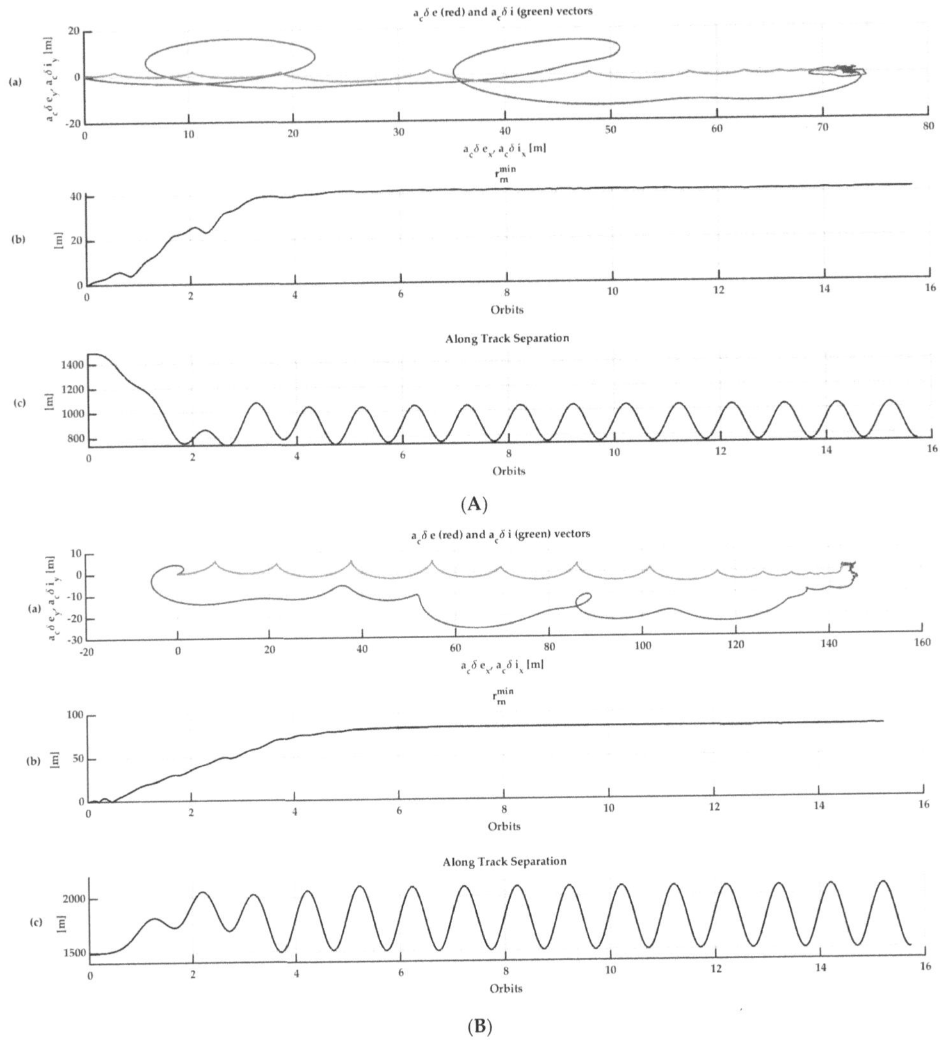

Figure 7. *(A) Simulation of a Robust Multibaseline AT-InSAR formation for follower 1. (a) Shows the components of $\delta \underline{e}$ and $\delta \underline{i}$, (**b**) shows the $\delta \underline{r}_{rn}^{\min}$ and (c) shows the along track separation between follower 1 and the chief. (B) Simulation of a Robust Multibaseline AT-InSAR formation for follower 2. (a) Shows the components of $\delta \underline{e}$ and $\delta \underline{i}$, (**b**) shows the $\delta \underline{r}_{rn}^{\min}$ and (c) shows the along track separation between follower 2 and the chief.*

The constellation objective O_C is stated considering the absolute orbit elements as generated by the constellation design, which is typically free of non-conservative forces, i.e., without drag and solar pressure effects, and with a certain reduced order model of the gravitational effects. On the other hand, to build the constellation control error, we need to estimate the center of mass of each formation F_i. This requires the knowledge of the centre of mass and the mass of each of the

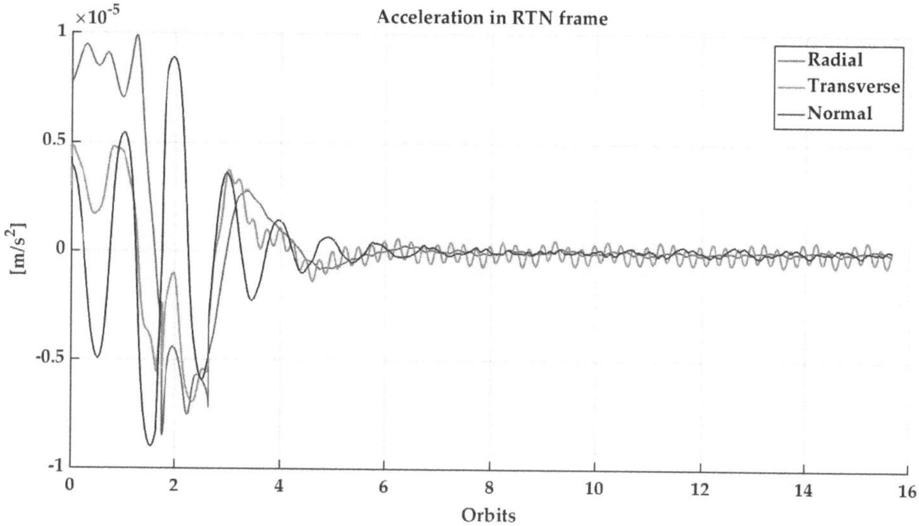

Figure 8. *Control acceleration for the follower 1—chief formation, where red is the Radial, green is the Transverse, and blue is the Normal component.*

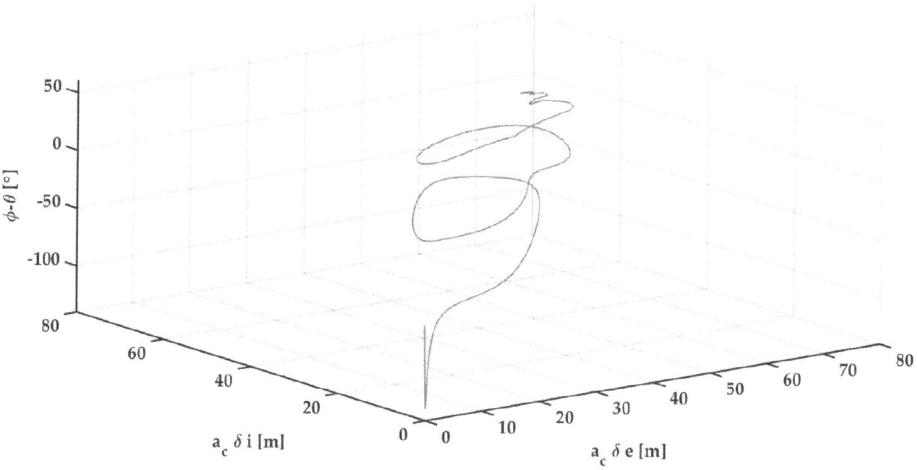

Figure 9. *Safe condition evaluation for follower 1-chief formation. 5. Constellation of Formations.*

satellites but has the benefit to enable a smoothing of the navigated orbit, as the center of mass is a weighted average. In particular, if all the satellites have the same mass and independent identical navigation error distributions, the navigated position and velocity of the center of mass will have their standard deviation reduced by a factor of $\sqrt{N_{Fi}}$. On the other hand, if there is a satellite dominant in mass, the centre of mass navigation error is dominated by the navigation error of this satellite. In this way, the constellation objective has been reduced to the problem

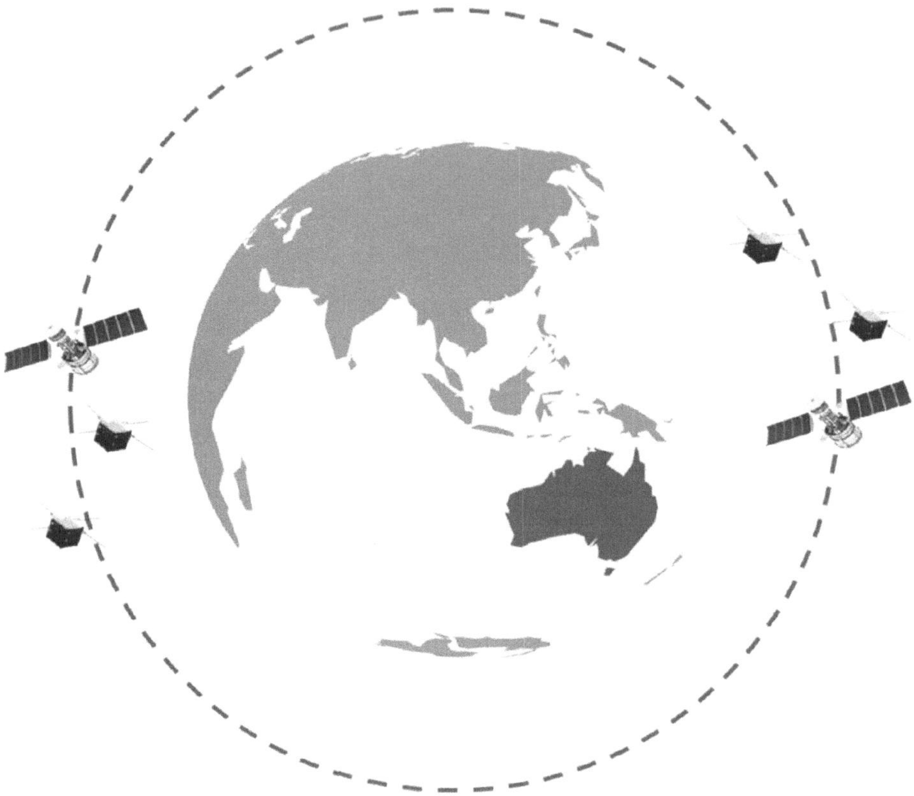

Figure 10. *A constellation of two formations with three satellites.*

of the control of the center of mass of the systems or particles determined by each formation. Once the control acceleration for the center of mass is computed, this is translated into the specific forces to be implemented on each satellite, considering their masses. Additional features of this proposal are developed in the following sections.

5.1. Dedicated Navigation

It is well-known that the relative navigation based on Global Navigation Satellite System (GNSS) receivers can be improved by using interferometry in the L-band using the carrier phase (see [33]), which is known as Real-Time Kinematic (RTK). Other sensors can also be used to improve the accuracy of relative navigation, which confirms the benefit of implementing specific feedback for the formation control separated from the absolute control. On the other hand, the maintenance of the absolute orbit within the constellation must use absolute information, which can be implemented by using Precise Point Positioning (PPP) as proposed in [34].

The Relative Orbit Elements for each of these objectives are computed using the Mean Orbit Elements based on the Ustinov parameters and the analytic formulas as shown in [31] and the works of literature. However, this could not be enough to attain the high accuracy needed for autonomous orbit control, even using the PPP and RTK methods. To this end, a nonlinear filter with finite time memory can directly smooth the control error given by the Relative Orbital Elements, which is compatible with low thrusts, as shown in [34]. This filter can be applied by storing all the implemented control accelerations and measured Relative Orbital Elements, during a certain time horizon, for instance, the last (moving) orbit period. The resulting smoothed control error has enough accuracy to enable autonomous orbit control with a feasible propellant consumption (i.e., the navigation noise is not translated into a permanent actuation and waste of propellant).

5.2. Dedicated Control

Every satellite on the formation implements the same control computed for the center of mass of this formation to preserve / achieve the constellation objective. This can be seen as a "common mode" control, using absolute orbit navigation of the center of mass. On the other hand, for each satellite on the formation, there is an additional term obtained as the necessary feedback to implement the relative orbit control within the formation with the restriction that the dynamics of the center of mass of the local formation is not perturbed. Following the previous analogy, this can be seen as a "differential mode", using relative navigation between the satellites on the same local formation. Consider the two followers and the chief in the previous section's application as a formation; thus, the control acceleration to achieve relative dynamics while preserving the formation's center of mass must be computed. Let $\underline{a}_{0R}, \underline{a}_{1R}$, and $\underline{a}_{2R}$ be the relative terms of the control accelerations for the chief (index 0), follower 1 (index 1) and follower 2 (index 2). Because the SAR interferometry requirements are written in terms of the error between the chief and each of the followers rather than the error between the followers, one can begin by stating the desired formation objectives:

$$-k_F \delta \underline{\alpha}_{01R} = B\left(\underline{a}_{1R} - \underline{a}_{0R}\right) \tag{2}$$

$$-k_F \delta \underline{\alpha}_{02R} = B\left(\underline{a}_{2R} - \underline{a}_{0R}\right) \tag{3}$$

where k_F is a proportional gain for the formation control and $\delta\underline{\alpha}_{01R}$ and $\delta\underline{\alpha}_{02R}$ are the relative orbital elements of each of the followers with respect to the desired Maneuver orbit (see [31]) written in both cases relative to the same chief:

$$\delta \underline{\alpha}_{01R} = T_0 \left(\underline{\xi}_1 - \partial \underline{\xi}_1 - \underline{\xi}_0\right) \tag{4}$$

$$\delta \underline{a}_{02R} = T_0 \left(\underline{\xi}_2 - \partial \underline{\xi}_2 - \underline{\xi}_0\right) \tag{5}$$

where $\underline{\xi}_0, \underline{\xi}_1, \underline{\xi}_2$ are the mean Ustinov parameters (see [14,15]) of the chief, follower 1 and follower 2 orbits respectively, while $\partial\underline{\xi}_1$ and $\partial\underline{\xi}_2$ are the desired deviation relative to the chief necessary to implement the mission orbit requirement for follower 1 and follower 2, respectively. The matrix T_0 is written in terms of the chief parameters and can be found in [30]. Finally, the matrix B in (2) and (3) is the control input matrix of these relative orbit elements dynamics, which is assumed equal for all the satellites in the formation. These relative orbital elements dynamics are given as (see [31,32]):

$$\frac{d\delta\underline{\alpha}_{01R}}{dt} = \underline{f}_{01R} + B\left(\underline{a}_{1R} - \underline{a}_{0R}\right) \tag{6}$$

$$\frac{d\delta\underline{\alpha}_{02R}}{dt} = \underline{f}_{02R} + B\left(\underline{a}_{2R} - \underline{a}_{0R}\right) \tag{7}$$

where $\underline{f}_{02R}$ and $\underline{f}_{02R}$ are considered very small disturbances, which can be partially compensated as a feed-forward term by the control law. The common input matrix B for a formation F_i will be determined by the orbit parameters of its centre of mass orbit, using its orbital elements as follows:

$$B = \frac{1}{\bar{a}\bar{n}}\begin{bmatrix} 0 & 2 & 0 \\ -2 & 0 & 0 \\ \sin(\bar{\lambda}) & 2\cos(\bar{\lambda}) & 0 \\ -\cos(\bar{\lambda}) & 2\sin(\bar{\lambda}) & 0 \\ 0 & 0 & \cos(\bar{\lambda}) \\ 0 & 0 & \cos(\bar{\lambda}) \end{bmatrix} \tag{8}$$

where $\bar{a}, \bar{n}$, and $\bar{\lambda}$ are the mean orbital elements of the centre of mass of the formation F_i, associated respectively to the semi-major axis, mean motion and mean argument of latitude. The columns of matrix B span the whole vector space R^6 every orbit, but locally only can generate a subspace of dimension 3. Therefore (2) and (3) cannot actually be met unless the left-hand sides belong to the column vector space of matrix B, but this can be solved in general by using the pseudo-inverse $B^+$of the input matrix B:

$$\underline{a}_{1R} = \underline{a}_{0R} - k_F \cdot B^+ \cdot \delta\underline{\alpha}_{01R} \tag{9}$$

$$\underline{a}_{2R} = \underline{a}_{0R} - k_F \cdot B^+ \cdot \delta\underline{\alpha}_{02R} \tag{10}$$

The center of mass constraint for the control accelerations is given as follows, for a chief with mass, and the followers with masses:

$$\underline{a}_{0R} \cdot m_0 + \underline{a}_{1R} \cdot m_1 + \underline{a}_{2R} \cdot m_2 = 0 \tag{11}$$

The linear Equations (9)-(11) can be solved for the control vectors a_{0R}, a_{1R}, and a_{2R} as follows:

$$\underline{a}_{1R} = -k_F \cdot B^+ \left(\frac{(m_0 + m_1)\,\delta\alpha_{01R} - m_2 \cdot \delta\alpha_{02R}}{m_0 + m_1 + m_2}\right) \tag{12}$$

$$a_{2R} = -k_F \cdot B^+ \left(\frac{(m_0 + m_2) \cdot \delta\alpha_{02R} - m_1\delta\alpha_{01R}}{m_0 + m_1 + m_2}\right) \tag{13}$$

$$a_{0R} = -k_F \cdot B^+ \left(\frac{m_1 \cdot (m_0 + m_1 - m_2) \cdot \delta\alpha_{01R} + m_2 \cdot (m_0 + m_2 - m_1) \cdot \delta\alpha_{02R}}{m_0 \cdot (m_0 + m_1 + m_2)} \right) \tag{14}$$

This method can incorporate disturbance rejection, control saturation, and fuel consumption management as in [31], but the emphasis on the linear combination of the relative orbital elements is maintained. Notice that the control acceleration shown in the previous section's example for each of the followers did not specify the implementation completely, as there were undefined degrees of freedom. For instance, one could define zero relative control acceleration for the chief, as can be found in a non-cooperative leader-follower approach. By exploiting these degrees of freedom more generally, allowing to preserve the center of mass for relative control, and on the other hand, one can compute the control acceleration for the center of mass in order to track the desired constellation objective as a common control acceleration $\underline{a}_C^{F_i}$ for a given formation F_i. Therefore, the total control accelerations to be implemented on each of the satellites of this formation F_i are as follows:

$$\underline{a}_0^{F_i} = \underline{a}_C^{F_i} + \underline{a}_{0R}^{F_i} \tag{15}$$

$$\underline{a}_1^{F_i} = \underline{a}_C^{F_i} + \underline{a}_{1R}^{F_i} \tag{16}$$

$$\underline{a}_2^{F_i} = \underline{a}_C^{F_i} + \underline{a}_{2R}^{F_i} \tag{17}$$

which is the control law for each formation F_i in the constellation of formations C. For a given objective for the centre of mass of the formation F_i in the constellation, it is defined as a relative orbital element $\delta\underline{\alpha}_C^{F_i}$ which determines the control term $\underline{a}_C^{F_i}$ as follows:

$$\underline{a}_C^{F_i} = -B_{Fi}^+ \left(k_C \cdot \delta\underline{\alpha}_C^{F_i} + \underline{f}_C^{F_i} \right) \tag{18}$$

where the input matrix corresponds to the formation F_i which is explicitly stated in the notation. The term $\underline{f}_C^{F_i}$ may be used for feed-forward compensation of non-conservative dynamics, as aerodynamic drag or solar radiation pressure, as a degree of freedom for the designer. In order to compute the constellation error, it is necessary to compute the desired orbit for the center of mass, which can be performed on-board with a suitable orbit propagator, which should be modified/initialized considering the mission needs. As both control objectives O_C and O_F have different accuracy limits, the proposed separation helps to optimize the application of each of the laws on the specific time periods on which they may be more effective.

Note on the control law: In [15], several relative orbit control laws are formulated in terms of the linearized Clohessy-Wiltshire equation as:

$$\frac{d\underline{x}}{dt} = A^{cw}\underline{x} + B^{cw}\underline{u} \tag{19}$$

and the control is obtained as:

$$\underline{u} = -K(\underline{x})\,(\underline{x} - \underline{x}_d) \tag{20}$$

for a given desired coordinate $\underline{x}_d$. As the relation between the control and the error $(\underline{x} - \underline{x}_d)$ can be considered linear as in (9) and (10), the same approach can be implemented to determine the relative control component associated with the formation objective, restricted to determine a null deviation of the center of the mass formation.

Note on the saturated control law: The thrust control authority must be selected so that the constellation objectives can be achieved with a large enough margin. In this way, it is always possible to select a small enough gain k_F for the formation control which achieves stabilisation of the formation objective. As there might be time and propellant consumption restrictions, this gain and the thrust and satellite masses allocation in the formation should be selected carefully (see [31,35]). In particular, the gain k_F could be selected specifically for each chief / follower pair in order to consider different features of each follower satellite and associated objective. However, to make the presentation simpler on (12)-(14), a unique gain k_F is chosen for all the followers.

5.3. Allocation of Satellite Masses on Each Formation

In the studies of a companion satellite for the L-band SAR Argentine MicroWave Observation Satellite: Satélite Argentino de Observación COn Microondas (SAOCOM) mission [36,37], the relation of masses between the chief and the follower was around ten times. It is reasonable to fix the same mass for the followers, i.e., $m_1 = m_2 = m_F$, and the chief mass is given as $m_0 = \beta m_F$ for $\beta \geq 1$. Moreover, it would be convenient to implement on the chief a thruster β times bigger in terms of force and propellant mass, for a given common propulsion technology and specific impulse. Under this mass model, the total mass of the constellation of formation would be $m_T = N_c(2+\beta)m_F$, where N_c. is the number of formations of three satellites (one chief and two followers). The following particular cases can be identified by inspection of Equations (12)-(14):

- $\beta \gg 1$: In this case, the required chief's control acceleration becomes negligible with respect to the control acceleration of the followers, which tends to be like a classical leader-follower topology on which the control is made by the follower only.
- $\beta = 1$: The required chief's control acceleration authority doubles the required control acceleration authority of each of the followers.
- $\beta = 2$: The required chief's control acceleration authority equals the required control acceleration authority of each of the followers.
- $\beta > 2$: The required chief's control acceleration authority is smaller than the required control acceleration authority of each of the followers.

In general, if there were N_{F_i} satellites on a formation F_i, the mass ratio for equal control acceleration authority for a chief with mass $m_0 = \beta m_F$ is given by $\beta = N_{F_i} - 1$. Moreover, notice that the case with $0 < \beta < 1$ would be feasible, but this case is not practical for a SAR formation, where the chief performs more tasks than the followers and thus requires more satellite mass.

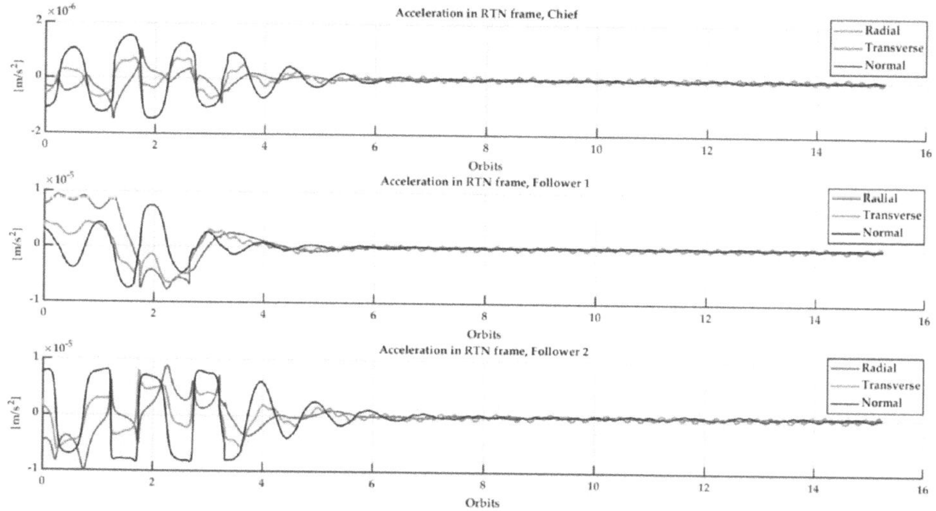

Figure 11. *Control acceleration results with chief/follower mass ratio* $\beta = 2$.

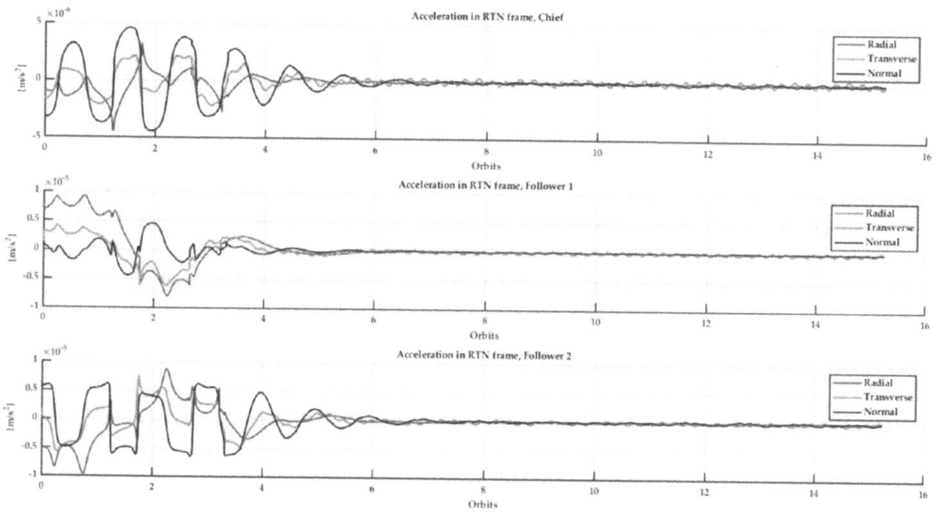

Figure 12. *Control acceleration results with chief/follower mass ratio* $\beta = 10$.

Figures 11 and 12 show the control acceleration evolution to implement the same formation reconfiguration as proposed for AT-InSAR, with $\beta = 2$ and $\beta = 10$ respectively. It is a verified fact that for a larger value of β, the control authority required on the chief becomes reduced in comparison with the followers'. This also has an impact on the DeltaV of each satellite, as shown in Figure 13. This could be taken for a trade-off on the specific constellation / formation system design under the particular restrictions and mission objectives, which is beyond the scope of this work.

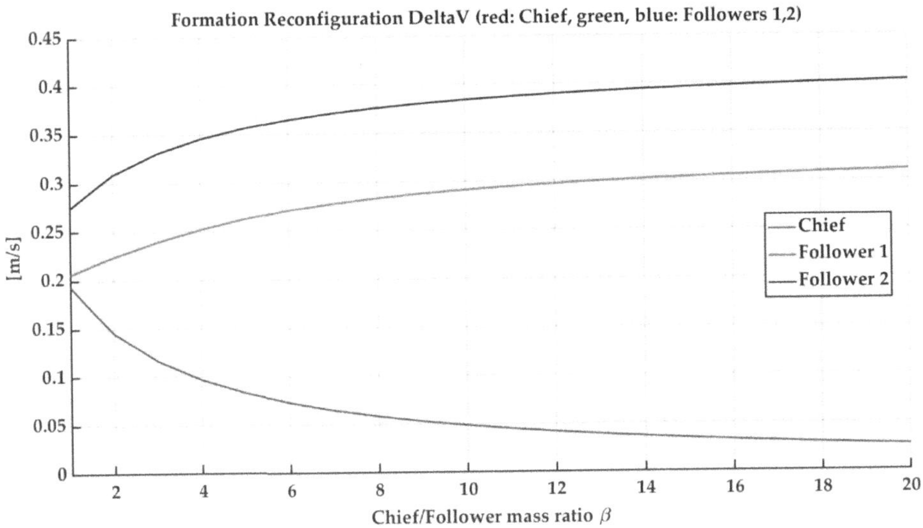

Figure 13. *DeltaV on a formation reconfiguration as a function of the mass ratio* β.

Finally, note that the implementation of this distributed control requires knowledge of the satellite masses, which are time-varying; one of the main uncertainties in knowing the satellite mass which is given by the propellant consumption. However, by using very high specific impulse electric propulsion of several thousands of seconds, this becomes negligible. The feasibility of this specific impulse level can be verified with the Field Emission Electric Propulsion (FEEP) technology, which is available now as Commercial Off-The-Shelf (COTS) products for small satellites (see [38,39]).

In order to simplify the implementation of the saturation, the saturation on the control acceleration differences is defined as follows:

$$\Delta \underline{a}_{1R} = \text{sat}\left(\underline{a}_{1R} - \underline{a}_{0R}\right) \tag{21}$$

$$\Delta \underline{a}_{2R} = \text{sat}\left(\underline{a}_{2R} - \underline{a}_{0R}\right) \tag{22}$$

Therefore, it can be found that under previous assumptions and two followers:

$$\underline{a}_{1R} = \Delta \underline{a}_{1R} \cdot \left(1 - \frac{1}{\beta + 2}\right) + \Delta \underline{a}_{2R} \cdot \left(-\frac{1}{\beta + 2}\right) \tag{23}$$

$$\underline{a}_{2R} = \Delta \underline{a}_{2R} \cdot \left(1 - \frac{1}{\beta + 2}\right) + \Delta \underline{a}_{1R} \cdot \left(-\frac{1}{\beta + 2}\right) \tag{24}$$

$$\underline{a}_{0R} = \frac{-1}{\beta + 2}\left(\Delta \underline{a}_{1R} + \Delta \underline{a}_{2R}\right) \tag{25}$$

There is no real actuator saturation in (21) and (22), as the saturation is applied here to a difference between control accelerations on different satellites. However, one could use an estimate of the upper bounds on these maximum available differences, considering the margin to guarantee that the real actuator on the full expressions (15)-(17) does not reach any saturation limit.

6. CONCLUSIONS AND FUTURE RESEARCH

Autonomous orbit maintenance paves the way for Trusted Autonomous Satellite Operations (TASO) to become a reality in Distributed Satellite Systems (DSS). Here it is shown that TASO is attainable with low-thrust electric propulsion for two main objectives: achieve and maintain the satellite orbit on the constellation, using absolute orbit navigation, and on the formation, using a more precise relative orbit navigation. In this way, there is a dedicated navigation type for each Autonomous Orbit Control objective.

As a case study, we considered a DSS mission for Maritime Domain Awareness (MDA) using Distributed Synthetic Aperture Radar (SAR) instruments. We showed a formation geometry capable of tracking ship movements using single-pass Along-Track SAR Interferometry (AT-InSAR). In particular, we have proposed a single pass and multi-baseline implementation using a safe three-satellite formation, which allows us to avoid temporal decorrelation and to have different velocity scales to track simultaneously. In order to improve the repeatability of these SAR products, we have proposed a constellation of these formations. As we demonstrated, this can be solved by autonomous orbit control using low thrusts compatible with electric propulsion. A particular formation mass distribution was analyzed on which there is a chief whose mass is equal or greater than the mass of the followers by certain common factor. It was shown that for the combined formation/constellation control, the relative importance of the control authority (in terms of maneuver total Delta V) of the chief decreases in relation to the equivalent figure for the followers, as this mass ratio increases.

A Constellation of Formations approach was proposed as a way to model the problem, and the solution's concept has been determined. The approach is based on the concept of a system of particles to describe each of the formations in the constellation in such a way that the relative control within the formation determines the Formation Flying, while there is a separate constellation control objective stated in terms of the center of mass of each formation, i.e., a constellation of formation's center of masses. Both objectives were solved with the same feedback control law structure using relative orbital elements obtained from the mean orbit elements of each of the spacecrafts. This requires an intersatellite communication link between the satellites on the same formation for the Formation Flying feedback computation and the knowledge of the constellation objective in terms of mean orbital elements for the constellation feedback computation, which may be obtained onboard by the desired orbit propagation.

Finally, notice that with the recent evolution of inter-satellite communications, it is possible to augment this Inter Satellite Link (ISL) capacity in order to share also the SAR data information in order to generate the interferogram onboard and, therefore, deliver it in near real-time to the user. In this way, as distributed satellite system solutions become more readily available, a concept of operation with onboard single pass multibaseline interferometry computation will make it possible to deliver high-quality SAR data products faster to the user for effective maritime monitoring.

Additional work must be addressed to perform more realistic simulations, including hardware in the loop to test Global Navigation Satellite System (GNSS) navigation hardware, control nonlinearities, and possible inter-satellite links topologies, in order to complete the mission concept at the flight segment system level.

Author Contributions

Conceptualization, K.T. and P.S.; methodology, K.T. and P.S.; formal analysis, K.T. and P.S.; investigation, K.T. and P.S.; validation, K.T. and P.S.; writing—original draft preparation, K.T. and P.S.; writing-review and editing, D.S., H.F., K.T., M.E., P.M., P.S., R.S. and S.H.C.; visualization, K.T. and P.S.; supervision, D.S., H.F., M.E., P.M., R.S. and S.H.C. All authors have read and agreed to the published version of the manuscript.

REFERENCES

1. Chintoan-Uta, M.; Silva, J.R. Global maritime domain awareness: A sustainable development perspective. WMU J. Marit. Aff. 2017, 16, 37-52.
2. Tetreault, B.J. Use of the Automatic Identification System (AIS) for maritime domain awareness (MDA). In Proceedings of the Oceans 2005 Mts/IEEE, Washington, DC, USA, 18-23 September 2005; pp. 1590-1594.
3. Rahman, N.Z.C. Maritime domain awareness in Australia and New Zealand. In Maritime Security; Routledge: New York, NY, USA, 2009; pp. 228-249.
4. Lauro, A.; Corrêa, C.R. Futures for the Maritime Domain: Signs and Trends That Shape Scenarios. In Power and the Maritime Domain; Routledge: New York, NY, USA, 2022; pp. 286-301.
5. Andrews, S. Four Oceans: An Integrated Approach To Australia's Maritime Domain. Ph.D. Thesis, UNSW Sydney, Sydney, Australia, 2021.
6. Australian Government. Australian Government Civil Maritime Security Strategy. 2022. Available online: www.homeaffairs.gov au/about-us/our-portfolios/national-security/civil-maritime-security (accessed on 31 October 2022).
7. Denton, G.L.; Harris, J.R. Maritime piracy, military capacity, and institutions in the Gulf of Guinea. Terror. Political Violence 2022, 34, 1-27.
8. Greenway, S.R.; Sipes, C.J. Maritime Domain Awareness In The South China Sea: An Operational Picture Design. Naval Postgraduate School Monterey United States, 2018. Available online: https://apps.dtic.mil/sti/pdfs/AD1052657.pdf (accessed on 9 January 2022).
9. Bouchard, C. Small Island Developing States and Maritime Security. In Routledge Handbook of Maritime Security; Routledge: New York, NY, USA, 2022; pp. 250-264.
10. Bigatti, G.; Signorelli, J. Marine invertebrate biodiversity from the Argentine Sea, South Western Atlantic. ZooKeys 2018, 791, 47.

11. Jugović, A.; Schiozzi, D. Comparative analysis of concessions on maritime domain in ports of regional significance in Croatia and Italy. Pomorstvo 2013, 27, 299-312.
12. Cordner, L. Rethinking maritime security in the Indian Ocean Region. J. Indian Ocean. Reg. 2010, 6, 67-85.
13. Fishy networks: Uncovering the Companies And Individuals Behind Illegal Fishing Globally. Financial Transparency Coalition 2022. Available online: https:// financialtransparency.org/reports/fishy-networks-uncovering-companies-individuals-behindillegal-fishing-globally / (accessed on 22 October 2022).
14. Behrens, J.; Lal, B. Exploring Trends in the Global Small Satellite Ecosystem. New Space 2019, 7, 126-136.
15. Mathavaraj, S.; Padhi, R. Satellite Formation Flying: High Precision Guidance Using Optimal and Adaptive Control Techniques; Springer Nature: Berlin/Heidelberg, Germany, 2021.
16. Enrico, L.; Samuel, H.; Andoh, A.; Alessandro, G.; Roberto, S. Autonomous Trajectory Optimisation for Intelligent Satellite Systems and Space Traffic Management. Acta Astronaut. 2022, 194, 185-201.
17. Thangavel, K.; Spiller, D.; Sabatini, R.; Marzocca, P.; Esposito, M. Near Real-time Wildfire Management Using Distributed Satellite System. IEEE Geosci. Remote Sens. Lett. 2022, 1.
18. Wang, D.; Wu, B.; Poh, E.K. Satellite Formation Flying Relative Dynamics, Formation Design, Fuel Optimal Maneuvers and Formation Maintenance; Springer: Berlin/Heidelberg, Germany, 2017. 19. Vasile, M.; Minisci, E.; Tang, K. Computational Intelligence in Aerospace Science and Engineering [Guest Editorial]. IEEE Comput. Intell. Mag. 2017, 12, 12-13.
19. Brown, W. Synthetic Aperture Radar. IEEE Trans. Aerosp. Electron. Syst. 1967, AES-3, 217-229.
20. Curlander, J.C.; Mcdonough, R.N. Synthetic Aperture Radar: Systems and Signal Processing; Wiley: New York, NY, USA, 1991.
21. Marrero, L.M.; Merlano-Duncan, J.C.; Querol, J.; Kumar, S.; Krivochiza, J.; Sharma, S.K.; Chatzinotas, S.; Camps, A.; Ottersten, B Architectures and Synchronization Techniques for Distributed Satellite Systems: A Survey. IEEE Access 2022, 10, 45375-45409.
22. Cerruti, S.H.; España, M.; Servidia, P. Software Architecture Design of Distributed Satellite Systems Test Bed. In Proceedings of the 2022 IEEE Biennial Congress of Argentina (ARGENCON), San Juan, Argentina, 5-6 September 2022; pp. 1-8.
23. Thangavel, K.; Spiller, D.; Sabatini, R.; Marzocca, P. On-board Data Processing of Earth Observation Data Using 1-D CNN. In Proceedings of the SmartSat CRC Conference 2022, New South Wales, Australia, 12-13 September 2022.
24. Spiller, D.; Thangavel, K.; Sasidharan, S.T.; Amici, S.; Sabatini, R. Wildfire segmentation analysis from edge computing for on-board real-time alerts using hyperspectral imagery. In Proceedings of the 2022 IEEE International Conference on Metrology for Extended Reality, Artificial Intelligence and Neural Engineering (MetroXRAINE), Rome, Italy, 26-28 October 2022.
25. Thangavel, K.; Spiller, D.; Sabatini, R.; Amici, S.; Sasidharan, S.T.; Fayek, H.; Marzocca, P. Autonomous Satellite Wildfire Detection Using Hyperspectral Imagery and Neural Networks: A Case Study on Australian Wildfire. Remote Sens. 2023, 15, 720.

26. Louet, J.; Bruzzi, S. ENVISAT mission and system. In Proceedings of the IEEE 1999 International Geoscience and Remote Sensing Symposium, IGARSS'99 (Cat. No. 99CH36293), Hamburg, Germany, 28 June-2 July 1999; IEEE: New York, NY, USA, 1999; Volume 3, pp. 1680-1682.
27. Krieger, G.; Moreira, A.; Fiedler, H.; Hajnsek, I.; Werner, M.; Younis, M.; Zink, M. TanDEM-X: A satellite formation for highresolution SAR interferometry. IEEE Trans. Geosci. Remote Sens. 2007, 45, 3317-3341.
28. Budillon, A.; Pascazio, V.; Schirinzi, G. Multichannel along-track interferometric SAR systems: Moving targets detection and velocity estimation. Int. J. Navig. Obs. 2008, 2008, 310656.
29. Gosh, K. Multi-Baseline Polinsar Inversion And Simulation Of Interferometric Wavenumber For Forest Height Retrieval Using Spaceborne Sar Data. Master's Thesis, University of Twente, Enschede, The Netherlands, 2018.
30. Servidia, P.A.; España, M. On Autonomous Reconfiguration of SAR Satellite Formation Flight With Continuous Control. IEEE Trans. Aerosp. Electron. Syst. 2021, 57, 3861-3873.
31. D'Amico, S. Autonomous Formation Flying In Low Earth Orbit. Ph.D. Thesis, TU Delft, Delft, The Netherlands, 2010.
32. Montenbruck, O.; Ebinuma, T.; Lightsey, E.G.; Leung, S. A real-time kinematic GPS sensor for spacecraft relative navigation. Aerosp. Sci. Technol. 2002, 6, 435-449.
33. Hauschild, A.; Tegedor, J.; Montenbruck, O.; Visser, H.; Markgraf, M. Precise onboard orbit determination for LEO satellites with real-time orbit and clock corrections. In Proceedings of the 29th International Technical Meeting of The Satellite Division of The Institute of Navigation (Ion Gnss+ 2016), Portland, OR, USA, 12-16 September 2016; pp. 3715-3723.
34. Burroni, T.; Servidia, P. Control Orbital Autónomo Restringido de Bajos Empujes y Filtrado de Elementos Orbitales. In Proceedings of the 2022 IEEE Biennial Congress of Argentina (ARGENCON), San Juan, Argentina, 7-9 September 2022; pp. 1-8.
35. Davidson, M. SAOCOM-CS Mission and ESA Airborne Campaign Data. 3rd Advanced Course on Radar Polarimetry, 2015. Available online: https://eo4society.esa.int/wp-content/uploads/2021/01/2015_3rdPolarimetry_SAOCOM_MDavidson.pdf (accessed on 9 January 2022).
36. Gebert, N.; Dominguez, B.C.; Davidson, M.W.; Martin, M.D.; Silvestrin, P. SAOCOM-CS-A passive companion to SAOCOM for single-pass L-band SAR interferometry. In Proceedings of the EUSAR 2014; 10th European Conference on Synthetic Aperture Radar, Berlin, Germany, 3-5 June 2014; pp. 1-4.
37. MorpheusSpace. Available online: https://morpheus-space.com/products/nanofeep/.morpheus-apace.com/products/nanofeep (accessed on 22 November 2022).
38. ENPULSION. Available online: www.enpulsion.com (accessed on 22 November 2022).

CHAPTER 2

Enhanced Communications on Satellite-Based IoT Systems to Support Maritime Transportation Services

Victor Monzon Baeza[1], Flor Ortiz[1], Samuel Herrero Garcia[2] and Eva Lagunas[1]

[1]*SIGCOM Group, Interdisciplinary Centre for Security Reliability and Trust (SnT), University of Luxembourg, 1855 Luxembourg, Luxembourg*
[2]*Telefonica Ronda de la Comunicacion, C4, 28050 Madrid, Spain*

ABSTRACT

Maritime transport has become important due to its ability to internationally unite all continents. In turn, during the last two years, we have observed that the increase of consumer goods has resulted in global shipping deadlocks. In addition, the future goes through the role of ports and efficiency in maritime transport to decarbonize its impact on the environment. In order to improve the economy and people's lives, in this work, we propose to enhance services offered in maritime logistics. To do this, a communications system is designed on the deck of ships to transmit data through a constellation of satellites using interconnected smart devices based on IoT. Among the services, we highlight the monitoring and tracking of refrigerated containers, the transmission of geolocation data from Global Positioning System (GPS), and security through the Automatic Identification System (AIS). This information will be used for a fleet of ships to make better decisions and help guarantee the status of the cargo and maritime safety on the routes. The system design, network dimensioning, and a communications protocol for decision-making will be presented.

Keywords: *zigbee; maritime transport and logistic services; satellite-based Internet of Things*

1. INTRODUCTION

1.1. Background

Today, emerging communications technologies are contributing to the evolution of cities toward Smart Cities. This favors the sustainability of the city in terms of resources such as relation between energy savings and well-being of citizens. Within cities, different economic sectors or verticals such as the energy sector, health, education, infrastructure, and transportation are being improved by the use of smart communications (Zdraveski et al. in [1]).

There are different smart applications as a solution to different services within the Smart City (Zafeiriou et al. in [2]). These smart solutions are articulated through connected modular and scalable systems that facilitate the control and management of different environments in a city. They can be cataloged in different blocks as follows:

- **Smart Environment:** focused on environmental management systems to improve energy efficiency and the quality of the environment in cities.
- **Smart Mobility:** providing intelligent monitoring systems for the transport and mobility of a city and its surroundings to be more efficient and reduce the carbon footprint.
- **Smart Living:** intelligent fire detection systems, video surveillance, and air conditioning to improve the quality of life of citizens.
- **Smart People:** smart solutions for citizens to integrate intelligent communication between the city and its citizens (see interactive maps, citizen apps, and social Wi – Fi).

Among the emerging technologies used in Smart Cities are the Internet of Thing (IoT) and Big Data (Talebkhah et al. in [3]). Through IoT, the city has an intelligent network of connected objects and machines that transmit data using wireless technology protocols such as Long Range (LoRa), Zigbee, and Bluetooth, among others, and Message Queuing Telemetry Transport (MQTT) and Web Application Messaging Protocol (WAMP) for the cloud environment. Supported by Big Data, Smart Cities use these data to implement measures that improve the quality of life of the inhabitants.

The technological transformation of cities is posing several technical challenges. On one hand, climate change is one of the main challenges to be addressed in the coming years within the evolution of Smart Cities. 11% of global greenhouse gas emissions come from the logistics sector, with 3% coming from maritime transport. For this reason, we must pay attention to coastal ports and efficiency in maritime transport, with the aim of decarbonizing their impact on the environment Ahwazi et al. in [4].

On the other hand, recently, the COVID-19 pandemic has opened another challenge that must be faced since it has led to a change in the consumption habit in society Koelle et al. in [5]. During these years, there has been greater demand from the transport sector, especially in the maritime case. About 80% of the things we consume are transported by sea. The transport of containers through shipping companies has fostered a change in consumption and purchasing patterns. As a result of the increase in the demand for goods and an unprecedented shortage of empty containers, maritime freight rates have skyrocketed by up to 328%, while the demand for air transport fell by 90% (Koelle et al. in [5]).

In just 11 months, the average shipping price of the standard container used in maritime transport, known as twenty-foot equivalent unit (TEU), and the most common in international trade has gone from 1103 dollars to 3785 dollars (Koelle et al. in [5]). This increase in maritime logistics transport clearly impacts the final consumer. The delays caused by the shortage of empty containers also affect the final costs, making container movements even more expensive and increasing prices. In the order of magnitude of this problem, it can be pointed out that the current level of freight costs is equivalent to values that oscillate between 0.35% of the retail value in the case of high-value clothing and 63.55% in the case of high volume and low value in furniture Koelle et al. in [5].

Entering the type of merchandise transported, we find the problem that certain products require special treatment or greater supervision, where any unforeseen event can spoil the merchandise, resulting in large economic losses or directly impacting the health of the citizen. This is the case of certain foods or medicines that need special temperature conditions during the shipping process. During the pandemic, the amount of transported medicines increased, 80% of the merchandise transported by sea. Faced with this situation of increasing maritime demand, monitoring the cold chain on the high seas and, consequently, acting in the face of problems that arise is a challenge to be tackled. To do this, satellite networks breakthrough with their innate global coverage capability as a key candidate to offer coverage on the high seas.

This work focuses on finding a solution for various services offered in maritime transport that resolve the problems introduced within this sector, responding to the challenges posed. Within the Smart Cities applications, this work will be included as solutions for Smart Ports (Smart Environment environment) and Smart Transport (Smart Mobility environment).

1.2. Related Work

In this section, we review several works related to intelligent transportation (IT), its evolution, as well as IoT technology and satellite communications (SatCom) within the maritime transport sector (MTS) as topics related to the proposal presented in this work.

In the IT sector, the focus has been mostly on ground transportation. Many systems and proposals for vehicular communications have been developed to advance autonomous driving. To provide a higher level of safety for both driving and the driver. These systems make cities more sustainable with the environment by also controlling emissions through traffic management (Cao et al. in [6]) as well as improve the infrastructure of the city by communicating these with the vehicles (Malik et al. in [7]). On the other hand, for the maritime case, the convergence of MTS and IoT has led to the promising IoT-improved MTS (IoT-MTS); however, the research is still under development and IT systems have not advanced so much to apply equally to ship cases like ground and air vehicles.

For example, Huang et al. in [8] propose a UAV-based system to smartly control and watch over real-time maritime areas in terms of complicated geographical, meteorological, and electromagnetic conditions. This control system offers functions such as automatic detection of ships violating rules, water area intelligent monitoring, water transportation order control, and water area pollution monitoring. However, the solution is exclusively for the safety of the ship, not for the cargo it carries.

Large unattended maritime areas are conducive to risks corresponding to piracy, interference attacks, or ransomware attacks which require ensuring the reliability and efficiency of information transmission in maritime areas. Therefore, one solution is to include blockchain technology that guarantees security during the journey. Yang et al. in [9] propose the keys to including this technology and operating in edge computing in an intelligent maritime transport system. The main keys in Yang et al. [9] are an IoT-enabled maritime transport communication system, which is a distributed system composed of base stations and offshore buoys and uses the unique structure of the blockchain to solve the problems of security and reliability in the network considering only the ships. While Zhang et al. in [10] propose an IoT-based collaborative processing system that unifies the modular structure and integrates multiple modules involved in modern maritime transportation systems. The proposal uses blockchain technology for transportation flow scheduling and management.

Despite advances in applications between IoT and maritime environments to improve safety on board, the trajectory of ships can have highly negative impacts on the management of IoT-MTS. Therefore, Hu et al. in [11] propose a transfer-learning-based trajectory anomaly detection strategy. However, we highlight that this detection is a posteriori; thus, any realtime correction can be made for the maritime route. The limited maneuverability in the trajectory anomaly detection, combined with the lack of efficient communication among ships and ports, maintain the risk of accidents and unwanted situations. In order to take action on the transport routes, we must have the position of the ships controlled at all times. The Global Navigation Satellite Systems (GNSS) system is currently the most attractive positioning system. The keys and challenges to using the GNSS system with autonomous vehicles within IT are presented in Jing et al. [12]. However, as mentioned at the beginning, this has been developed for the ground vehicle but is still in development for the maritime sector. In addition, for monitoring purposes, positioning and navigation are critical functions to determine the absolute and relative positions in the environment in which ships operate.

On the other hand, much interest has arisen worldwide in SatCom as the only means to globally cover the entire land and sea surface. Many operators and manufacturers have gradually built Low Earth Orbit (LEO) satellite constellations. Thanks to their characteristics of small delay and large coverage using constellations, this type of satellite provides a possible solution for the full coverage of the ground communication network. Traditionally, the satellite has also been preferred to provide maritime coverage. Instead, it was not until a few years ago that it gained greater importance thanks to the union with IoT. The work presented by Jin et al. in [13] provides a solution for the implementation of an IoT network supported by LEO satellite constellations, opening the way to the architecture of the IoT-SatCom. In addition, it provides solutions and research directions for the in-depth integration of the satellite-based devices.

We need to effectively allocate relatively limited onboard resources for meeting massive access demand when the satellite is integrated into IoT networks. Therefore, a global traffic model needs to be considered. The work performed by Qu et al. in [14] analyzes the characteristics of satellite-based IoT devices using geographic properties and applications where the design leads to the global traffic model of an LEO-based IoT network. This model has to be extended for maritime traffic.

In addition, from point of view of user traffic, the transmission of huge amounts of data generated from various sensors is the basis and prerequisite for the analysis and prediction of the marine ecosystem. The special geographical conditions of the ocean determine that its data transmission is different from the common terrestrial data transmission. Zong et al. in [15] present a study about marine environmental monitoring network architecture in order to collect environmental data from marine ecosystems using a heterogeneous network constructed by a multihop network and a satellite network. Considering the characteristics of satellite networks, such as high bandwidth-delay product, long latency, and high bit error rate, Zong et al. [15] proposes a transmission scheme that improves the transmission performance of Internet networks over satellite networks based on Transmission Control Protocol/Internet Protocol. Hence, providing an useful reference for the network model and data transmission for ocean monitoring.

Considering the global demands on IoT, the work presented by Yan et al. in [16] proposes a design for a satellite-terrestrial integrated architecture for the IoT relying on the software-defined network. Moreover, based on this architecture, they further propose a dynamic channel resource allocation algorithm to control the access of the IoT terminals with different priorities.

1.3. Our Contribution

We can see how the integration of the satellite with IoT networks is recommended for the transport sector, specially for maritime transport it is a potential candidate. Instead, there are several challenges that need to be addressed to include this architecture. The IoT-satellite union has been proposed for solutions on ships and their surrounding environments, however, it has not been proposed in detail for the cargo carried by ships. There are solutions for anomalies caused by external factors, but not in the cargo itself that the trajectories are not valid. The lack of maneuverability in the event of any anomaly or unforeseen event would cause the loss of the ship's cargo. In addition, the inefficiency of communications with the ports prevents real-time route updates, causing delays in vessel departures and arrivals.

Against this background, we present two types of maritime transport services: monitoring the cold chain inside the containers and integrating data in Global

Positioning System (GPS) and Automatic Identification System (AIS) protocols to improve communications with the ground segment through a MEO mega-constellation. A communications protocol and a proprietary messaging system based on IoT are defined for these services. Thanks to the proposed satellite network, the data are sent to the control center on land to order the update of the route of the ship in case of any problem in the load.

For SatCom's estimation, we evaluated the communication parameters supported by a constellation of satellites in the MEO orbit that seeks to maintain the balance between delay and depot cost while guaranteeing 100% connection availability.

This proposal is framed on the one hand in the MEGALEO Project funded by Luxembourg National Research Fund (FNR) together with GOMSPACE as a partner. On the other hand, the 5G-Space Lab Project funded by the Luxembourg Space Agency (LSA) and the Ministry of Economy (MECO) of the Government of the Grand Duchy of Luxembourg.

The research hypothesis of "MegaLEO" is that a large (MEO and LEO) satellite constellation can operate semiautonomously by deciding and executing satellite and network operation configurations in space. In this direction, MEGALEO will focus on use cases where distributed control of the satellites' orbital positions and radio resources would be imperative, matching the heterogeneous traffic demand across the globe (maritime in this case). 5G-Space Lab has provided the software for testing the constellation design.

The rest of the paper is organized as follows. The system model is presented in Section 2, the architecture for a satellite-based IoT system proposed for maritime transportation communications is presented in Section 3. The proposed services are defined in Section 4. The network dimensioning on the platform on the ship is shown in Section 5. The communication protocol are proposed in Section 6. Finally, we conclude in Section 7.

2. SYSTEM MODEL: MARITIME COMMUNICATIONS AND INTERNET-OF-SHIP

Communications for the maritime sector have historically used radio waves for the exclusive transmission of voice and data. Its main objective from the beginning has been to communicate distress and safety signals between coastal stations and ships. Secondly, with the advancement of telephony, they began to be able to communicate by voice with other vessels and receive weather forecasts by text. Subsequently, maritime radio communications experienced a great boom with the incorporation of satellites. With these systems there are radio beacons with which disasters or distress calls are located, likewise, there are satellite network systems that allow the use of telephony and data. Inmarsat is the operator that has traditionally offered maritime communications services via satellite. In addition, the satellite brought with it navigation systems, introducing systems such as Global Positioning System (GPS) and Automatic Identification System (AIS). In the context of this work, we are interested in the AIS protocol through the SatCom link. The AIS system is used to identify ships, their speed, and course at a specific time.

This system, in maritime navigation, is a considerable element in the safety of ships, since throughout the navigation it makes the vessel visible, in front of large merchant ships. It also allows you to identify other ships that have AIS systems and see their course. Today, the AIS system is one of the most important systems in maritime communications since the radar was introduced. This system was designed with the intention that ships can be seen in any weather condition and thus avoid collisions and is widely used in the merchant marine. Two kinds of AIS systems are available:

- **Class A:** It is a more complex and quite expensive system that has different very high-frequency transmitters and receivers that are usually used by larger ships.
  - Frequencies: 156.025MHz – 162.025MHz
  - Sending information: Continuous
  - Emission power: 12.5 W
  - Range: 50Nm
- **Class B:** It is a lighter and more accessible system in terms of price. This class B AIS is the one used in recreational navigation.
  - Frequencies: 161.500MHz – 162.025MHz
  - Sending of information: Every 3 min
  - Emission power: 2 W

The recent emergence of IoT technologies in mission-critical applications in the maritime industry has led to the introduction of the Internet-of-Ships (IoS) concept. IoS is a novel application domain of IoT that refers to the network of smart interconnected maritime objects, which can be any physical device or infrastructure associated with a ship, a port, or the transportation itself, with the goal of significantly boosting the shipping industry toward improved safety, efficiency, and environmental sustainability. Aslam et al. in [17] propose the architecture, key elements, and main characteristics for IoS in modern maritime communications era. The Aslam et al's. work includes a literature review for IoS. At the present time, there are new digitalization and data exchange initiatives, providing new opportunities by better connecting operators, ports, people, and infrastructure through technologies such as IoS.

3. PROPOSED ARCHITECTURE

The architecture to provide services in the transport of refrigerated containers through SatCom is designed and configured depending on the characteristics of the services provided on the ship. Hence, it is composed of three wireless networks with different technologies as shown in the Figure 1. These subsystems or subnetworks are:

- **WSN (Wireless Sensor Network):** On-board cargo monitoring subsystem. It is a network of sensors inside container ships based on Zigbee as a representative of the IoT family protocols. It is responsible for providing the data collected from the monitoring of the refrigerated container sensors. -

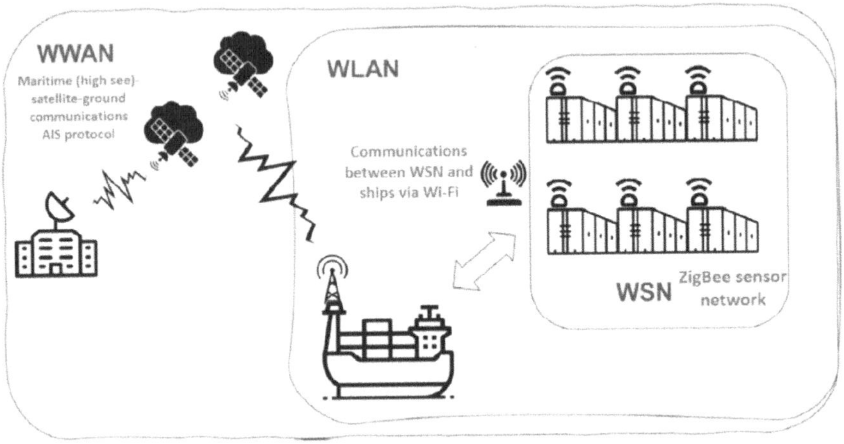

Figure 1. *Proposed architecture scheme.*

- **WLAN (Wireless Local Area Network):** On-ship communications subsystem. It is responsible for supporting communications on the ship (via WIFI), without being affected by corrosion of the wiring due to weather conditions. They are cheaper than wiring container ships as well as being more adaptable to the environment (mobility on the ship).
- **WWAN (Wireless Wide Area Network):** Global communications subsystem. It is provided by a constellation of MEO satellites to communicate the container ships with the maritime control center on the ground segment. This link uses the AIS protocol over satellite.

3.1. IoT (Zigbee) Based WSN Subsystem

In the new IoS paradigm, sensor networks allow the monitoring of different parameters of refrigerated containers in maritime logistics, such as temperature and relative humidity, the object of the solution proposed in this work. To this end, the proposed WSN consists of a wireless sensor-based network that would comply with ISO 10368:2006 standard, which establishes a series of requirements regarding communication between devices, data records, and other standards. The proposed technology to be able to communicate these sensors will be through Zigbee due to:

1. The Zigbee standard operates in license-free bands that facilitate the integration of services in the transport sector in the 2.4GHz, 900MHz, and 868MHz bands. It offers speeds of up to 250kbps and 65,000 connected devices (enough to monitor the temperature and relative humidity of all containers on the ship).
2. Zigbee's ability to support mesh networks that, due to their decentralized nature, are ideal for the structure of containers carried by ships. In addition, each node is capable of self-discovery on the network. When nodes leave the network, the mesh topology (not available in Bluetooth) allows nodes to reconfigure routing paths based on the new network structure. Mesh topology and ad-hoc routing features provide greater stability in changing conditions

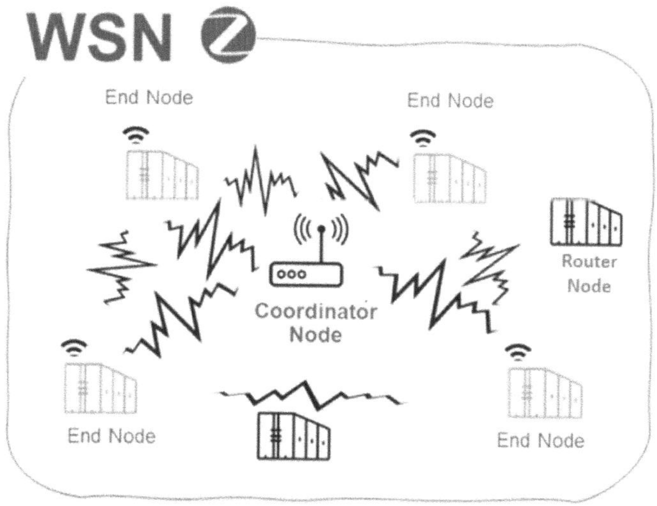

Figure 2. *Distribution of node for the WSN subsystem.*

or in the event of a single node failure. This feature can be beneficial in transportation and container logistics since it would frequently vary from place to place when used in intermodal transport.

3. Lower power consumption compared to that required by Bluetooth. When it is transmitting, Zigbee offers consumption of 30 mA and at the rest of the time just 3μA. On the other hand, Bluetooth consumes about 40 mA transmitting and 0.2 mA at rest. This is because the Zigbee system spends most of its time in a "sleep" state, while Bluetooth is always in a transmit and/or receive state. In this way, it is optimal in terms of energy savings and battery life, something necessary in long-term transoceanic trips.

For WSN, according to the Zigbee standard defined in IEEE Standard [18], we find three types of devices or nodes as shown in Figure 2.

These nodes are:

- Coordinator:
- There can only be one per network.
- Starts the formation of the network.
- Router:
- It is associated with the network coordinator or with another ZigBee router.
- It may act as coordinator.
- It is in charge of multihop routing of messages.
- End device:
- Basic element of the network.
- It does not perform routing tasks.

In order to calculate the distribution of the IoT sensors on the ship, it is necessary to have the stowage arrangement (placement of the cargo) in a container

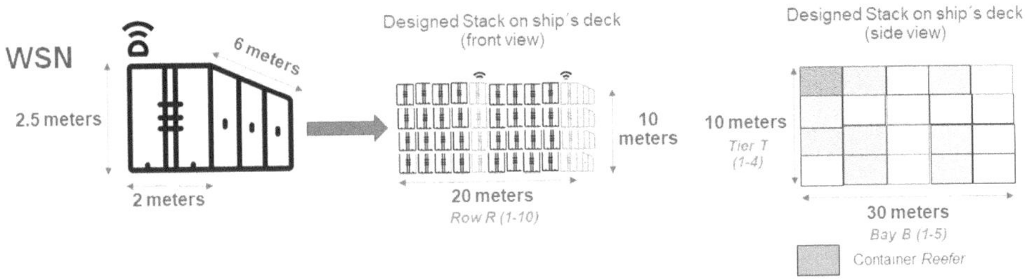

Figure 3. *WSN subsystem scheme.*

ship. An example of container layout is shown in Figure 3. It is common to separate cargo both on deck or ship's hold, into different stacks (or blocks) which are differentiated into bays, rows, and tiers according to nomenclature defined by D. Steenken et al. in [19]. Each stack can contain both normal containers and "refer" containers (referred to as refrigerated containers). The measures of the standard TEU container (twenty-foot equivalent unit) are 20feet, corresponding to 6 m long, 2 m wide, and 2.5 m high as shown in Figure 3 . Based on these measurements, 200 containers are grouped on deck in each stack, storing 4 tiers high, 10 rows wide, and 5 bays deep. The layout of the containers is observed from a frontal perspective (10 rows and 4 tiers of containers). The side view forms a cube of containers (5 bays of containers). In this way, the layout of a container-type stack (on deck) and the corresponding measurements are displayed.

In summary, the container type stack on deck is made up of 10 rows of containers placed horizontally, 4 tiers of containers placed vertically, and 5 bays of containers placed deeply.

In the ship's hold, the design changes as half of the containers can be accommodated by each stack or block, since the height is reduced by half, each stack can hold about 100 containers. In total, the design of the stacks placed along the ship can accommodate 1200 containers. In order to generate a Zigbee network, the previously shown grouping type will be taken as the type network. In no case, the distance of 75-100 m (allowed by the Zigbee network) between end nodes and coordinator node is exceeded.

In this case, the design of the WSN architecture requires the following conditions:

- Being able to accommodate a maximum of 200 reefer containers in each Zigbee network.
- A decentralized mesh network is proposed so that each node is capable of selfdiscovery in the network (this will facilitate the transit of monitored reefer containers). This topology allows the nodes to reconfigure the routing routes each time new ones are included or they disappear (in the case of transit or failure of a reefer).
- A Zigbee coordinator node is included for each cluster with a maximum of 100 (ship 's hold)/200 (deck) containers, in charge of forming the Zigbee network. It establishes a communications channel and a network identifier (PAN ID).

Table 1. *Zigbee parametrization.*

IEEE Standard	Bit Rate	Latency	Frequency	Range	Number Dev	Topology	Battery Life	Security	Cost	Complexity
802.15.4	20–250 kbps	30 ms	2.4 GHz	10–75 m	2–65,000	Mesh Star	>1 year	128 bits AES	Low	Simple

Table 2. *WIFI5 parametrization.*

Standard	Release	Frequency	Bit Rate	Indoor Range	Outdoor Range	Bandwidth
WIFI5 802.11ac	14	5 GHz	3.5 Gbps	100 m	50 m	20, 40, 80 + 80, 160 MHz

Handles packet routing of the final nodes. In addition, the same coordinator node will serve as a gateway in the next WLAN subsystem.

- A Zigbee end node is implemented for each refrigerated container in each cluster.
- There is a sensor capable of measuring the relative humidity and the temperature of the container.

The main characteristics taken from the Zigbee protocol in IEEE Standard [18] for our proposal of the WSN subsystem are listed in Table 1.

3.2. Wi-Fi-Based WLAN Subsystem

The second subsystem, WLAN, is shown in Figure 4. This subnetwork consists of WIFI5 technology according to IEEE IEEE Standard [20], whose elements are detailed as follows:

- Gateways or coordinating nodes (in the WSN network) for each group of containers (they share the 2.4GHz frequency of the Zigbee-based WSN network as the aforementioned 5GHz for WIFI5). These devices will be redundant.
- Automatic Identification System and geolocation antenna (AIS + GPS transponder) class A based on WIFI5 technology with a redundant access point (AP). Commer- cial solution found: Ray-Marine with the AIS4000 product, which is a class A AIS transponder with built-in WiFi and GPS.
- Control system based on PDAs for ship crew with WIFI5 in which to monitor the previously described WSN sensor network.
- Concentrator router (redundant) of the rest of the previously described elements located in the command bridge.
- The ship's electrical power line serves to provide sufficient energy to all the elements of the network. There is a redundant element through a UPS (uninterruptible power supply) to guarantee the power supply.

The main characteristics of WiFi for the WLAN subsystem taken from IEEE Standard [20] are listed in Table 2.

3.3. Satellite Based WWAN Subsystem

A communication link based on a MEO satellites constellation has been chosen to provide coverage on the ship itself, as well as the ship with any data monitoring point on land, in order to find the balance between delay and cost of deployment for use in this WWAN (Wide Wireless Area Network).

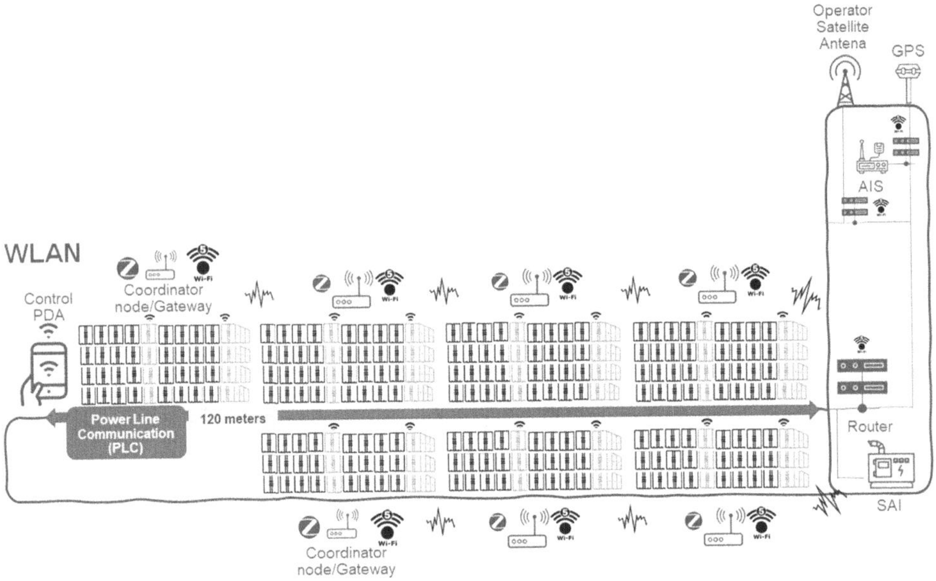

Figure 4. *WLAN subsystem scheme.*

As shown in Figure 5:

- The ship has a satellite antenna that allows both transmitting and receiving in the frequency bands used by this satellite, which will be connected to the communications rack established in the wheelhouse.
- Shore-based satellite base stations that function as gateways for ship-to-shore-to-ship communications. - These base stations are connected to the backbone network to provide an IP connection.
- Once a connection to the data backbone is provided, data is sent to a cloud server to secure the data transmitted by the ship (WSN data monitoring, AIS and ship geolocation, and any other necessary ship communications -Voice and Data-).
- An application is provided to visualize the data received by the ship.

To find a balance between satellite link delay and the cost required for system deployment, a medium earth orbit (MEO) satellite constellation designed to provide low-latency broadband connectivity to remote locations for mobile network operators and maritime service providers is chosen.

A total of 20 satellites are deployed in a circular orbit along the equator at an altitude of 8063 km at a speed of 11,755 mph (18,918 km/h), similar to that used by SES with O3b in [21,22]. The deployed constellation is shown in Figure 6.

Each satellite is equipped with fully steerable Ka-band antennas at 19.7GHz for the downlink and 24GHz for the uplink. Round-trip latency is 140 ms for data services.

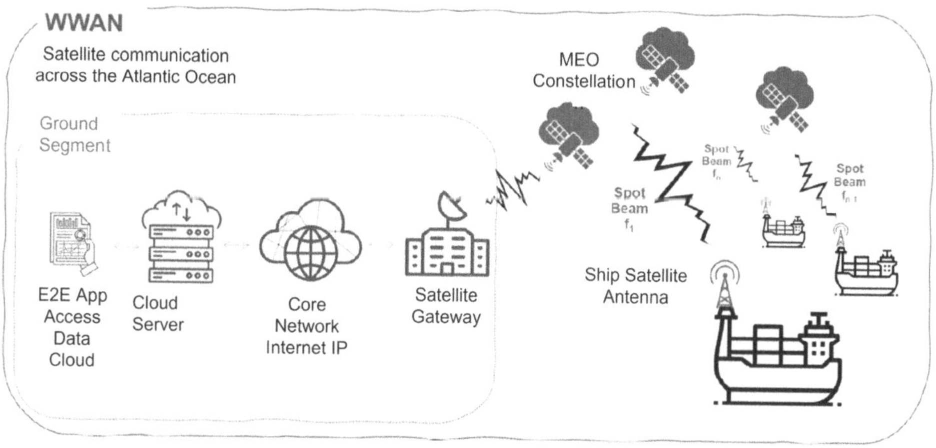

Figure 5. *WWAN subsystem scheme.*

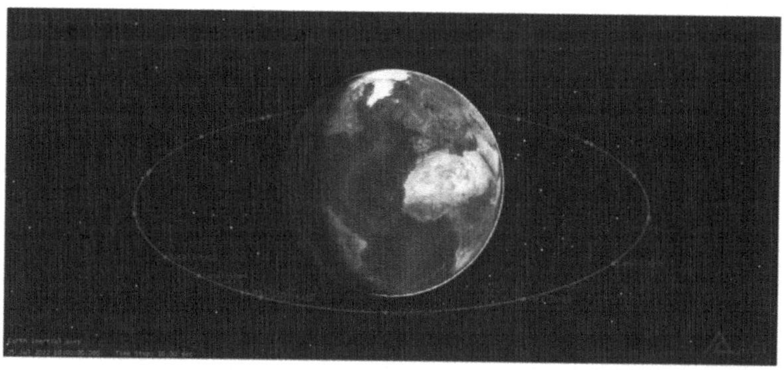

Figure 6. *As part of the constellation, a total of 20 satellites are deployed in a circular orbit along the equator at an altitude of* 8063 km *in MEO orbit.*

The shipping route crosses the Atlantic from Southern Europe to the Caribbean area. Regarding the service area, as shown in Figure 7, the sea route is always within the standard service area, thus ensuring its connectivity.

4. MARITIME TRANSPORTATION SERVICES

The evolution of IoT in the field of maritime transport ("Internet of Ships") has gained special importance, as argued in Section 1, and this has generated new areas of application in said industry. In this context, this work presents two new services supported by IoT and a satellite network for logistics transport at sea that has benefited society.

4.1. Monitoring Service of Transported Cargo

This service consists of creating a communications network based on IoT and a constellation of MEO satellites to monitor the cold levels of cargo transported by containers on ships to guarantee its conservation by maintaining the cold chain and

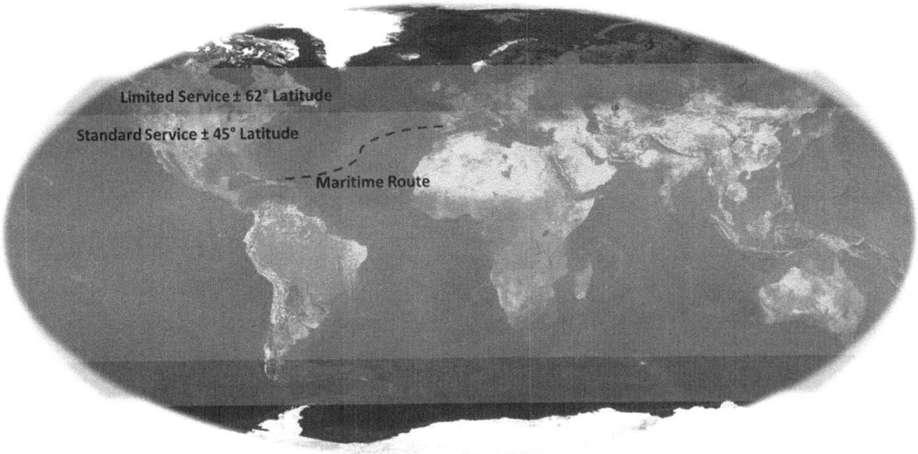

Figure 7. *Maritime route and service area.*

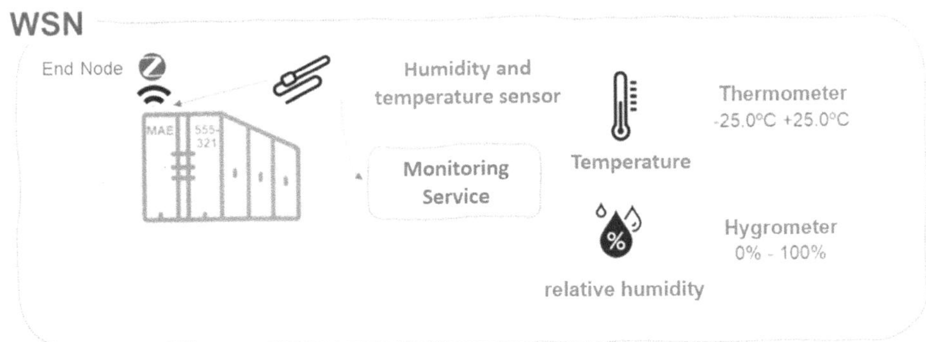

Figure 8. *Monitoring of temperature and relative humidity in the container.*

optimal relative humidity conditions (necessary in the transportation of medicines and perishable food).

Figure 8 presents a scheme for the proposed service. In this case, a thermometer sensor and a hygrometer capable of monitoring the temperature and relative humidity of the load are detailed. In addition, as can be seen, this scheme is included within the wireless sensor network WSN, which will be detailed in the following chapter:

This service must take into account the following:

- Have sensors capable of measuring the following parameters:
  - Thermometer sensor, capable of measuring temperature ranges between −25.0°C and +25.0°C (with a precision of 1 decimal).
  - Hygrometer sensor, capable of measuring relative humidity ranges between 0% absolute dry air and 100% air completely saturated with water vapor.

- Regulations to be met by the sensors placed according to the Agreement on the International Carriage of Perishable Food Products (ATP) (UNCTAD) [17],
  - Standard UNE EN 13846-Temperature recorders and thermometers for the transport, storage, and distribution of refrigerated, frozen, and deep-frozen foods, and ice cream. Periodic verification-.
  - Standard UNE EN 12830-Temperature recorders for the transport, storage, and distribution of temperature-sensitive products. Tests, operation, aptitude for use-.

4.2. Enhanced Positioning Service

The power lines installed on container ships (Power Line Communications, PLC) suffer from deterioration caused by corrosion on the high seas, by handling the loading/unloading of containers in ports (stowage), and have limitations both in speeds of transmission, as well as how little immune they are to noise. In this way, a wireless communications service capable of covering both the monitoring service through sensors and other services that this type of transport can use (such as GPS and AIS) is offered. For this, the solution must take into account the regulations that must be complied with in wireless communications on the ship that define the exchange of information:

- ISO 10368. The main summary lists the interfaces and protocols necessary to achieve this certification.
- Standardized architecture, correct dimensioning, and messaging protocol that defines correct monitoring of the cargo transported, as well as the communication of other services on the ship.

Therefore, the solution to respond to this service is based on designing and developing the technological evolution that responds to the aforementioned problems generated by wired communications.

5. NETWORK DIMENSIONING ON THE PLATFORM ON THE SHIP

The necessary dimensioning within the ship's communications is shown in Figure 9. On the one hand, how the available spectrum is distributed both for the WSN subsystem and for the WLAN subsystem is considered, and what channels will be used to transmit both from the endpoints located in the containers and the channel used by other possible services such as AIS, and GPS. The data are centralized in a central node (communications rack) which groups all the information before being sent to the satellite.

5.1. WSN Subsystem

For the wireless sensor area network WSN is dedicated to monitoring through Zigbee, a different communication channel assuming Frequency Division Multiplexing (FDM) is used for each type of container grouping (200 on deck and 100 in the ship's hold). In this way, the IEEE 802.15.4 standard according to IEEE Standard in [18] that allows users in the 2.4GHz band is distributed as follows in this solution:

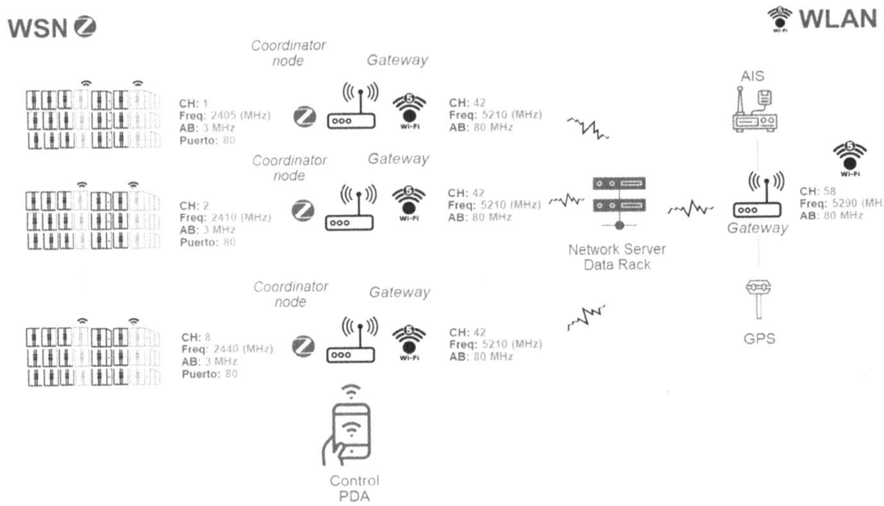

Figure 9. *WSN/WLAN subsystems dimensioning across the ship.*

- Number of channels available: 16 channels separated in their central frequency (f_c) by 5MHz (80 MHz of available spectrum). Thus, each coordinating node is assigned one of 16 channels. The 16 channels are never exceeded since there is a limitation of space on the ship for each group of containers.
- Frequency bandwidth of about 3MHz for each channel.
- Each node has a spectrum of 1.23kHz. According to the standard in the total 80MHz spectrum, 65,000 endpoints can be arranged.
- A maximum number of endpoints per channel: 4065 nodes (not exceeded since each group type allows a maximum of 200 containers).

Here is how the Zigbee spectrum is used to cover clusters of containers along the ship with different channels: Example of channels for 1200 containers placed in stacks:

1. Zigbee channels from 1 to 4 for 4 stacks of 200 containers on deck (800 containers in total).

- Channel 1: $f_c = 2405$MHz.
- Channel 2: $f_c = 2410$MHz. - Channel 3: $f_c = 2415$MHz
- Channel 4: $f_c = 2420$MHz.

2. Zigbee channels 5 to 8 for 4 stacks of 100 containers in the warehouse 400 containers in total).

- Channel 5: $f_c = 2425$MHz.
- Channel 6: $f_c = 2430$MHz
- Channel 7: $f_c = 2435$MHz.
- Channel 8: $f_c = 2440$MHz.

The protocol used is that established by the Zigbee standard, where each end node (the one used for monitoring each reefer container) transmits on its assigned channel. To identify each node, the license plate (inscribed on the container door) which is marked on each reefer and its spatial reference is used, so that the onboard crew can configure it in each stowage both in its Channel (CH), and assign it an identifier (reefer registration) at the application level (see explanation later in the chapter of communication protocols).

5.2. WiFi Subsystem

On the other hand, for the WLAN subsystem, technology has been established that allows communication within the area of the container ship itself that mainly does not interfere in the Zigbee band (2.4GHz) and that provides sufficient coverage for each stack of containers designed (let us remember 200 containers on deck and 100 in the hold) and other multiplexed services such as AIS and GPS.

In this way, the choice of the WIFI5 standard (802.11ac), allows us to solve the problem of interference since by operating in the 5GHz band the signals emitted do not interfere with those of the end nodes installed in each reefer.

Therefore, by broadcasting in the 5GHz frequency, the gateways that are going to be used (remember that they will also be used as coordinating nodes in the Zigbee network), have new communication channels and available spectrum. When using FDM, this spectrum will be distributed in bands of 80MHz per channel (remember that more spectrum can be dedicated per channel thanks to the use of this standard), in such a way that, of the 160MHz possible usable, the spectrum will be divided into two channels of 80MHz each. Thus:

1. A channel is guaranteed to collect information from each of the channels provided in the Zigbee network (WSN). Since 80MHz is available, 40MHz is used in the design example provided for the WSN communication protocol. If new reefer-type pools are added to the container ship, more spectrum is still available to collect the new monitoring channels.
2. Another 80MHz spectrum is guaranteed on a different channel for communications within the ship that collects information received from both the AIS and the GPS coordinates of the ship.

6. COMMUNICATION PROTOCOLS

In order to establish a communication protocol, we start from the base of the proposed architecture referring to the ship (WSN and WLAN) - it will be detailed from the smallest to the largest wireless network. Access to the medium will be based on CSMA/CA (Carrier Sense Multiple Access with Collision Avoidance) in both and will be defined in each of the wireless networks. Lastly, a message protocol will be established to define the communication between services (monitoring, identification, and position of the container on the ship and information related to AIS and GPS).

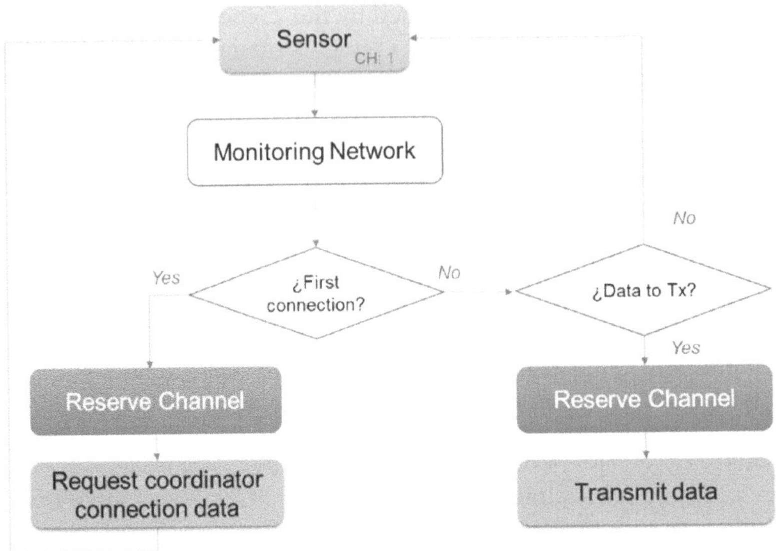

Figure 10. *Sequence for the reservation of the medium by a sensor in the Zigbee network.*

6.1. Channel Reservation Protocol

The protocol is based on the IEEE 802.15.4 standard [18], which defines CSMA/CA (Carrier Sense Multiple Access with Collision Avoidance) as the Media Access method for each grouping type designed (200 on the deck of the ship or 100 containers in the ship's hold in each case). It is a random waiting mechanism that avoids collisions in the transmission of each Zigbee node.

Detailing step by step shown in Figure 10, the following flow is followed:

1. Each container pool broadcasts on a different channel.
2. Within each channel, the end nodes that monitor each reefer listen to the medium.
3. If the medium is free, they reserve the channel to request connection data from their coordinator, if not, they wait a random time and try again.
4. If they have data to transmit and have not yet reserved the channel, they listen to the medium again and when it is free they ask for a temporary space to be able to reserve it.
5. If the end node already has the reserved channel and has data to transmit, it waits for a random time and then starts the data transfer.

Below, the association procedure of the sensor with its coordinating node is detailed, as well as the following procedure that will be chosen for sending data.

6.1.1. Sensor Association Procedure with Its Coordinator

In the following Figure 11 it is observed how before sending data, within the 802.15.4 standard there is the association procedure in which the final node (in our case the Zigbee device placed in the reefer), requests the data of the PAN ID

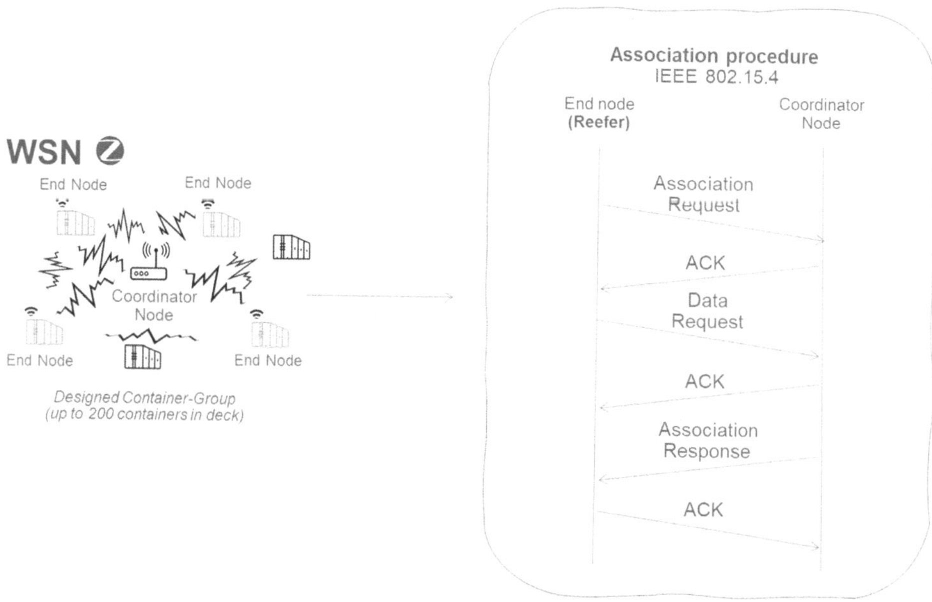

Figure 11. *IEEE 802.15.4 Sensor association procedure with its coordinator node.*

(Personal Area Network Identification) to the coordinating node. Later, it will be seen how once associated, the sending of data from the final node to the coordinator will continue and how this communication is carried out.

The CSMA/CA method used for the WSN network uses control frames to able to coordinate the sending of data on the channel, where:

- **SIFS:** Short Inter Frame Space. The time interval between transmission of packets (mandatory period of idle time on the transmission medium).
- **DIFS:** Distributed Inter Frame Space. Minimum delay for asynchronous traffic competing for access. An end node waits for a DIFS before transmitting any monitored data when the channel is free.
- **RTS:** Request to Send. Control packet for data transmission request.
- **CTS:** Clear to Send. Control packet indicating that the receiving node is free to send.

In this way, by using the control frames explained previously, it is established a procedure -after association- able to send data from end nodes to the coordinator in the WSN.

6.1.2. Sensor Data Submission Procedure with Its Coordinator

As we have seen previously, Figure 11 details the procedure for associating an endpoint (final node) placed in the container with its coordinator node. Now, Figure 12 details the channel reservation process to send data (in this case those that come from monitoring, as well as those that identify the container). In addition, a designed stack of containers located on the deck is observed in detail. They

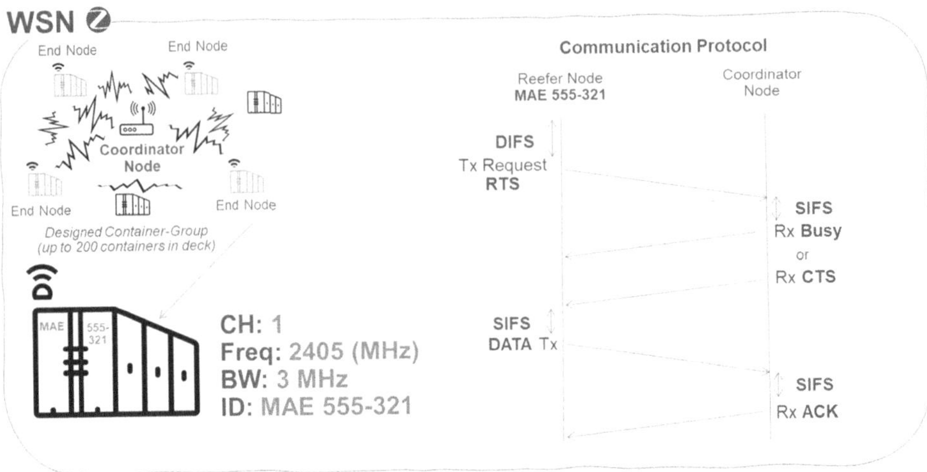

Figure 12. *Access to the Medium for data transmission in the WSN Reefer-Coordinator network.*

use the same channel (in this case Channel 1), with a bandwidth of 3MHz (separation between central frequencies of each channel of 5MHz), and a reefer identifier assigned as ID (registration of the container) to be able to recognize it within the Zigbee mesh.

6.1.3. Procedure for Sending Data in WLAN of the Services Provided

Once the channel has been reserved and the useful data of the sensor network has been transmitted to its coordinators, it is time to see how the channel reservation is made in the following wireless local area subsystem -WLAN-. WIFI5 is the technology chosen, using the previous subsystem detailed with the same access to the medium (CSMA/CA). Unlike the WSN network, it uses beaconing (virtual carrier) through a vector of reserve (NAV). In this way, all the gateways placed on the ship follow the next steps as shown in Figure 13:

1. They listen if the medium is free and if it is not, they wait. The gateways update their reservation vector of the NAV network (it works as a countdown, from which moment you can try to send data again).
2. Once the medium is free, they wait for an IFS time (Inter Frame Space, or frame time).
3. After this IFS time, they listen to the medium again, and if they find it still busy, they wait for another IFS time.
4. If, on the other hand, when listening to the medium after the IFS time, it is free and its reservation vector of the NAV network is 0 , the sending of data begins.
5. Once the data is sent, the end node waits for a data confirmation ACK received by the concentrator router of all ship communications. Buffer is released.
6. If this ACK is not received, the data frame is forwarded.

The gateways (coordinating nodes in the sensor network) placed throughout the ship, encapsulate the received frames (both from the sensor network and AIS

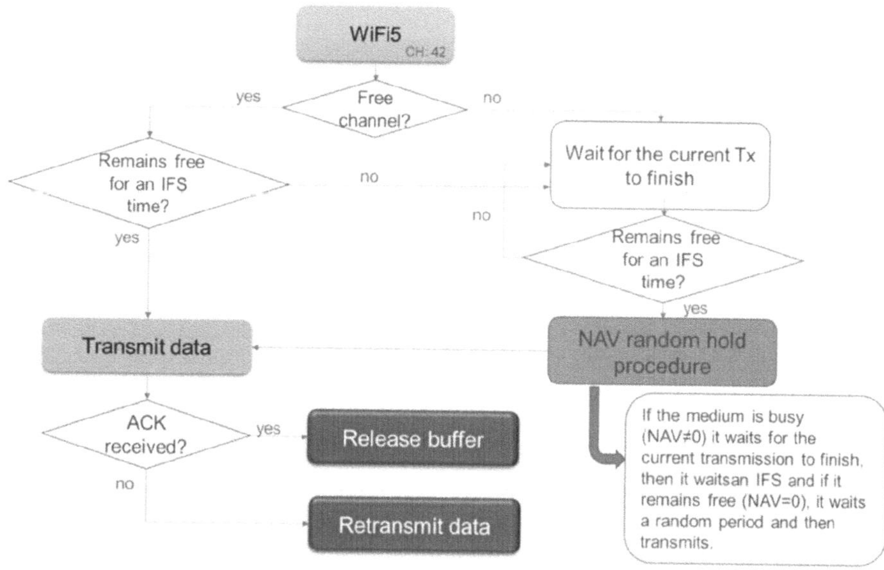

Figure 13. *Distributed Coordination function in WLAN transmission gateways.*

and GPS). In the MAC layer frame headers in the IEEE 802.11 standard, a duration field is marked that specifies the transmission time required for the frame, in short, the time that the medium will be busy. Gateways in the ship's WLAN mesh listening on the wireless medium read the duration field and set its NAV (Network Allocation Vector), which is an indicator to a gateway of how long to defer access to the medium.

A WLAN network gateway (WIFI5) listens to the free medium, and then sends data as shown in Figure 14. Meanwhile, the rest of the Gateways wait for the channel to be free and after a random time, they send the received data (sensors/AIS/GPS). It is also observed, how the data sent from the sensors is received by different gateways (mesh network) with the destination address of the router, and how the AIS/GPS data is sent directly to the same router, before being sent to the satellite (next WWAN subsystem).

6.2. Messaging Protocol

For the architecture proposed, also we define a message protocol within the ship. This messaging protocol will allow us to:

1. Identify the data monitored by the sensor related to temperature and relative humidity.
2. Locate the position of the monitored container inside the ship, so that action can be taken quickly in the event of failures.
3. Identify with a reefer container ID that allows monitoring (in a cloud application) the sensor data (humidity and temperature).
4. Identify what data comes from each different service.

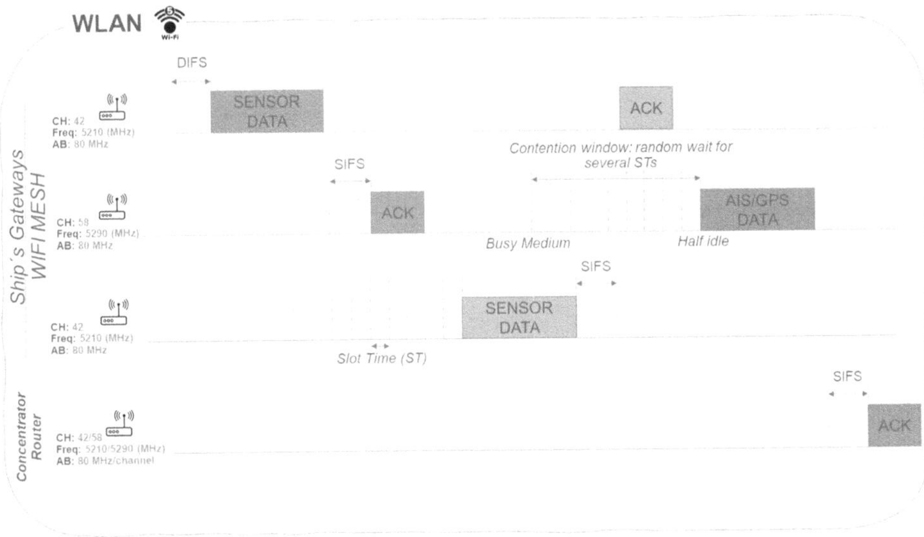

Figure 14. *Sensor data sending detail, AIS, GPS in WLAN.*

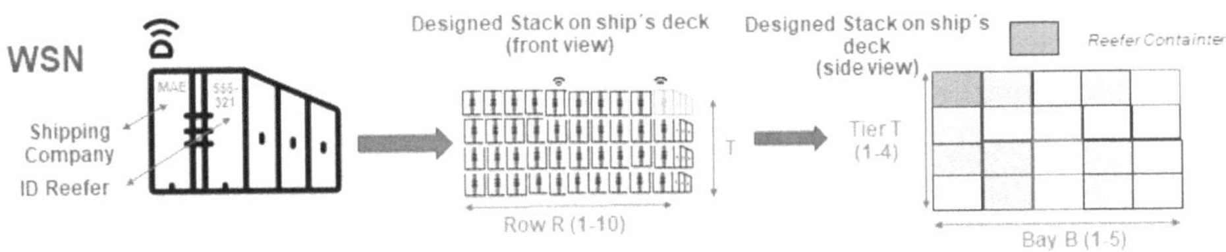

Figure 15. *Design of a container stack with the spatial location of the monitored reefer container.*

A stack of containers is designed as shown in Figure 15, in which 200 containers are arranged on the ship's deck using the spatial form of a cube. Figure 15 shows the placement and spatial form definition of the container stack:

- 10 rows of containers placed horizontally named " R ".
- 4 tiers of containers placed vertically named " T ".
- 5 bays of containers placed deeply named "B".

At the time of stowage (loading and connection of the reefer to the power line), the manual configuration at the application level of the Zigbee sensor placed in the container is carried out. To do this, in the different layers of the 802.15.4 standard, there is a payload and different headers (control and addressing) that are added to each defined layer, as can be seen in Figure 16:

The payload at the Application Sublayer level (ASDU -Application Service Data Unit-) has a variable capacity depending on the use or not of headers in other layers. Through the application and the monitoring of the sensor itself, a minimum space will be reserved so that the information shown in Figure 17 can be transmitted:

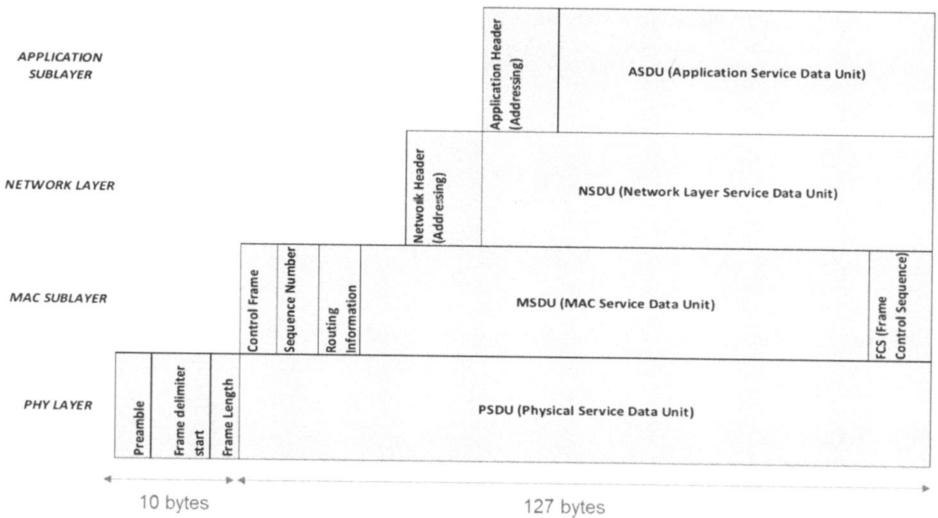

Figure 16. *Data Units in the layer structure.*

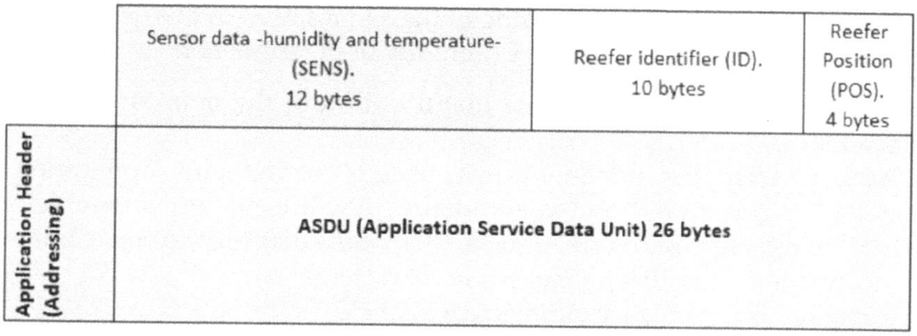

Figure 17. *Zigbee Application protocol frame bit reservation for container monitoring.*

1. Information received from the sensor referring to its humidity and temperature (SENS): A minimum of 12 octets (bytes) reserved in the ASDU payload, will be given by the information gathered from the sensors placed in the monitored reefer container. Although only 3 bytes in binary encoding are used for monitoring, a further 9 bytes are reserved for future use.

- Thermometer sensor (2 bytes):
- 1 byte: integer part of the temperature (range $-25°C, +25°C$). The most significant bit of the 8 bits will tell us if the temperature is negative or positive:
  - 1: positive value.
  - 0 : negative value.
- 1 byte: decimal part of the temperature.
- Hygrometer sensor (1 byte):
- 1 byte: the relative humidity (number from 0-100).

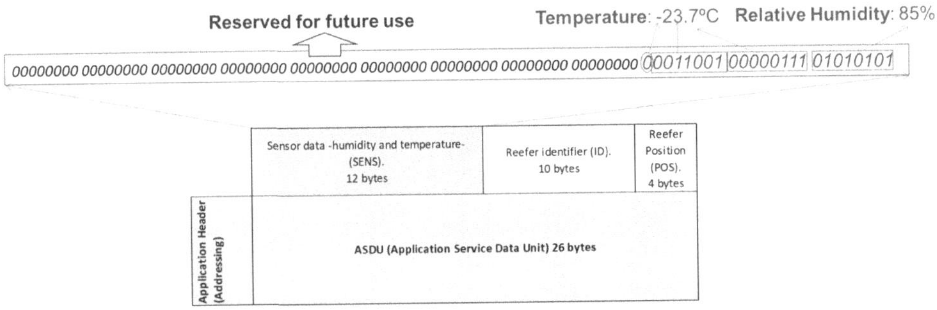

Figure 18. *Detail of the message with the temperature and relative humidity of the container.*

Example:

- Temperature: -23.7°C
  - $1byte$: $(-) : 0 \rightarrow 23$: 0011001
  - $1byte$: 7: 00000111
- Relative humidity: 85
  - 1 byte: 01010101

Figure 18 below shows the coding of the 3 bytes used in this example of monitoring the temperature and relative humidity of the container:

2. Information referring to the identification of the monitored reefer container (ID):

As seen before, the container license plate is entered in the application when the storage is carried out and is encoded in ASCII code (alphanumeric characters), devoting a maximum of 10 bytes (the plates that identify the container do not occupy more than 10 alphanumesric characters).

Container license plate example: MAE 555-321

- MAE: referring to the shipping company that owns the container (Maersk).
- 555-321: identification of the container.

In this way, Figure 19 shows the detail of the information referring to the identification of the container (ID) of the example exposed with the corresponding ASCII encoding:

3. Information referring to the spatial position of the monitored reefer container within a container type grouping:

To identify the position of the container within the different stacks of containers along the ship's deck/hold, 4 binary coded bytes are dedicated to helping an operator to locate it. For it:

- 1 byte: location of the stack of containers in order to identify it on the ship (up to 255 possible locations).
- 1 byte: row number of the stack of containers (R).
- 1 byte: bay number of the stack of containers (B).
- 1 byte: tier number of the stack of containers (T).

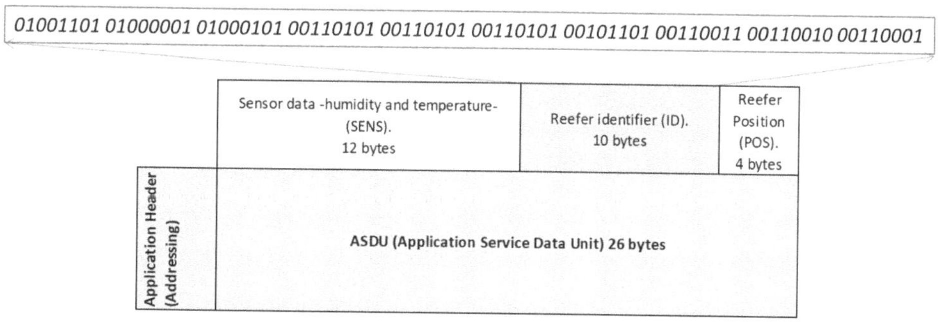

Figure 19. *Container ID Message Detail.*

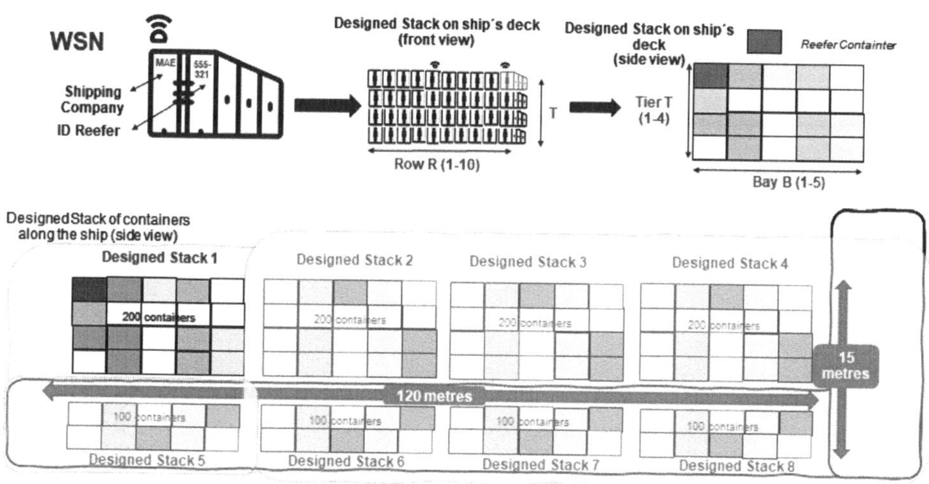

Figure 20. *Manual configuration in the storage of the container location.*

An example of the spatial location of the container is shown in Figure 20, and the following Figure 21 shows how the message is manually encoded in the ASDU during stowage to choose its position within its grouping.

In this example the following information travels:

- Number of a stack of containers location (1 byte): Stack 1: 00000001
- Row identification (R) (1 byte): Row 10: 00001010
- Bay identification (B) (1 byte): Column 1: 00000001
- Tier identification (T) (1 byte): Height 4: 00000100

Therefore, the container position message that will go in the ASDU payload will be as follows:

4. Information referring to identifying the services in the communication:

Once the communication protocol of each of the nodes that make up the WSN network has been defined, it will be explained below how this network will communicate with the WLAN network through WIFI5. It is established that, in each

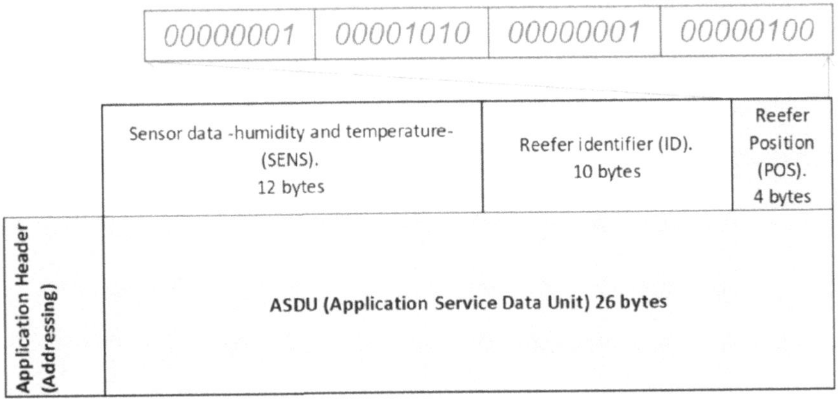

Figure 21. *Detail of the message with the position of the container on the ship.*

Frame information ("Application Service Data Unit")	
LSB encoding in payload	***Definition of data to transmit***
00	Information received from the sensor network
01	Information received from AIS
10	Information received from GPS

Figure 22. *Coding in IEEE 802.11ac payload the services provided.*

Gateway placed along the ship, depending on whether it receives information from the sensors or from the GPS or AIS service, this information is encapsulated and encoded according to the 802.11ac (WIFI5) standard, the first bits of the message data frame. In the case of encapsulating the sensor network data, it will be done by the specification of the WiFi standard such as Zigbee, adapting the network cards of the Gateways. Once encapsulated, an encoding is defined in the payload of the message at the application level. In this way, it will be possible to know what data from which services are being collected and issued, as illustrated in Figure 22:

6.3. Satellite Link

The satellite constellation has been tested using STK and MatLab software. We obtain different parameters such as visibility time and communication time between ship and satellite and carrier-to-noise ratio (CNR). With these results, we validate the requirements for services in the design.

Table 3 represents the main parameters of the satellite communications system that is used to provide round-the-clock coverage of the sea route.

Based on the MEO orbit constellation proposed in Section 3.3, Figure 23 depicts the 24-h access time to each of the constellation's satellites from any point along the maritime route. It can be seen that with the current configuration, access to at least one satellite is always available throughout the 24 h, ensuring 100% availability.

On the other hand, Figure 24 represents the histogram of the carrier to noise ratio (C/N) obtained during the visibility time for Satellite 11. It is observed that

Table 3. *Communications Satellite System Parameters.*

Orbit	MEO
Number of Satellites	20
Altitude	8063 km
Downlink Frequency	19.7 GHz
TX power	30 dBW
TX Gain	43.69 dB
Uplink Frequency	24 GHz
Latency	140 ms

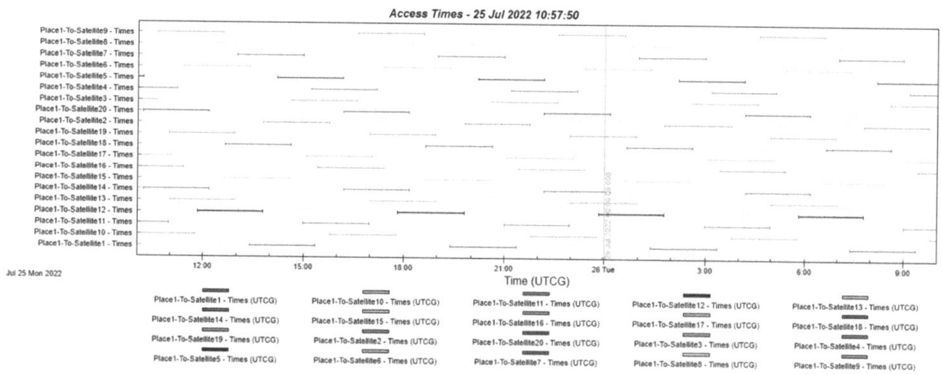

Figure 23. *Availability of service at any point along the maritime route.*

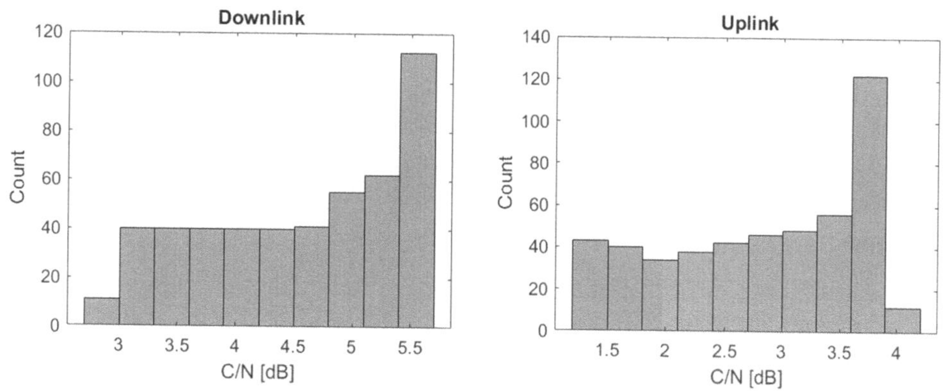

Figure 24. *Carrier to Noise Ratio histogram in the maritime route.*

in the downlink the C/N can vary from 2.4 to 5.7 dB, while in the uplink the range is from 0.4 to 4.2 dB.

7. CONCLUSIONS

In this work, an architecture based on three types of wireless networks has been proposed to solve several current problems in maritime transport, such as the transport of merchandise under special conditions (products under specific cold temperatures). The solution will be valid to provide a cold chain monitoring service through an IoT network supported by a constellation of MEO satellites.

The networks that make up the architecture are WSN, WLAN, and WWAN which use Zigbee, WiFi, and SatCom technologies respectively. In the proposed solution, the communication between the sensors within the WSN network has been detailed, forming an IoT network. In turn, the network of sensors communicates with the base station located on the deck of the ship via WiFi to subsequently communicate through the MEO satellite constellation with the control center on land. This communication with the ground segment is done through the classic maritime communications protocol, the AIS protocol. The proposed design presents the advantage in the throughput of the system by unifying existing protocols in maritime communications such as AIS with a new, more simplified protocol. In addition, we highlight the use of MEO satellites to reduce the latency that current systems based on GEO satellites produce.

On the other hand, the communications protocol and the messaging system have been defined to send the information from the sensors to the control center. This information within the service is the temperature and relative humidity of the containers. The messages defined in this solution are integrated into the AIS protocol messages to forward them through the satellite link. Unlike other works of the state of the art, the architecture proposed together with the protocol and communications messages allows updating the route of the ships so as not to lose the merchandise in the event that the cold chain is broken, updating the destination port.

The messaging system also offers the option of extending the messages in the future to include other physical parameters of the refrigerated containers, such as the levels of O_2/CO_2 in the container, and volumetric levels of the cargo transported. Therefore, Zigbee frames have space reserved for future expansion.

Future work will include further analysis and optimization of the satellite communications system, a more extensive study of LEO, MEO and GEO orbit compensation, and a more detailed study of rain attenuation.

Author Contributions

All authors contributed material to the paper based on their different research directions and projects. All authors contributed to the ideas that generated the original papers referenced herein. All authors contributed to writing and reviewing the manuscript. All authors have read and agreed to the published version of the manuscript.

Acknowledgments

The authors thank the project MegaLEO (C20/IS/14767486) and The STK-based simulations for the sizing of the MEO constellation have been possible thanks to the 5G-SpaceLab (https://5gspacelab.uni.lu/ (accessed on 15 July 2022)) funded by the Luxembourg Space Agency (LSA) and the Ministry of Economy (MECO) of the Government of the Grand Duchy of Luxembourg.

Abbreviations

The following abbreviations are used in this manuscript:

The following abbreviations are used in this manuscript:

AIS	Automatic Identification System
ASDU	Application Service Data Unit
C/N	carrier to noise ratio
CSMA/CA	Carrier sense multiple access with collision avoidance
CTS	Clear to Send
DIFS	Distributed Inter Frame Space
EMSA	Agencia Europea de Seguridad Marítima.
FDM	Frequency-division multiplexing
FEU	Forty-feet equivalent unit
GEO	Geostationary Earth Orbit
GNSS	Global Navigation Satellite Systems
GPS	Global Positioning System
HTS	High Throughput Satellite
IMO	Organización Marítima Internacional
INCOTERMS	International Commercial Terms
IoS	Internet of Ships
IoT	Internet of Things
IoT-MTS	IoT-improved MTS
ISO	International Standard Organization
IT	intelligent transportation
LEO	Low Earth Orbit
LoRa	Long Range
LPWAN	Low Power Wide Area
MEO	Medium Earth Orbit
MTS	maritime transport sector
NB-IoT	Narrow Band Internet of Thing
NNTT	Nuevas tecnologías aplicables al transporte
PAN	Personal Area Network Identification)
PLC	Power Line Communications
Ro-Ro	Roll on–Roll Off
RTS	Request to Send
SatCom	satellite communications
SIFS	Short Inter Frame Space
SOLAS	Safety of Life at Sea
TEU	Twenty-feet equivalent unit
TLA	Three letter acronym
UE	Unión Europea
UNCTAD	Conferencia de las Naciones Unidas sobre Desarrollo y Comercio
WLAN	Wireless Local Area Network
WSN	Wireless Sensor Network
WWAN	Wireless Wide Area Network

REFERENCES

1. Zdraveski, V.; Mishev, K.; Trajanov, D.; Kocarev, L. ISO-Standardized Smart City Platform Architecture and Dashboard. IEEE Pervasive Comput. 2017, 16, 35-43.
2. Zafeiriou, I. IoT and Mobility in Smart Cities. In Proceedings of the 3rd World Symposium on Communication Engineering (WSCE), Greece, Thessaloniki, 9-11 October 2020; pp. 91-95.
3. Talebkhah, M.; Sali, A.; Marjani, M.; Gordan, M.; Hashim, S.J.; Rokhani, F.Z. IoT and Big Data Applications in Smart Cities: Recent Advances, Challenges, and Critical Issues. IEEE Access 2021, 9, 55465-55484.
4. Ahwazi, A.Z.; Bordin, C.; Mishra, S.; Horsch, A.; Ha, P.H. Sustainable and decarbonized data-center facilities: A socio-technoeconomic discussion. In Proceedings of the IEEE PES Innovative Smart Grid Technologies-Asia (ISGT Asia), Brisbane, Australia, 5-8 December 2021; pp. 1-5.
5. Koelle, R.; Barbosa, F.L.C. Assessing the Global COVID-19 Impact on Air Transport with Open Data. In Proceedings of the IEEE / AIAA 40th Digital Avionics Systems Conference (DASC), San Antonio, TX, USA, 3-7 October 2021; pp. 1-8.
6. Cao, Y.; Xu, S.; Liu, J.; Kato, N. Toward Smart and Secure V2X Communication in 5G and Beyond: A UAV-Enabled Aerial Intelligent Reflecting Surface Solution. IEEE Veh. Technol. Mag. 2022, 17, 66-73.
7. Malik, R.Q.; Ramli, K.N.; Kareem, Z.H.; Habelalmatee, M.I.; Abbas, H. A Review on Vehicle-to-Infrastructure Communication System: Requirement and Applications. In Proceedings of the 3rd International Conference on Engineering Technology and its Applications (IICETA), Najaf, Iraq, 6-7 September 2020; pp. 159-163.
8. Huang, R. Maritime Intelligent Real-Time Control System Based on UAV. In Proceedings of the International Conference on Robots & Intelligent System (ICRIS), Amsterdam, The Netherlands, 21-23 February 2018; pp. 10-12.
9. Yang, T.; Cui, Z.; Alshehri, A.H.; Wang, M.; Gao, K.; Yu, K. Distributed Maritime Transport Communication System with Reliability and Safety Based on Blockchain and Edge Computing. IEEE Trans. Intell. Transp. Syst. 2022, 1-11.
10. Zhang, P.; Wang, Y.; Aujla, G.S.; Jindal, A.; Al-Otaibi, Y.D. A Blockchain-Based Authentication Scheme and Secure Architecture for IoT-Enabled Maritime Transportation Systems. IEEE Trans. Intell. Transp. Syst. 2022, 1-10.
11. Hu, J.; Kaur, K.; Lin, H.; Wang, X.; Hassan, M.M.; Razzak, I.; Hammoudeh, M. Intelligent Anomaly Detection of Trajectories for IoT Empowered Maritime Transportation Systems. IEEE Trans. Intell. Transp. Syst. 2022, 1-10.
12. Jing, H.; Gao, Y.; Shahbeigi, S.; Dianati, M. Integrity Monitoring of GNSS / INS Based Positioning Systems for Autonomous Vehicles: State-of-the-Art and Open Challenges. IEEE Trans. Intell. Transp. Syst. 2022, 1-22.
13. Jin, L.; Wang, L.; Jin, X.; Zhu, J.; Duan, K.; Li, Z. Research on the Application of LEO Satellite in IOT. In Proceedings of the IEEE 2nd International Conference on Electronic Technology, Communication and Information (ICETCI), Changchun, China, 27-29 May 2022; pp. 739-741.
14. Qu, Z.; Cheng, Y.; Zhang, G. Global Aggregated Traffic Model for LEO Satellite Constellation IoT Network. In Proceedings of the International Symposium on Advanced

Electrical and Communication Technologies (ISAECT), Rome, Italy, 27-29 November 2019; pp. 1-6.

15. Zong, L.; Wang, H.; Luo, G. Transmission Control Over Satellite Network for Marine Environmental Monitoring System. IEEE Trans. Intell. Transp. Syst. 2022, 1-8.
16. Yan, L.; Ding, X.; Zhang, G. Dynamic Channel Allocation Aided Random Access for SDN-Enabled LEO Satellite IoT. J. Commun . Inf. Netw. 2021, 6, 134-141.
17. Aslam, S.; Michaelides, M.P.; Herodotou, H. Internet of Ships: A Survey on Architectures, Emerging Applications, and Challenges IEEE Internet Things J. 2020, 7, 9714-9727.
18. IEEE 802.15.4-2006; IEEE Standard for Information Technology-Telecommunications and Information Exchange between Systems-Local and Metropolitan Area Networks—Specific Requirements Part 15.4: Wireless Medium Access Control (MAC) and Physical Layer (PHY) Specifications for Low Rate Wireless Personal Area Networks (LR-WPANs). IEEE Standard for Information Technology: Piscataway, NJ, USA, 2006.
19. Steenken, D.; Voß, S.; Stahlbock, R. Container terminal operation and operations research-A classification and literature review In Container Terminals and Automated Transport Systems; Springer: Berlin/Heidelberg, Germany, 2014; pp. 22 – 24.
20. IEEE Std 802.11ac-2013 (Amendment to IEEE Std 802.11-2012, as Amended by IEEE Std 802.11ae-2012, IEEE Std 802.11aa-2012, and IEEE Std 802.11ad-2012); IEEE Standard for Information Technology-Telecommunications and Information Exchange between Systems_Local and Metropolitan Area Networks-Specific Requirements-Part 11: Wireless LAN Medium Access Control (MAC) and Physical Layer (PHY) Specifications-Amendment 4: Enhancements for Very High Throughput for Operation in Bands below 6 GHz. IEEE Standard for Information Technology: Piscataway, NJ, USA, 2013; pp. 1-425.
21. ITU. BR IFIC (Space Services). Available online: https://www.itu.int/en/ITU-R/space/Pages/brificMain.aspx (accessed on 1 July 2022).
22. SES. IProven High-Performance NGSO Connectivity. Available online: https://www.ses.com/our-coverage/o3b-meo (accessed on 1 July 2022).

CHAPTER 3

Design and Experiment of Satellite-Terrestrial Integrated Gateway with Dynamic Traffic Steering Capabilities for Maritime Communication

Hyounhee Koo[1,2], Jungho Chae[2] and Wooseong Kim[1]

[1]*Department of Computer Engineering, Graduate School of Gachon University, 1342 Seongnamdaero, Sujeong-gu, Seongnam 13120, Republic of Korea*
[2]*SyncTechno Inc., Rm 914/918, 311 Gangnamdaero, Seocho-gu, Seoul 06628, Republic of Korea*

ABSTRACT

This study presents the architectural design and implementation of a multi-RAT gateway (MRGW) supporting dual satellite and terrestrial connectivity that enables moving maritime vessels, such as autonomous surface ships, to be connected to multiple radio access networks in the maritime communication environment. We developed an MRGW combining LTE and very-small-aperture terminal (VSAT) access networks to realize access traffic steering, switching, and splitting functionalities between them. In addition, we developed communication interfaces between the MRGW and end-devices connecting to their corresponding radio access networks, as well as between the MRGW and the digital bridge system of an autonomous surface ship, enabling the MRGW to collect wireless channel information from each RAT end-device and provide the collected data to the digital bridge system to determine the optimal navigation route for the autonomous surface ship. Experiments on the MRGW with LTE and VSAT end-devices are conducted at sea near Ulsan city and the Kumsan satellite service center in Korea. Through validation experiments on a real maritime communication testbed, we demonstrate the feasibility of future maritime communication technologies capable of providing the minimum performance necessary for autonomous surface ships or digitized aids to navigation (A to N) systems.

Keywords: *multi-rat; gateway; autonomous surface ship; maritime; MRGW; 5G-Advanced; 6G*

1. INTRODUCTION

Maritime communications using maritime frequency bands within the very high frequency (VHF) spectrum became the most widely-used communication system for vessels after Marconi's success in transatlantic communications with a transmission distance of over 3000 km in 1901 [1]. Currently, commercial maritime communication uses cellular communication based on LTE or 5G technology for vessels along the coast, where vessels are within the coverage of the cellular network. This is also applicable to onboard communication on luxury cruise ships or state-of-the-art ships such as small-size electric ships or autonomous surface ships [2].

Traditional maritime communication technologies may coexist with state-of-the-art technologies under two conditions: the long life expectancy of ships, which is usually $>$ 20 or 30 years, and the difference in countrywide communication infrastructures, which requires ships to be equipped with multiple communication interfaces for access networks permitted by a country whose ships visit during their navigation [3-5]. In addition, the international maritime organization (IMO) has recently completed a regulatory scoping exercise on Maritime Autonomous Surface Ships (MASS) [6]. The International Association of Marine Aids to Navigation and Lighthouse Authorities (IALA) is developing guidelines for the future development of maritime radio communication, including IMT-2020, which is applicable to their digital transformation, such as autonomous surface ships [7, 8]. Since the maritime sector was included in the standardization scope of 3rd Generation Partnership Project (3GPP) in December 2018, use cases and requirements related to the maritime sector have been considered in addition to other use cases and requirements from several vertical industries [9-12].

In such a maritime communication environment with heterogeneous communication technologies, the support of multiple radio access technologies (multi-RATs) is essential from the perspective of the upcoming digital transformation era in the maritime sector [13-18]. There are research works on the support of multi-RAT including LTE and Wi – Fi in terrestrial environments (e.g., in ultra-dense urban area, in smart cities, in vehicular or air-to-ground communications), but their outcomes were mainly based on the simulations and did not provide results based on the real experimental validation with the implemented equipment [19-23].

In this study, we design and implement a multi-RAT gateway (MRGW) with the capabilities of steering, switching, and splitting the access traffic, and provide experimental validation of the proposed MRGW prototype connected to LTE / VAT end-devices in the testbed of the Korean autonomous surface ship (KASS) at sea, near Ulsan city and the Kumsan satellite service center, Republic of Korea.

The key contributions of this study are summarized as follows:

- This study contributes to the literature by proposing the design of an overall system model applicable for the maritime communication environment where

an autonomous surface ship can exchange information with an on-land control center through an MRGW that enables intelligent routing among multi-RATs.
- Another contribution is the proposition of a detailed design of an MRGW with several key features adopted from open-source software and newly developed to monitor the radio channel status of each RAT end-device.
- Finally, this study demonstrates that the MRGW can enable the interworking between two heterogeneous RAT end-devices and support real-time video streaming through experiments on a real maritime testbed in three test scenarios.

This paper is organized as follows. Section 2 describe the related works and Section 3 introduces the overall multi-RAT supporting communication system model applicable to the maritime communication environment where autonomous surface ships exchange information with an on-land control center and neighboring ships or digital transform aids to navigation equipment using multi-RATs. Section 4 describes the design of the MRGW with an interface for each RAT end-device. Section 5 describes the system configurations for the field test environment and three test deployment scenarios, each of which comprises four or five steps for the experiment execution process. Section 6 analyzes the experimental outcome on the basis of test results. Finally, the conclusions and future research directions are discussed in Section 6.

2. RELATED WORKS

For the modernization of maritime communications, the transition to digital technologies has begun in the maritime sector and the existing systems are being enhanced and new technologies continue to emerge [4]. The VHF data exchange system (VDES) and Navigational Data (NAVDAT) were developed as maritime communication systems supporting the e-navigation concept and the modernization of the Global Maritime Distress and Safety System (GMDSS) [24-26].

The maritime autonomous surface ship (MASS) is also one of the main research streams for the digital transformation of the maritime sector [27-29] and there are ongoing works to enable advanced communication technologies including satellites, Unmanned Aerial Vehicles (UAVs), and IMT systems (e.g., 5G, 6G) with improved coverage and data rates to be applied for the MASS [30-34].

Satellite communication has been widely employed in the maritime sector because of the communication coverage and the accessible network infrastructure in the ocean, and traditionally provides limited applications due to the long propagation delay and affordable data rate. However, the recent technological advancement is ongoing for the integration of satellites into the 5G or 6G networks [35-38].

The support of the multi-RATs is one of the key requirements of several vertical industries, including the maritime sector, because of the heterogeneous networks. The multi-connectivity combining different access technologies (e.g., cellular, satellite, Wi-Fi) continue to be enhanced for seamless connection in diverse service scenarios over 5G or 6G systems [39-42].

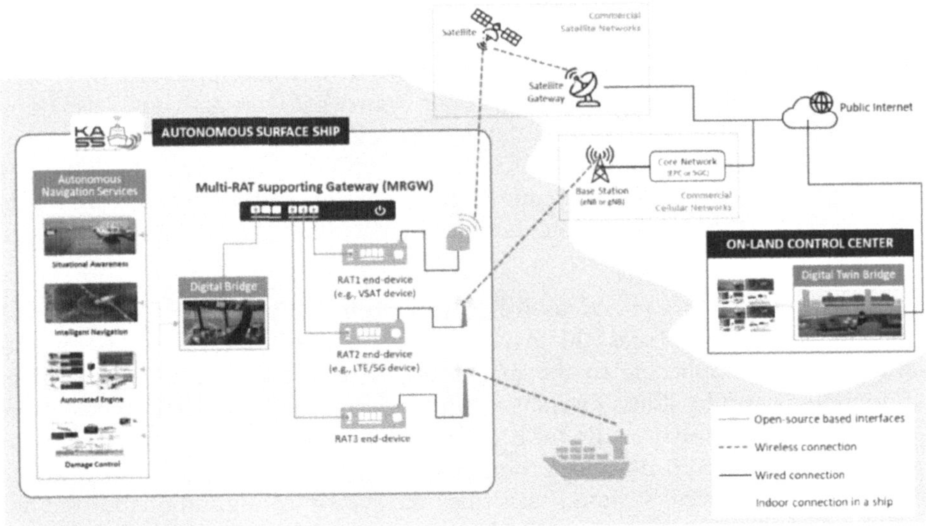

Figure 1. *System Model.*

3. SYSTEM MODEL

Our system model for the maritime communication environment is depicted in Figure 1. It enables an autonomous surface ship to exchange the information with an on-land control center via external communication systems accessible from a ship near the shore or at sea.

A commercial satellite access network (e.g., very-small-aperture terminal (VSAT) system) and a cellular access network (e.g., LTE system) are assumed to be accessible by each subscribed RAT end-device connected to the MRGW. In addition, the digital twin bridge of the on-land control center communicates with the digital bridge of the autonomous surface ship through the commercial satellite or cellular access network. This communication channel can be established through public Internet or a closed network (e.g., 5G private network) if available commercially or proprietarily deployed near a shore or at port.

The MRGW handles the acquisition of wireless channel information from each RAT end-device and the periodic provision of that information to the digital bridge. It provides the digital bridge of the autonomous surface ship with information about the best RAT at the current location that an autonomous surface ship is navigating. The MRGW provides two types of interfaces:

- An interface between the digital bridge and the MRGW to provide the wireless channel status measurement of each RAT end-device to the digital bridge, and to enable the MRGW to obtain the navigation information of the autonomous ship managed by the digital bridge;
- An interface between the MRGW and each RAT end-device, which depends on the communication protocol supported by each RAT end-device. As 5G-Advanced or 6G communication systems are expected to provide seamless

connection anywhere (e.g., in either the maritime or terrestrial sector), with the support of multiRATs, the current commercial communication networks will evolve accordingly [43-45]. Therefore, the MRGW is designed to be capable of providing interfaces to future RAT end-devices of evolved communication networks for its sustainability in the maritime communication environment where various types of ships exist (e.g., traditional ships with legacy radio communication systems such as VHF systems and autonomous surface ships with state-of-the-art radio communication systems such as 5G-Advanced or 6G system). The cyber security is assumed to be guaranteed by the firewall and VPN, etc., for the on-board communication as well as ship-to-shore communication.

This study focuses on the design of an MRGW supporting flexible and scalable interfaces with RAT end-devices using commercial communication networks (e.g., VSAT system or LTE system) and the public Internet.

4. DESIGN OF AN MRGW

4.1. Functional Architecture

An MRGW architecture consisting of a hardware system, an operating system and the MRGW platform is functionally designed, as shown in Figure 2. The open source Unix-like operating system, known as FreeBSD, provides various networking features [46]. For instance, a routing table is dynamically configurable using the MRGW platform through the template of the routing configuration. The MRGW platform implemented on top of the open source software supports the provision of basic networking services and the addition of any user-defined functionality as a plugin.

To support multi-RATs, two functionalities are realized at the MRGW platform. The first functionality is the wireless channel monitoring, developed as a plugin added to the operating system, consisting of following functional components:

- Status information collector of the wireless channel;
- Status information controller;
- Application programming interfaces (APIs) for interworking with third-party applications.

The second functionality is the routing functionality developed to support multi-RATs and the optimal routing selection among accessible RATs, consisting of the following functional components:

- User interface APIs (UI APIs) for MRGW control;
- Routing configuration controller.

In addition, a human machine interface (HMI) controller is developed for the display control in the MRGW hardware system. In addition, APIs for the interworking with third-party applications and UI APIs for MRGW control are developed, through which third-party applications (e.g., the user interface of the digital bridge of the autonomous surface ship) can control the MRGW to provide the

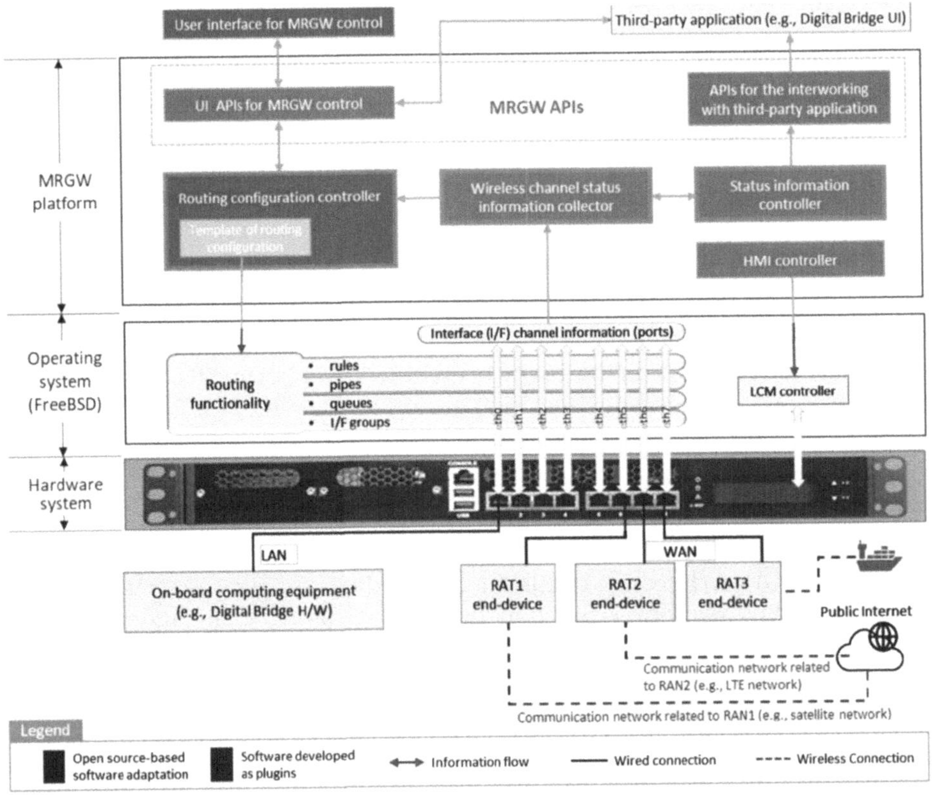

Figure 2. *MRGW functional architecture.*

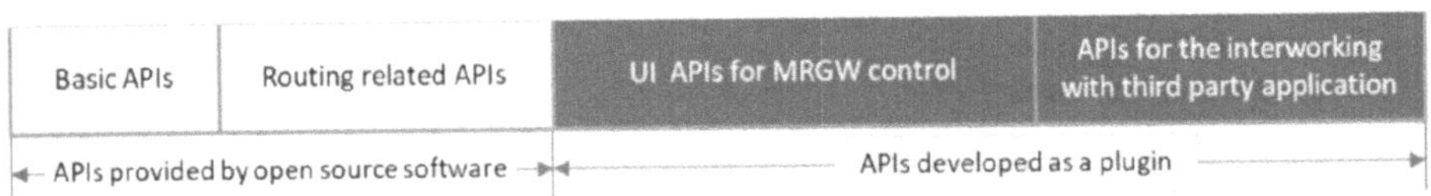

Figure 3. *API structure for MRGW access and control.*

wireless channel information of each RAT end-device and select the optimal RAT among the accessible RATs.

4.1.1. MRGW APIs

MRGW APIs provide programmable interfaces for accessing and controlling the MRGW to users and third-party applications. In addition to the various APIs provided by the open source software, the UI APIs for MRGW-control, based on the web interface and interworking APIs for external third-party applications, are provided. The overall API structure for the MRGW access and control is depicted in Figure 3 and is classified into two parts: APIs provided by the open source software and APIs developed as plugins.

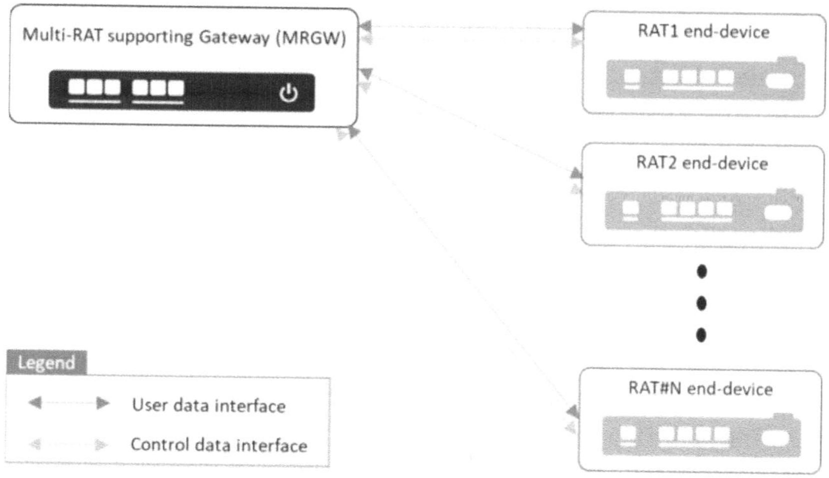

Figure 4. *Interfaces between the MRGW and each RAT end-device.*

4.1.2. Wireless Channel Monitoring Functionality

The wireless channel monitoring functionality is realized by two functional components; namely, the wireless channel status information collector and the status information controller in the MRGW platform, as depicted in Figure 2. Accordingly, the wireless channel status information is collected from each RAT end-device and the collected information is provided to other functional components, such as the routing configuration controller or a third-party application (e.g., digital bridge UI), as shown in Figure 2. The interfaces between the MRGW and each RAT end-device are illustrated in Figure 4. The representational state transfer (REST) API and simple network management protocol (SNMP) are used for exchanging the control data between the MRGW and an LTE end-device, as well as between the MRGW and a VSAT end-device, respectively.

4.1.3. Routing Functionality

The MRGW platform adopts routing functionalities supported by the FreeBSD operating system. In addition to those routing functionalities supported by the open source software, the routing configuration controller is implemented as a functional component with several features, such as traffic shaping and traffic steering operations, as shown in Figure 5, to manage user data traffic.

The routing configuration controller receives the wireless channel status information of each RAT and processes this information to select the optimal RAT in the maritime communication environment. The data analysis result provided by the routing configuration controller is delivered to the UI APIs for MRGW control. Furthermore, the basic routing functionality supported by the FreeBSD operating system can be controlled on the MRGW platform by updating the routing configuration information of the FreeBSD operating system through the routing configuration controller's routing configuration template.

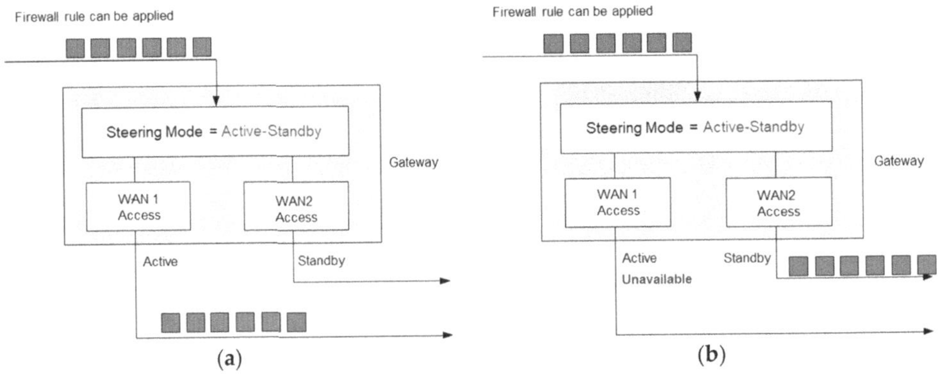

Figure 5. *An example of traffic steering operation on Active Standby (a) Active access available; (b) Active access unavailable.*

4.2. Environment for System Design Verification

The system design for the verification of the MRGW is depicted in Figure 6. The MRGW's operation and performance are verified using the DX Ocean TM, remote monitoring software, and packet capturing software.

To verify the MRGW's operation and performance, a remote monitoring procedure is implemented using the DX Ocean TM, which enables the information collected from, or created by, the MRGW (e.g., wireless channel status information measured from each RAT end-device or MRGW status information) to be stored in the onboard system as well as a Cloud-based on-land data hub when the communication network is connected between an autonomous surface ship and on-land systems. According to the service flow for remote monitoring, shown in Figure 6, the operation and performance of the MRGW and each RAT end-device can be verified through the visualized data using the remote monitoring software.

To verify the data transmission and reception between the digital bridge of the autonomous surface ship and the digital twin bridge of the on-land control center, the packet capture software sniffs the data exchanged between the digital bridge of the autonomous surface ship and the digital twin bridge of the on-land control center. The packet capture software executes always-on sniffing packets which continuously forward to the DX Ocean TM (Cloud-based on-land data hub). In cases of the stored data being transmitted from the DX Ocean TM (onboard equipment), 30 s is configured as the period of the data transmission from the DX ocean TM (onboard equipment) to the DX ocean TM (Cloud-based on-land data hub). According to the service flow for data verification, shown in Figure 6, whether all of the packets are received is determined by comparing the number of transmitted packets and received (sniffed) packets, and it is also possible to determine whether the MRGW enables the data to be exchanged between the digital bridge and the digital twin bridge using the remote monitoring software, which compares the well-known training data sent from the digital bridge and the sniffed data received by the digital twin bridge for data verification.

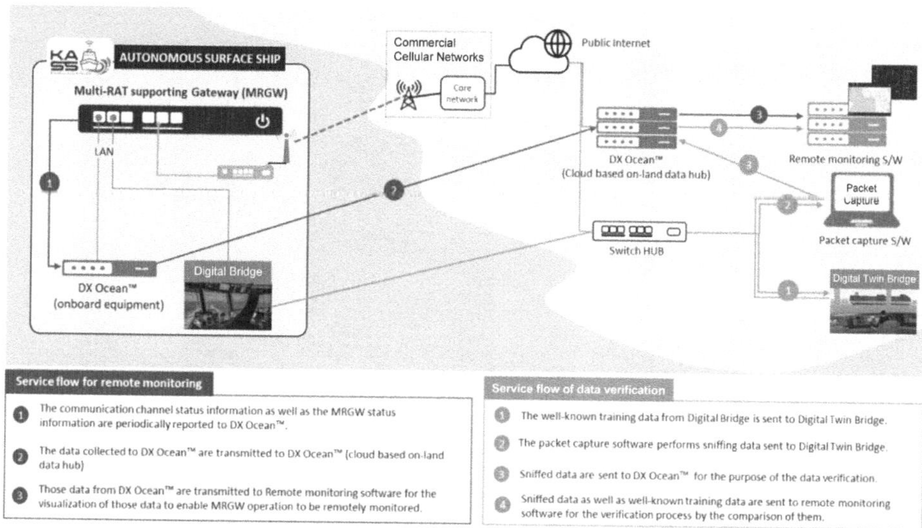

Figure 6. *System design for verifying the MRGW's operation and performance.*

5. EXPERIMENTAL VALIDATION

5.1. Field Experiment Configuration

A field experiment was conducted to verify the operation and performance of the MRGW connected with an LTE end-device and an LTE device for testing in terms of traffic steering between different RATs in the testbed of the autonomous surface ship, known as KASS, plying along the navigation route next to the JangSaengPo port in Ulsan, Korea, as depicted in Figure 7.

Another field experiment was conducted to verify the operation and performance of the MRGW connected with a VSAT end-device and an LTE end-device in terms of the latency between the different RATs at the KTSAT Kumsan satellite service center in Kumsan, Republic of Korea, as shown in Figure 8 [47]. The VSAT end-device is connected to Republic of Koreasat-6, which is a geostationary satellite operated by KTSAT, with an orbit height of 35,786 km. There is a limitation on the available bandwidth (i.e., 1Mbps) due to the high cost of the satellite service during the field experiment, so the main purpose of the field experiment is to check the operation of the MRGW and the latency of the satellite communication link.

For the field experiment, an LTE end-device developed using the LTE and Nvidia Jetson Nano modules is used to enable the MRGW to extract wireless channel status information (e.g., Reference Signal Received Power (RSRP) and Reference Signal Received Quality (RSRQ)) from the LTE end-device. Figure 9 shows the specification of the LTE module and the main board of the LTE end-device.

5.2. Test Scenarios and Experiment Outcome Analysis

Three system configurations are considered to test the operation and performance of the MRGW. For the experimental validation, the expected potential operation errors or the breakdown of the MRGW are considered when each experiment

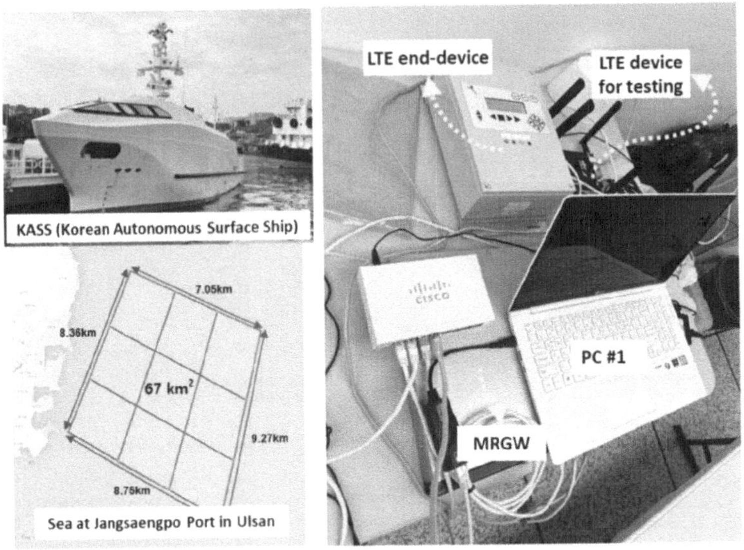

Figure 7. *Experimental environment for testing MRGW and LTE end-device in the testbed of KASS at Jangsaengpo port in Ulsan, Republic of Korea.*

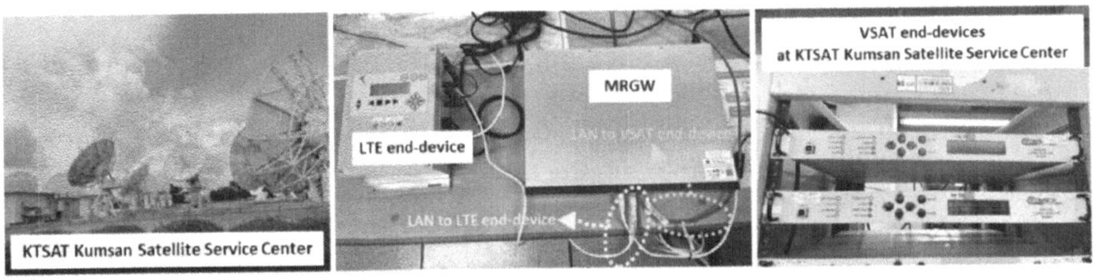

Figure 8. *Experimental environment of MRGW interworking between LTE end-device and VSAT end-device at the KTSAT Kumsan satellite service center, in Kumsan, Republic of Korea.*

execution process for three test scenarios is developed to enable such errors to be identifiable during the experiment by several methods (e.g., testing tools with ping or iperf to identify a hardware malfunction or a configuration error, testing with the DX Ocean TM for the data integrity check). The overall system configuration and experiment outcome analysis for the three test scenarios are provided in following subsections.

5.2.1. Test Scenario 1 for Testing the Switchover Performance of MRGW

Test scenario 1 for testing the switchover performance of the MRGW is to determine whether the MRGW can switch from an active RAT end-device to another standby RAT end-device within 1 s once it detects that the active RAT end-device is disconnected.

The overall system configuration for test scenario 1, described in Figure 10, shows the allocated IP addresses used for the validation experiment. The data used

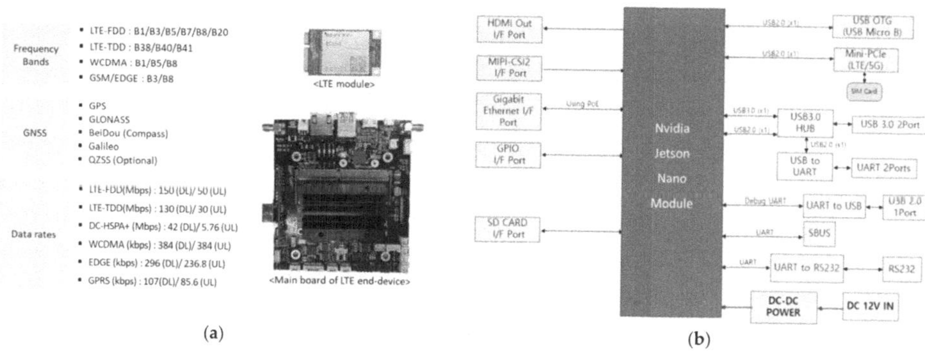

Figure 9. *LTE end-device-(a) LTE module specification and the main board of LTE end-device; (b) Block-diagram of LTE end-device.*

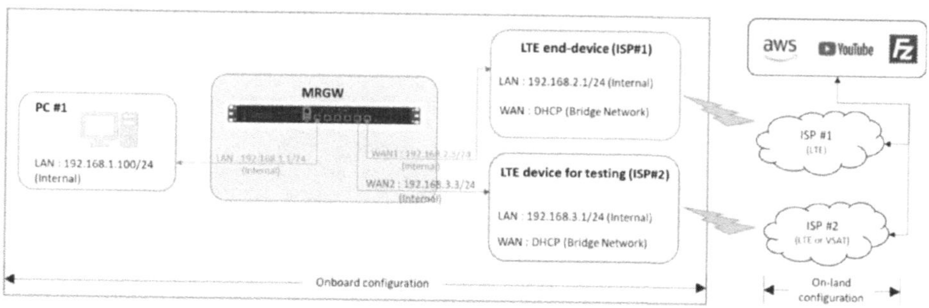

Figure 10. *System configuration for the test scenario 1 for evaluating the switchover performance of MRGW.*

for the test scenario are ICMP (Ping) messages generated at PC #1 every second and a public DNS address (i.e., Google public DNS: 8.8.8.8) used as the targeted IP address of the ICMP message.

The overall system configuration for test scenario 2, illustrated in Figure 12, shows the allocated IP addresses used for the experimental validation. The Chrome browser is used to access the video room application of the Web-RTC server (Janus) at PC #1 with a camera, and PC #2 with a camera and real-time video streaming data produced from the camera are sent out and received by PC #1 and PC #2, respectively, via the Web-RTC server (Janus).

The switchover performance of the MRGW is verified by performing the following experiment execution processes.

Step 1: The active-standby mode, shown in Figure 5, is configured through the MRGW, and the two RAT end-devices connected to the MRGW are the LTE end-device (ISP#1) and LTE end-device for testing (ISP#2).

Step 2: PC #1 executes the ping program to generate an ICMP message, which is sent to the public DNS address (8.8.8.8), and the information containing the start time of the transmission of the ICMP message is identified on

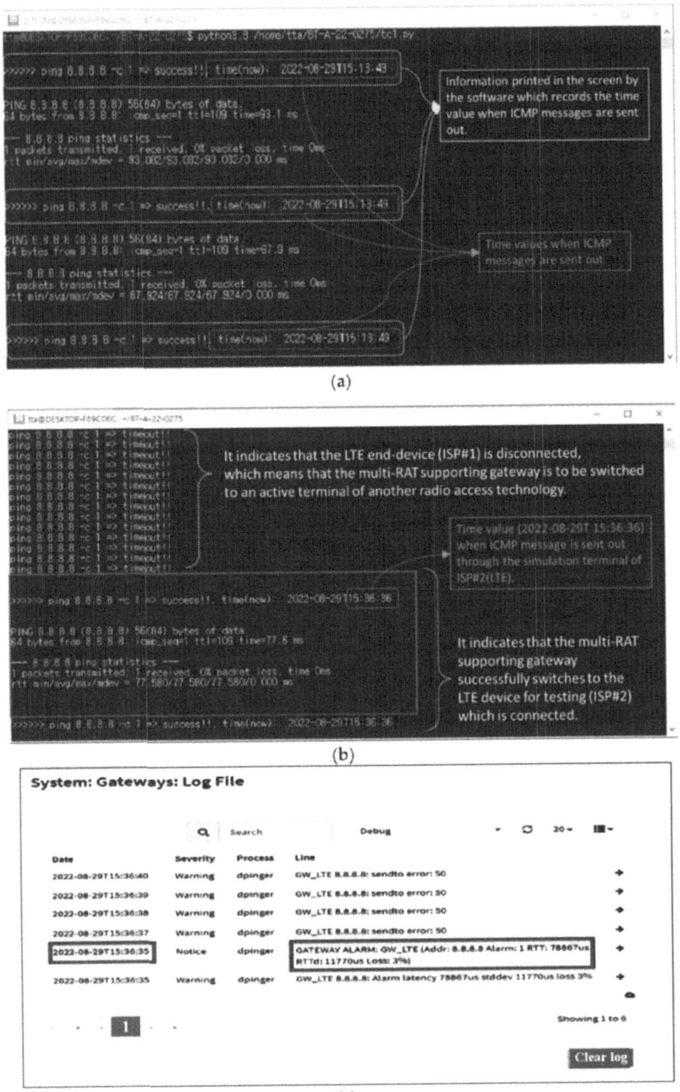

(c)

Figure 11. *Experiment outcome of test scenario 1-(a) Screenshot related to Step 2 of the experimental execution process; (b) Screenshot related to Step 4 of the experimental execution process; (c) Time value identified from the configuration menu of the MRGW when 'GATEWAY ALARM occurs because of the disconnection of LTE end-device (ISP#1).*

the screen of PC #1 by the software that records the time when the ICMP messages are transmitted (Figure 11a).

Step 3: : The power of the LTE end-device (ISP#1) is turned off to create a field experimental environment, where the LTE end-device (ISP#1) is disconnected.

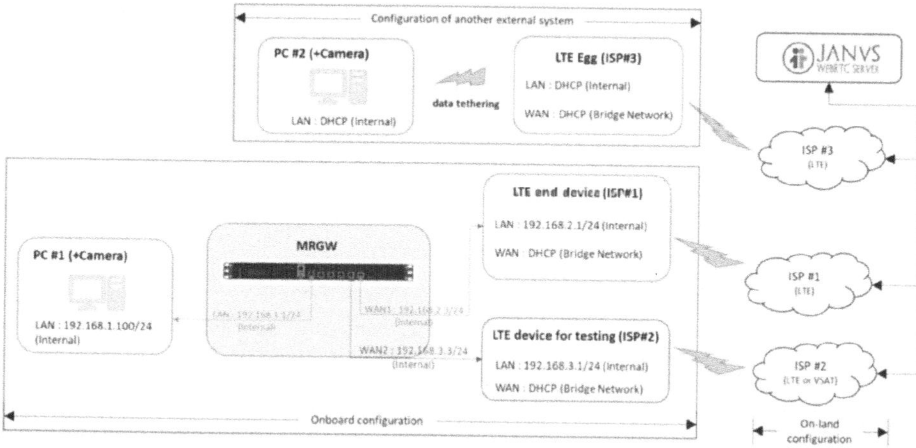

Figure 12. *System configuration for test scenario 2 for evaluating the support for real-time video streaming data provided by the MRGW's operation.*

Figure 13. *Experiment outcome of test scenario 2 (Chrome browser screenshot of the video room through Web-RTC server (Janus)).*

Step 4: The start time of the transmission of the ICMP messages is recorded and identified on the screen of PC #1 once the connection is re-established through the LTE device for testing (ISP#2), instead of the inaccessible LTE end-device (ISP#1) (Figure 11b).

Step 5: The time when the 'GATEWAY ALARM' is generated is recorded. This time is identified from the system/gateway/logfile of the user interface for MRGW control, shown in Figure 2 (192.168.1.1) (Figure 11c).

The MRGW's operation for supporting real-time video streaming data is verified by performing the following experiment execution processes.

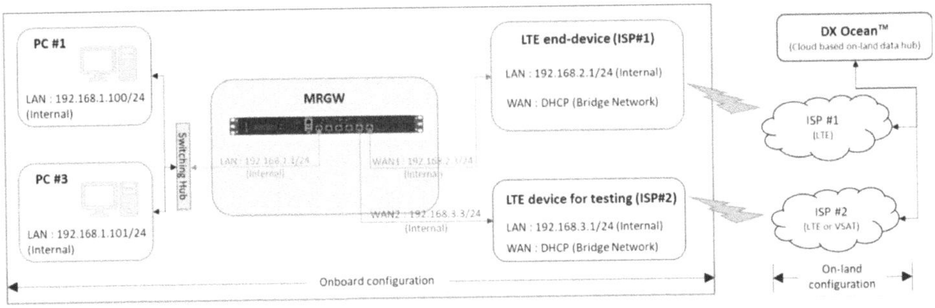

Figure 14. *System configuration for test scenario 3 for evaluating the operation of MRGW in interworking with DX Ocean* ™.

Step 1: The active-standby mode, shown in Figure 5, is configured using the MRGW, and the two RAT end-devices connected to the MRGW are the LTE end-device (ISP#1) and the LTE device for testing (ISP#2) for the interworking test.

Step 2: PC #1 accesses the Web-RTC server (Janus) through the Chrome browser and activates the camera and microphone to produce real-time video streaming.

Step 3: PC #2 accesses the Web-RTC server (Janus) through the Chrome browser and activates the camera and microphone to produce real-time video streaming.

Step 4: PC #1 and PC #2 determine whether the real-time video streaming produced from the activated camera and microphone is displayed through the video room of the Web-RTC server (Janus) with no service interruption.

The outcome of the experiment for evaluating the MRGW's operation for supporting real-time video streaming data is depicted in Figure 13, which includes two video streaming images of the video room application of the Web-RTC server (Janus). The left image is captured from the camera of PC #1 and the right one is captured from the camera of PC #2. They are displayed from the video room of the Web-RTC server (Janus) at PC #1 because the MRGW successfully exchanges the video streaming data via the LTE end-device (ISP#1). Based on the experiment outcome, we verified that the MRGW successfully ensured that the real-time video streaming was provided through the Web-RTC server (Janus).

The outcome of the experiment to evaluate the switchover performance of MRGW is presented in Figure 11. The time is printed every time an ICMP message is sent, and Figure 11a shows the examples of the times (blue dotted lines) that were printed every time an ICMP message was sent via the LTE end-device (ISP#1). As soon as the ISP#1 (LTE) end device is disconnected, the MRGW conducts the switchover from the ISP#1 (LTE) end device to the LTE device for testing (ISP#2). Once the MRGW is connected with the LTE device for testing (ISP#2), the time for the successfully sent ICMP message starts to be printed and the first time of an ICMP message sent via the LTE device for testing (ISP#2) is 36 s36 min15 h

(a)

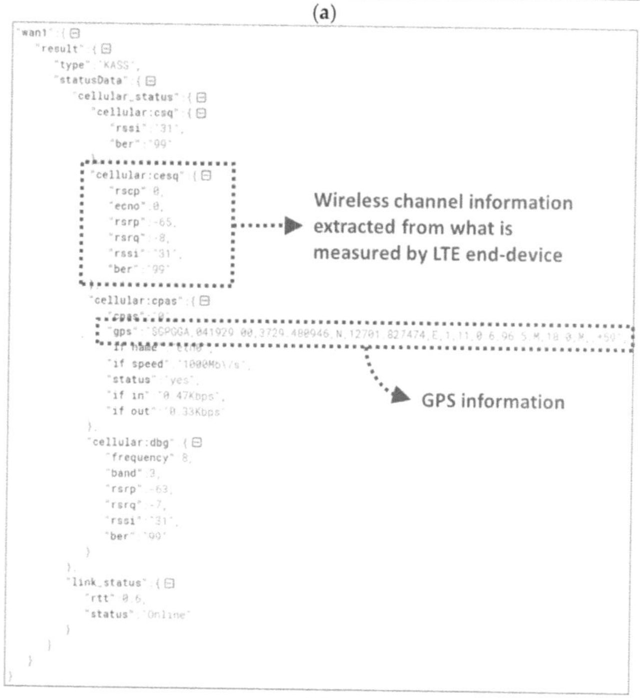

(b)

Figure 15. *Experiment outcome of test scenario 3 (a) DX Ocean* ™ *software screenshot (b) Sample of ISP#1(LTE) wireless channel status information collected from the MRGW and stored at DX Ocean* ™ *(onboard equipment and cloud-based on-land data hub).*

on 29 August 2022, as shown in Figure 11b. The time at which the LTE end-device (ISP#1) is disconnected can be identified only from the user interface for the control of the MRGW by the unit of the second. The time of the disconnection from the LTE end-device (ISP#1) is 35 s36 min15 h on 29 August 2022, identified as 2022-08-29T15:36:35, from the user interface for the MRGW control, as shown in Figure 11c. As the MRGW lost the connection from the LTE end-device (ISP#1) at 35 s36 min15 h and the MRGW was again connected to the LTE device for testing (ISP#2) at 36 s36 min15 h, we verified that the MRGW successfully switched the

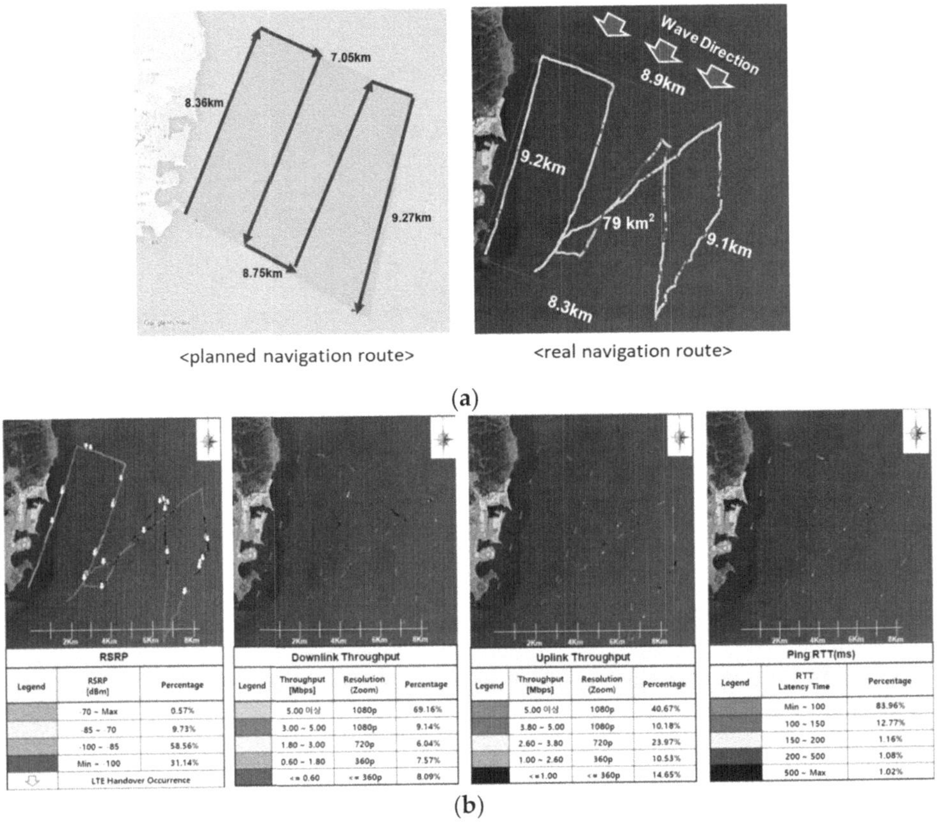

Figure 16. *Experiment outcome of LTE end-device (a) navigation route in front of JangSaengPo port in Ulsan, Republic of Korea; (b) measured LTE performance.*

connection from the LTE end-device (ISP#1) to the LTE device for testing (ISP#2) within 1 s.

5.2.2. Test Scenario 2 for Evaluating the Support for Real-Time Video Streaming Data

Provided by MRGW OperationTest scenario 2 for evaluating the support for the real-time video streaming data provided by the MRGW's operation is to determine whether the MRGW can establish a connection with an accessible targeted RAT end-device and transmit and receive real-time video streams through the web-RTC server (Janus).

5.2.3. Test Scenario 3 for Evaluating the Operation of MRGW in Interworking with the DX Ocean ™

Test scenario 3 for evaluating the operation of the MRGW in interworking with the DX Ocean ™ is to determine whether the MRGW interworks with the DX Ocean ™; the information collected by the DX Ocean ™ (onboard equipment)

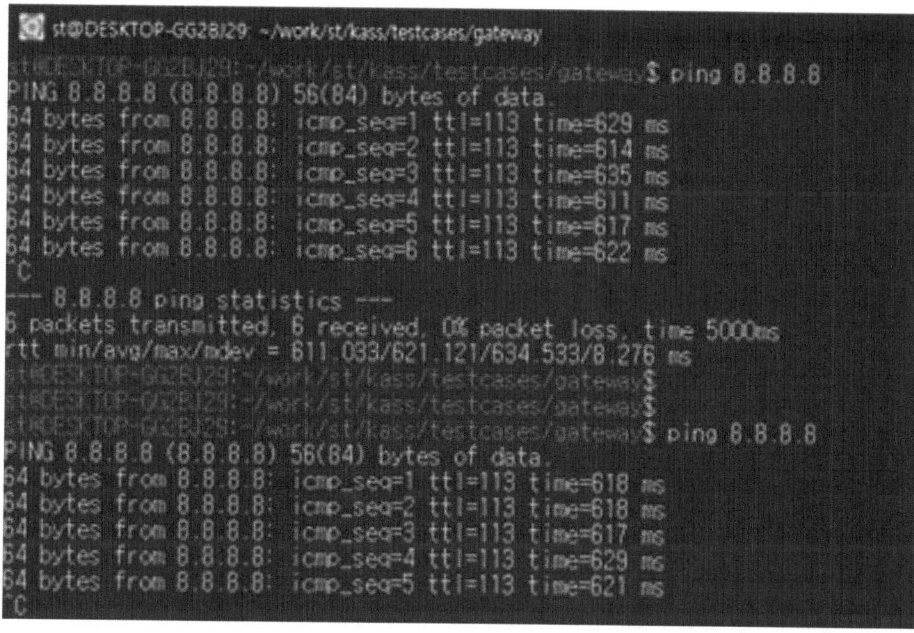

Figure 17. *Round trip time measured from VSAT end-device at KTSAT Kumsan satellite service center, in Kumsan, Republic of Korea.*

deployed in the KASS is sent out and delivered to the DX Ocean TM (Cloud-based on-land data hub) through the MRGW and each RAT end-device. The overall system configuration for test scenario 3, illustrated in Figure 14, shows the allocated IP addresses used for the experimental validation. The API related to the wireless channel status information collector, shown in Figure 2, is assumed to be identified and the DX Ocean TM software is installed and executed at PC #1.

The operation of the MRGW in interworking with the DX Ocean TM is verified by performing the following experiment execution processes.

Step 1: The active-standby mode, shown in Figure 5, is configured using the MRGW, and the two RAT end-devices connected to the MRGW are the LTE end-device (ISP#1) and the LTE device for testing (ISP#2) for the interworking test.

Step 2: The information related to the configuration of the MRGW is identified from the menu of the user interface for MRGW control, depicted in Figure 2.

Step 3: The DX Ocean TM software at PC #1 stores the information collected from the API related to the wireless channel status information collector, shown in Figure 2, and the information regarding a ship's location obtained from the positioning system deployed in the ship.

Step 4: The information stored in the DX Ocean TM (onboard equipment) is transmitted to the DX Ocean TM (cloud-based on-land data hub) through

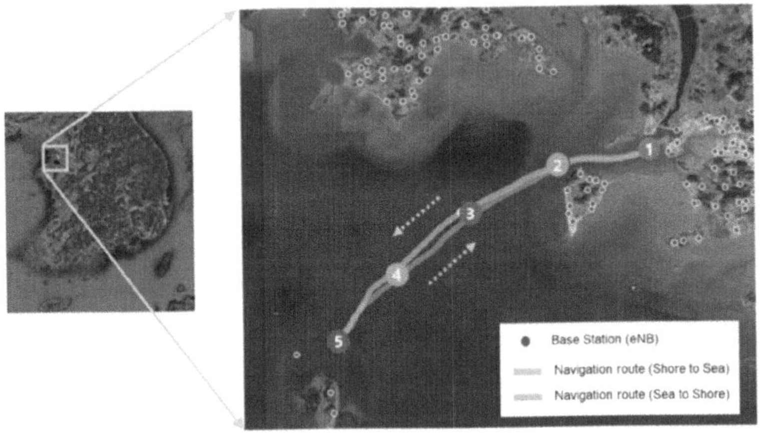

(a)

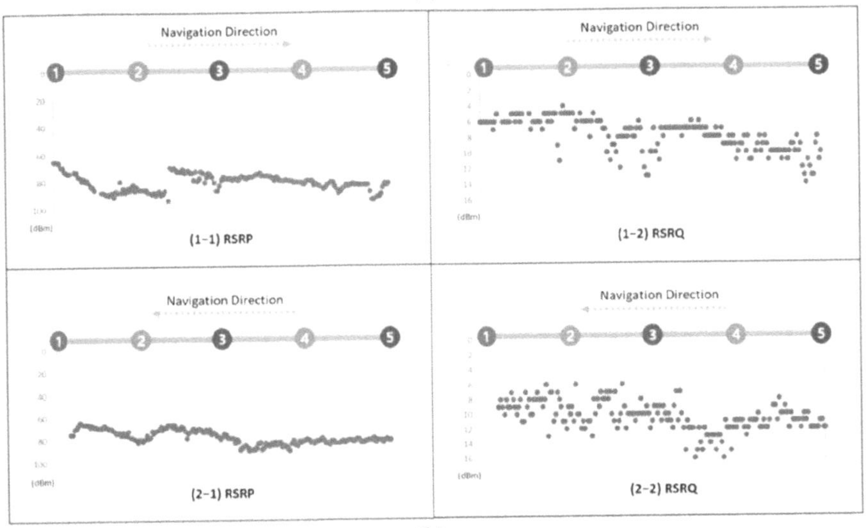

(b)

Figure 18. *Experimental outcome of LTE end-device (a) navigation route in front Jeongok Port in Hwasung city, Republic of Korea; (b) measured LTE performance (RSRP, RSRQ).*

the MRGW via either the LTE end-device (ISP#1), if accessible, or the LTE device for testing (ISP#2), if accessible.

The experiment outcome for test scenario 3 is shown in Figure 15. As shown in Figure 15a, the DX ocean TM shows the ship location based on the GPS information which the MRGW obtains from the LTE end-device (ISP#1) and subsequently sends it to the DX Ocean TM (onboard equipment). The DX Ocean TM (onboard equipment) sends the stored data out to the DX Ocean TM (cloud-based on-land data hub) every 30 s. The period of the data sent out from the DX Ocean TM (onboard equipment) to the DX Ocean TM (cloud-based on-land data hub) is configurable. Figure 15 b is the sample of collected data related to the wireless channel status information from the MRGW. Based on the experiment outcome, we found that the information collected from the MRGW is successfully stored in the DX Ocean ^{TM} (onboard equipment and cloud-based on-land data hub).

5.3. Experimental Outcome of Wireless Performance

The wireless performance of the LTE end-device is shown in Figure 16. The real navigation route was revised compared to the planned navigation route because of the wave direction due to the worse weather condition at sea for the reduction in the safety risk during the testing. In terms of the RSRP, 68.86% of the total navigation route shows values ≥ -100dBm. In terms of the downlink throughput, 92% of the total navigation route shows values $\geq$ 0.6Mbps. In terms of the uplink throughput, 85% of the total navigation route shows values $\geq$ 1Mbps. In terms of the latency, 98% of the total navigation route shows values $\leq$ 200 ms.

The wireless performance of the LTE end-device was inferior compared to that of the wireless performance of the LTE end-devices in urban areas. However, the performance may have been hampered by the operator policy in terms of the difference in network operation between urban and maritime/rural areas.

The round trip time measured from the VSAT end-device is shown in Figure 17. The worst case round trip time propagation delay of a GEO satellite with transparent payload is 541.46 ms [48], but the measured round trip time of the satellite was more than 610 ms during the testing at the KTSAT Kumsan satellite service center in Kumsan, Republic of Korea.

An additional experiment was performed using the fishing boat in front of the Jeongok Port at the west sea of Republic of Korea and the MRGW collected the wireless channel status information (i.e., RSRP, RSRQ, GPS information) measured from the LTE end-device. The wireless performance measured from the LTE end-device is shown in Figure 18. As shown in Figure 18a, the navigation route is identically selected for the shore-to-sea direction and the sea-to-shore direction to have the effect of the repeated experiments. The weather condition was good, and it was a sunny day, which means that waves did not cause any impact to degrade the wireless performance at sea. As shown in Figure 18b, two of the graphs (i.e., (1-1) and (2-1)) have a similar RSRP pattern and another two graphs (i.e., (1-2) and (2-2)) also have a similar RSRQ pattern. Based on the experiment outcome, we can assume that little change occurs to the wireless channel status in cases of a sunny day, which means that our experimental validation may be meaningful despite the limited number of the experiments.

6. CONCLUSIONS

Herein, we proposed a design for a novel MRGW and conducted corresponding experiments to verify the operation and performance of the MRGW based on three test scenarios with different system configurations. The experiment outcomes indicate that the MRGW has the capability of providing the seamless connection for an autonomous surface ship through multiple RAT end-devices in the maritime communication environment. Based on the MRGW's capability, proven by the experimental validation in the real maritime communication environment, it is important to continue to enhance the MRGW to support new upcoming end-devices (e.g., 5G end-device supporting both cellular access and satellite access) based on more evolved RATs, such as 5G-Advanced and 6G. Our research is expected to be continuously carried out during the second step of the KASS project [49], from the year 2023 to the year 2025, to enhance the functions of the MRGW and address any abnormal potential errors by applying assurance framework, such as a FMECA (Failure modes, effects, and criticality analysis) mechanism, and performing more experimental validations for three years in the testbed of autonomous surface ships in Ulsan, Republic of Korea. IMT-based technologies continue to be evolved to support the requirements of the vertical industries, including the maritime sector. The service requirements related to the digital transformation of the maritime sector, including the autonomous surface ship, are important to be provided during the global standardization of 5G-Advanced and beyond in 3GPP. Subsequent future works, as well as the research outcome of this paper, are expected to be leveraged to identify the gap analysis of existing commercial communication technologies and develop maritime requirements that should be satisfied by 5G or 6G systems that are possible to be employed as the fundamental communication platform for the digitalization and mobilization of the maritime sector.

In the near future, an end-device with both satellite and cellular access (e.g., 5G or LTE) is expected to be launched in markets as a 5G satellite was first specified to be integrated into 5G systems in 3GPP specifications in March 2022. However, further research is needed to enhance the capabilities of the current 5G satellite, particularly in terms of support for low earth orbit satellites, which are always in motion, not stationary. New opportunities and challenges for improving the MRGW can be found in research areas related to the support of multiple radio accesses, including low earth orbit satellites integrated into the 5G system.

The digital bridge system controlling the navigation of an autonomous surface ship and its counterpart on-land control center are to be deployed and operated 24/7/365 in the near future and the quantity of data is expected to increase rapidly. The data generated for the experimental verification are not the same as the data expected to be generated during the real operation of the digital bridge of the autonomous surface ship and the on-land control center. Additional research on the enhancement of the MRGW needs to be conducted to address how to flexibly manage the uplink and downlink bandwidth because the bandwidth allocated for the uplink transmission is generally lower than that of the downlink transmission in commercial cellular networks.

Author Contributions

Concept design, H.K. and J.C.; formal analysis, H.K. and W.K.; investigation, H.K. and J.C.; writing-original draft preparation, H.K.; writing-review and editing, H.K. and W.K.; visualization, H.K.; supervision, W.K.; project administration, H.K. All authors have read and agreed to the published version of the manuscript.

Acknowledgments

This research was supported by the "Development of Autonomous Ship Technology (20200615)" funded by the Ministry of Oceans and Fisheries (MOF, Republic of Korea).

REFERENCES

1. Marconi, G. Radio telegraphy. J. Am. Inst. Electr. Eng. 1922, 41, 561-570.
2. Mashino, J.; Tateishi, K.; Muraoka, K.; Kurita, D.; Suyama, S.; Kishiyama, Y. Maritime 5 G experiment in windsurfing World Cup by using 28GHz band massive MIMO. In Proceedings of the IEEE International Symposium on Personal, Indoor and Mobile Radio Communications, Bologna, Italy, 9-12 September 2018; pp. 1134-1135.
3. Modern Maritime Communications, ITU WRS 2016. Available online: https://www.itu.int/dms_pub/itu-r/md/15/wrs16/sp/R15-WRS16-SP-0026!!PDF-E.pdf (accessed on 1 December 2022).
4. Modern Maritime Communications, ITUWRS 2020. Available online: https://www.itu.int/en/ITU-R/seminars/wrs/2020/Plenary%20Sessions%20%20Presentations/03.%20Terrestrial%20Services%20-%201%20to%203%20Dec%202020/WRS20%20Modern%20Maritime%20Communications.pdf (accessed on 1 December 2022).
5. International Convention for the Safety of Life at Sea (SOLAS). 1974. Available online: http://www.imo.org/About/Conventions/ListOfConventions/Pages/International-Convention-for-the-Safety-of-Life-at-Sea-(SOLAS),-1974.aspx (accessed on 1 December 2022).
6. IMO. Outcome of the Regulatory Scoping Exercise for the Use of Maritime Autonomous Surface Ships (MASS). IMO MSC#103.May 2021. Available online: https://wwwcdn.imo.org/localresources/en/MediaCentre/PressBriefings/Documents/MSC.1Circ.1638-OutcomeOfTheRegulatoryScopingExerciseForTheUse-OfMaritimeAutonomous%20Surface%20Ships...%20(Secretariat).pdf (accessed on 1 December 2022).
7. IALA. Maritime Radio Communications Manual (MARCOM MANUAL). IALA ENAV30-5.1.2.5.1. September 2022. Available online: https://www.iala-aism.org (accessed on 1 December 2022).
8. IALA. Guideline on Developments and Implications of Maritime Autonomous Surface Ships for Coastal Authorities. IALA ENAV30- MTF04-8.1. September 2022. Available online: https://nextcloud.iala-aism.org/index.php/apps/files/?dir=-/Committees/ENAV/ENAV30/WG2/Task%204.1.1_4.1.2_4.2.1MASS&fileid=189934. (accessed on 1 December 2022).
9. 3GPP TR 22.819 Feasibility Study on Maritime Communication Services over 3GPP System; Stage 1. Available online: https://www.3gpp.org/ftp//Specs/archive/22series/22.819/22819-g20.zip (accessed on 1 December 2022).
10. 3GPP Release 16 TS 22.119 Maritime Communication Services over 3GPP System; Stage 1. Available online: https://www.3gpp.org/ftp/Specs/archive/22\series/22.119/22119-h00.zip (accessed on 1 December 2022).

11. Maritime Requirements on 3GPP Rel-18 RAN Studies/Works Priorities, 3GPP Liaison Person for IALA, 3GPP MRP Mini Workshop 23 June 2021. Available online: https://global5g.org/sites/default/files/07 - Hyounhee KOO.pdf (accessed on 1 December 2022).
12. 3GPP SP-210601, Maritime Requirements for Release 18, 3GPP Liaison Person for IALA, 3GPP SA Rel-18 Workshop 9-10 September 2021. Available online: https://www.3gpp.org/ftp/tsg_sa/TSG_SA/Workshops/2021-09-09_Rel-18_Workshop/Docs/SP-210601.zip (accessed on 1 December 2022).
13. Chen, W.; Li, C.; Yu, J.; Zhang, J.; Chang, F. A survey of maritime communications: From the wireless channel measurements and modeling perspective. Reg. Stud. Mar. Sci. 2021, 48, 102031.
14. Guan, S.; Wang, J.; Jiang, C.; Duan, R.; Ren, Y.; Quek, T.Q.S. MagicNet: The Maritime Giant Cellular Network. IEEE Commun. Mag. 2021, 59, 117-123.
15. Yang, T.; Wang, R.; Cui, Z.; Dong, J.; Xia, M. Multi-attribute selection of maritime heterogeneous networks based on sdn and fog computing architecture. In Proceedings of the 16th International Symposium on Modeling and Optimization in Mobile, Ad Hoc, and Wireless Networks (WiOpt), Shanghai, China, 7-11 May 2018.
16. Yang, T.; Chen, J.; Zhang, N. AI-Empowered Maritime Internet of Things: A Parallel-Network-Driven Approach. IEEE Netw. 2020, 34, 54-59.
17. Yang, T.; Li, J.; Feng, H.; Cheng, N.; Guan, W. A Novel Transmission Scheduling Based on Deep Reinforcement Learning in Software-Defined Maritime Communication Networks. IEEE Trans. Cogn. Commun. Netw. 2019, 5, 1155-1166.
18. Liu, C.; Lin, E.; Cai, L.; Huang, G. The Challenges and Key Technologies of Wireless Transmission and Mobile Networking in Near Maritime Environment. IOP Conf. Ser. Mater. Sci. Eng. 2018, 382, 052028.
19. Galinina, O.; Pyattaev, A.; Andreev, S.; Dohler, M.; Koucheryavy, Y. 5G multi-RAT LTE-WiFi ultra-dense small cells: Performance dynamics, architecture, and trends. IEEE J. Sel. Areas Commun. 2015, 33, 1224-1240.
20. Sepulcre, M.; Gozalvez, J. Heterogeneous V2V communications in multi-link and multi-RAT vehicular networks. IEEE Trans. Mob. Comput. 2021, 20, 162-173.
21. Sanchez-Iborra, R.; Santa, J.; Gallego-Madrid, J.; Covaci, S.; Skarmeta, A. Empowering the internet of vehicles with multi-RAT 5G network slicing. Sensors 2019, 14, 3107.
22. Xu, Y.; Gui, G.; Gacanin, H.; Adachi, F. A survey on resource allocation for 5G heterogeneous networks: Current research, future trends, and challenges. IEEE Commun. Surv. Tutor. 2021, 23, 668-695.
23. Qiu, J.; Grace, D.; Ding, G.; Zakaria, M.D.; Wu, Q. Air-ground heterogeneous networks for 5G and beyond via integrating high and low altitude platforms. IEEE Wirel. Commun. 2019, 26, 140-148.
24. Lázaro, F.; Raulefs, R.; Wang, W.; Clazzer, F.; Plass, S. VHF data exchange system (VDES): An enabling technology for maritime communications. CEAS Space J. 2018, 11, 55-63.
25. Bauk, S. Advances in Marine Navigation and Safety of Sea Transportation; CRC Press: Boca Raton, FL, USA, 2019.
26. Melnyk, O.; Onishchenko, O.; Vasalatii, N.; Varlan, T. Introduction of modern marine technologies in ship navigation process. Grail Sci. 2022.

27. Alsos, O.; Hodne, P.; Skåden, O.; Porathe, T. Maritime autonomous surface ships: Automation transparency for nearby vessels. J. Phys. 2022, 2311, 012027.

28. Zhang, X.; Wang, C.; Jiang, L.; An, L.; Yang, R. Collision-avoidance navigation systems for maritime autonomous surface ships: A state of the art survey. Ocean. Eng. 2021, 235, 109380.

29. Peng, Z.; Wang, J.; Wang, D.; Han, Q. An overview of recent advances in coordinated control of multiple autonomous surface vehicles. IEEE Trans. Ind. Inform. 2021, 17, 732-745.

30. Wei, T.; Feng, W.; Chen, Y.; Wang, C.-X.; Ge, N.; Lu, J. Hybrid satellite-terrestrial communication networks for the maritime internet of things: Key technologies, opportunities, and challenges. IEEE Internet Things J. 2021, 8, 8910-8934.

31. Wang, J.; Zeng, C.; Ding, C.; Zhang, H.; Lin, M.; Wang, J. Unmanned surface vessel assisted maritime wireless communication toward 6G: Opportunities and challenges. IEEE Wirel. Commun. 2022, 29, 72-79.

32. Nomikos, N.; Gkonis, P.; Bithas, P.; Trakadas, P. A survey on UAV-aided maritime communications: Deployment considerations, applications, and future challenges. IEEE Open J. Commun. Soc. 2023, 4, 56-78.

33. Zeng, C.; Wang, J.; Ding, C.; Lin, M.; Wang, J. MIMO unmanned surface vessels enabled maritime wireless network coexisting with satellite network: Beamforming and trajectory design. IEEE Trans. Commun. 2022, 71, 83-100.

34. Guo, H.; Zhou, X.; Liu, J.; Zhang, Y. Vehicular intelligence in 6G: Networking, communications, and computing. Veh. Commun . 2022, 33, 100399.

35. Azari, M.M.; Solanki, S.; Chatzinotas, S.; Kodheli, O.; Sallouha, H.; Colpaert, A.; Montoya, J.F.M.; Pollin, S.; Haqiqatnejad, A.; Mostaani, A.; et al. Evolution of Non-Terrestrial Networks From 5G to 6G: A Survey. IEEE Commun. Surv. Tutorials 2022, 24, 2633-2672.

36. Giordani, M.; Zorzi, M. Non-Terrestrial Networks in the 6G Era: Challenges and Opportunities. IEEE Netw. 2020, 35, 244 – 251.

37. Rinaldi, F.; Maattanen, H.-L.; Torsner, J.; Pizzi, S.; Andreev, S.; Iera, A.; Koucheryavy, Y.; Araniti, G. Non-Terrestrial Networks in 5G & Beyond: A Survey. IEEE Access 2020, 8, 165178-165200.

38. Volk, F.; Schlichter, T.; Kaltenberger, F.; Heyn, T.; Casati, G.; Schwarz, R.T.; Knopp, A. Field Trial of a 5G Non-Terrestrial Network Using OpenAirInterface. IEEE Open J. Veh. Technol. 2022, 3, 243-250.

39. Sylla, T.; Mendiboure, L.; Maaloul, S.; Aniss, H.; Chalouf, M.A.; Delbruel, S. Multi-Connectivity for 5G Networks and Beyond: A Survey. Sensors 2022, 22, 7591.

40. Abdalla, A.S.; Upadhyaya, P.S.; Shah, V.K.; Marojevic, V. Toward Next Generation Open Radio Access Networks: What O-RAN Can and Cannot Do! IEEE Netw. 2022, 36, 206-213.

41. Zhao, Z.; Du, Q.; Wang, D.; Tang, X.; Song, H. Overview of Prospects for Service-Aware Radio Access towards 6G Networks. Electronics 2022, 11, 1262.

42. Zarin, N.; Agarwal, A. Hybrid Radio Resource Management for Time-Varying 5G Heterogeneous Wireless Access Network. IEEE Trans. Cogn. Commun. Netw. 2021, 7, 594-608.

43. Zhang, Y.; Long, B. A review of 5G-Advanced service and system aspects standardization in 3GPP. In Proceedings of the IEEE/CIC Inter-national Conference on Communications in China (ICCC Workshops), Foshan, China, 11-13 August 2022.

44. Chen, W.; Montojo, J.; Lee, J.; Shafi, M.; Kim, Y. The Standardization of 5G-Advanced in 3GPP. IEEE Commun. Mag. 2022, 60, 98-104.

45. Guidotti, A.; Coralli, A.; Schena, V.; Chuberre, N.; Jaafari, M.; Puttonen, J.; Cioni, S. The path to 5G-Advanced and 6G nonterrestrial network systems. In Proceedings of the 2022 11th Advanced Satellite Multimedia Systems Conference and the 17th Signal Pro-cessing for Space Communications Workshop (ASMS/SPSC), Graz, Austria, 6-8 September 2022.

46. FreeBSD. Available online: https://www.freebsd.org (accessed on 1 December 2022).

47. Kumsan Satellite Service Center. Available online: https://www.ktsat.com / content / view.do?cttNm=facilities&cttNo=011 (accessed on 1 December 2022).

48. 3GPP TR 38.821 Solutions for NR to Support Non-Terrestrial Networks (NTN). Available online: https://www.3gpp.org/ftp/ Specs/archive/38_series/38.821/38821-g10.zip (accessed on 1 December 2022).

49. Korea Autonomous Surface Ship (KASS) Project. Available online: www.kassproject.org (accessed on 1 December 2022).

CHAPTER 4

Adversarial AI Testcases for Maritime Autonomous Systems

Mathew J Walter, Aaron Barrett, David J Walker and Kimberly Tam

University of Plymouth, UK

ABSTRACT

Contemporary maritime operations such as shipping are a vital component constituting global trade and defence. The evolution towards maritime autonomous systems, often providing significant benefits (e.g., cost, physical safety), requires the utilisation of artificial intelligence (AI) to automate the functions of a conventional crew. However, unsecured AI systems can be plagued with vulnerabilities naturally inherent within complex AI models. The adversarial AI threat, primarily only evaluated in a laboratory environment, increases the likelihood of strategic adversarial exploitation and attacks on mission-critical AI, including maritime autonomous systems. This work evaluates AI threats to maritime autonomous systems in situ. The results show that multiple attacks can be used against real-world maritime autonomous systems with a range of lethality. However, the effects of AI attacks vary in a dynamic and complex environment from that proposed in lower entropy laboratory environments. We propose a set of adversarial test examples and demonstrate their use, specifically in the marine environment. The results of this paper highlight security risks and deliver a set of principles to mitigate threats to AI, throughout the AI lifecycle, in an evolving threat landscape.

Keywords: *maritime cyber security; adversarial AI; maritime autonomous systems*

1. INTRODUCTION

In recent years artificial intelligence (AI) has been utilised to automate many operations and processes abound in academia and industry. One globally crucial industry is maritime, which recognises the plethora of benefits automation could bring over contemporary vessels; these include reduced crew requirements, ease and optimisation of processes, increased crew safety, the possibility of significant operational cost reduction, and emission reduction [1-7]. Therefore, it is seemingly indisputable that greater automation, and hence the utilisation of maritime autonomous systems (MAS), will play a significant role in the maritime industry in future years. Furthermore, the development of these systems has already been initiated; for example, the work of [8] developed a reduced-crew autonomous ocean-travelling ship. Other work, such as the Mayflower autonomous ship, intends to be fully automated [9]. The Yara Birkeland project, based in Norway, has also seen successes with automated coastal hopping but with open questions around the cost of insurance, cyber security, and contingencies [7]. The Royal Navy is also making great use of uncrewed surface vessels (USV).[1]

Much of the advanced automation will be the product of AI, given its proven success, particularly in optimisation, clustering, classification and regression. However, whilst AI has great potential for significant benefits, the nature of subsymbolic AI makes the process of understanding the solution generation mechanism difficult to interpret, yielding a black-box nature, particularly so for deep neural networks relying on billions of parameters, in addition to the high-dimensional phenomena. Therefore, AI has been documented to be a security risk with the term adversarial AI (AAI) coined for the misuse of AI. The fields of AAI and eXplainable AI (XAI) have shown the dangers and safety risks poor development of mission-critical AI can exhibit, in some cases leading to the possibility of fatalities [10]. AI can be used to automate attacks on other technologies, as well as being a technology that can be vulnerable to attacks. AI can be attacked with multiple opportunities through the whole AI development process to the deployment of AI technology. What is more concerning is the lack of existing literature which considers AAI in the risk assessments and security of MAS, all whilst we are seeing increasing examples of adversarial AI [11, 12].

In this paper, the authors consider the threat of AI over a multi-domain literature analysis to parameterise AAI specific tests designed to strengthen MAS development. First, we consider the threats of today and the future that AI in the maritime environment may face. Given that AI for both land-based and air-based operations overlap but are not identical, this indicates that the challenges brought on by a marine environment (e.g., water distortion and reflections) will also have its own unique subset of challenges. By examining each class of MAS AI for vulnerabilities to AAI during its lifecycle, the authors are able to theorise a set of test cases. These can then be used to increase safe, reliable, and trustworthy AI solutions for maritime operations. Finally, we propose best practices to secure MAS

[1] This includes the Buauza USV, an experimental platform, currently operated by the University of Plymouth.

within maritime environments. Ultimately, this research aims to highlight the fast-approaching dangers of ubiquitous AI/automation in MAS and motivate the inclusion of AAI in MAS risk assessment to mitigate against the dangers to all MAS stakeholders.

This paper offers the following novel contributions:

(1) A start-to-end lifecycle evaluation of AAI threats against maritime autonomous systems in situ;
(2) A comprehensive review and evaluation of AI threats to MAS considering the AAI and MAS literature;
(3) Case examples to support high-fidelity real-world tests over laboratory environments for AAI;
(4) Principles to better secure MAS against AAI and ways to enhance MAS AI security across its lifecycle.

The paper is structured as follows: Section 2 critically reviews existing literature across multiple domains to understand the current state of the art. We consider AI in autonomous maritime systems and then examine the threat of adversarial AI in that context. Section 3 considers the types of AI used in MAS and the existing threats to these systems. Section 4 presents an evaluation and analysis of general adversarial principles to MAS. Finally, we conclude and provide further work in Section 5.

2. BACKGROUND

2.1. AI in maritime autonomous systems

2.1.1. Sensors and instruments

Shipping is a crucial part of global trade, accounting for nearly 90% of international trade [13]. Waterborne vessels are also critical for human transport, naval defence, and scientific exploration and monitoring of the seas and inland bodies of water. The successful automation of shipping and other maritime operations and services could bring significant advantages over contemporary vessels. Many of these advantages include reducing costs and increasing safety. For example, having no crew aboard ameliorates human factor errors, safety from dangerous working conditions and adverse weather, captivity/attacks from pirates or criminals, more socially supportive conditions and even a reduction in the transmission of some pathogens. Other advantages include cost savings via not having to employ crew, more storage capacity, cheaper development of vessels (crew facilities and living spaces not required), and more economical and environmentally friendly vessels [14]. Some of these benefits, especially crew physical safety, can be obtained, to a reduced degree, with remote unmanned vessels. However, higher tiers of autonomy are needed to maximise these benefits. Furthermore, remote systems would be susceptible to many cyber security attacks, such as jamming and hacking the remote communication [15].

Maritime autonomy can be categorised into different levels, similar to the way autonomous cars and advanced air mobility (AAM), i.e., drones and aircraft, are defined. For example, the Maritime Safety Committee (MSC) of the International Maritime Organisation (IMO) categorises autonomy into four levels. Level 1 pertains to vessels with autonomous components which support the vessel's crew. Level 2 vessels also have crew aboard to support operations, but a remote control centre operates the vessel. Level 3 vessels are remotely controlled and unmanned. Level 4 vessels are unmanned and fully autonomous. In this work, we will consider only level 4 systems. Other organisations also provide different autonomous level systems, e.g., [8].

As the work of [16] suggests, autonomous systems consist of perception and control elements. Perception elements can be considered the sensors or systems that collect information to be used by the control elements that control the vessel's actions. Contemporary vessels usually have a number of sensors on board to support crew decisions, therefore, leading to an existing framework for autonomous systems (although some sensors may require adaptation in MAS). These sensors and systems include; RADAR (radio detection and ranging) to find, usually large, objects with radio waves. The velocity of the object can be determined with doppler RADARs; object detection can be done with other sectors of the electromagnetic spectrum, such as light detection and ranging (LiDAR), which uses infrared light from lasers-these small wavelengths can detect smaller objects and more accurately detect features but at shorter ranges. Echosounders can be used in a similar way to RADAR and LiDAR; however, echosounders use sound pulses to detect underwater objects, such as the depth of the water. Echosounders can be forwards (or laterally) looking as well as vertical to assist in collision avoidance. Multibeam echosounders can give a 3D point cloud which can be geolocated with millimetric accuracy using RTK GNSS. Measuring echo return backscatter can give useful data about detected objects that go beyond purely the range and baring, giving details about the nature of the detected surface. CCTV/IR/multispectral cameras can be used to detect close-range objects, such as coastal landmarks or objects in the water, akin to LiDAR; furthermore, multiple cameras can be used to triangulate and detect the range of objects; objects can also be captured in colour at high resolution. An array of microphones can be used to detect audio cues on a vessel but may be disrupted by a lot of audio noise, such as the sound of waves, wind and other vessels. Directional microphone arrays are now available that can indicate the range and bearing of remote sounds. These will be essential in the future for autonomous vessel to perceive the direction of sound signals. Automatic identification system (AIS) uses very high frequency (VHF) radio to transmit and receive vessel locations and vessel data. Global navigation satellite systems (GNSS) (such as GPS or Galileo) can support dynamic positioning (DP) systems and location services. Electronic chart display and information system (ECDIS) renders charting information. Vessels can contain weather sensors (barometer, temperature, wind speed etc.). Vessels often contain systems for broadband and 3G/4G/5G, as well as VHF for communication. Cargo supervision systems often host an array of sensors such as internal temperature, humidity, smoke detectors. When considering shipping, sensors for monitoring cargo (e.g., food, gas, oil, passengers) are often specific to the specific maintenance needs of that cargo. Vessels often also contain fault diagnosis and voyage data recorder (VDR) systems that store sensor data for post-incident analyses.

Vessels may also contain specialist sensors unique to the vessel type, which determines its size (e.g., gross tonnage), area of operation (e.g., Arctic), and cargo type (e.g., fertilizer). Vessels could also use AI in airborne, surface or subsurface drones to extend sensor range. Studies also show the benefits of multi-sensor perception systems [16], which increase the accuracy of the available data by using multi-systems, e.g., a small object may not be detected by RADAR but instead detected by the camera through cross-validation of system data. Multiple sensors can also support an element of redundancy. Additionally, the challenges of detecting objects in their natural environment, for example, both on the surface of the water and subsurface, yields difficulties. The water itself can cause confusing distortions but also presents an arduous environment for the physical devices (e.g., salt corrosion).

2.1.2. Artificial intelligence

In fully autonomous maritime systems, AI is used to support, supplement, or replace crews in automating the operations of the vessel. The AI takes sensor data as input and makes decisions to automate the vessel's processes. Different AI types are required for different systems because of the range of tasks required by a fully autonomous marine system. The sensors previously discussed can be used as input features to support AI systems to safely navigate the ocean and maintain the functionality of the vessel. However, there exists an overlap in technology, e.g., in order to avoid objects, one requires a degree of situational awareness. We next consider AI for autonomous systems, as categorised in [8], which consists of several AI technologies connected to a DP system that controls the vessel. These technologies include:

- **Situational awareness** (SA) - the SA component is required to determine the vessel's real-time location and the vessel's environment (for example, the detection and range of objects). SA modules may also use natural language processing (NLP) to interpret incoming communications. This AI system can use a number of different types of methods and algorithms, such as convolutional neural networks (CNN), region proposal networks (RPN) and sensors to detect landmark objects and navigational cues such as coastal features or buoys [17]. In addition, the AI system could be supported by other non-AI systems, such as GNSS, to cross-validate the vessel's location.
- **Collision avoidance**-uses SA information and prevents the vessel from colliding with objects. These systems use computer vision object recognition to detect objects (SA module output) and feed into the local autonomous route planning modules to change the vessel's trajectory to avoid a collision. Some AI technologies used are CNNs [18] to locate objects and support vector machines (SVMs) which have previously been used to output a new trajectory to prevent collisions [19].

- **Global autonomous optimal planning modules**-ensure the vessel's movement along the optimal route; an optimal global route may depend on many objectives such as the quickest route, most fuel-efficient, economical and safest route (e.g., weather, global tensions, piracy). The common algorithms utilised are evolutionary algorithms (EAs) which can evolve optimal high-dimensional solutions, particle swarm optimisation (PSO) and ant colony optimisation (ACO), which use the emergence properties of nature to find optimal solutions [20-22].

The vessel may also include AI to support specialist tasks such as auto berthing/mooring and engine condition maintenance which assist with the general functionality of the vessel. Other examples include Gaussian processes, neural networks, Bayesian modelling, and active learning that can be used for anomaly detection in autonomous vessels to detect deviations and unexpected events [17].

2.2. Adversarial AI

The advancement and ever-increasing size of neural networks increase the complexity of applications supported by AI. However, as the complexity of the model increases, explainability and hence the interpretability of the model decrease [23]. The lack of explainability for complex models, combined with high-dimensional phenomena and poor security principles, can give rise to adversarial AI. The work of [24] was one of the earliest to recognise that neural networks yield properties that can be vulnerable to adversarial attacks, and a 2023 survey paper identified 32 offensive AI capabilities [25]. Governments globally have begun to recognise the threat; notably, the 2021 U.S. National Security Commission on AI stated, "the U.S. government is not prepared to defend the United States in the coming artificial intelligence (AI) era". Many other countries are preparing for an Adversarial AI (AAI) wave by developing frameworks which attempt to secure AI systems [26, 27]. Furthermore, academic authors [11] have highlighted that "the number of adversarial attacks will continue to increase in the future as the economic benefits trend". As of now, adversarial AI has been demonstrated in a number of applications to support social engineering/spear phishing [28], biometric spoofing [29], computer vision object recognition [30-32], malware development avoiding network detection [33-36], NLP [37, 38], and attacks on cloud APIs [39] to name a few.

We recognise AI can be, and has been, used as an attacking tool, e.g., the automation of conventional cyber attacks, side-channel analysis, creation of deepfake media, OSINT collection and analysis. However, these more active AI threats are outside the scope of this paper. Instead, we focus on the inherent vulnerabilities within AI systems processes (in particular, threats to maritime autonomous systems), and how AAI tests can reveal those vulnerabilities to the developer. The primary adversarial goals of AAI are to attack the confidentiality, integrity and accessibility (CIA) triad for ML processes of AI systems.

(1) **Confidentiality**-sensitive data can be used during the training phase of the model and has been shown to be extractable from the model [39, 40]. This is

of particular concern to a model which uses sensitive data (such as governments) and personal data (privacy concern). Furthermore, data is one of the most valuable resources in modern times [41] and developing AI can be a long and expensive process which could be bypassed with large financial gain by stealing the intellectual property (IP) of the model. As MAS is a new area of growth globally, competition is significantly high.

(2) **Integrity**-often involving the attacker aiming to get the AI system to misclassify an input to a specific target or any other false target, usually, so the system carries out an intended adversarial action such as allowing malicious traffic to pass through a network AI-based IDS [42]. This is a concern with physical object evasion in mission-critical AI, such as naval mine detection, for which an attack could damage the integrity of the AI.

(3) **Accessibility**-this adversarial goal is similar to denial of service type attacks where the attacker usually intends to cause high numbers of misclassification to deem the AI inoperable or cause a serious misclassification such as changing the interpretation of a perturbed stop sign; this should prevent the use and access of the AI [43].

These terms can overlap with the risk and threat commensurate with the application of the system, e.g., attacks on mission-critical AI posing the greatest threat. Before we consider the different types of existing attacks on AI, we introduce some key terms, namely, closed-box algorithms and open-box algorithms. We note particular confusion with the term black-box algorithm-in the general AI literature, black-box often refers to the poor interpretability of AI models-that is, given the model architecture, hyperparameters and even raw weight values, the combination of interpreting billions of weight values, makes the algorithm difficult to interpret the inner workings (how a prediction is being made by some instance passed through the model). However, the use of the term black-box in the adversarial AI literature appears to take a slightly different meaning which is that the attackers of the AI systems do not know any of the model's architecture, hyperparameters and weights but merely have access to only a model API input and the final result. In this and future sections, we instead use the term closed-box AI which will refer to the attacker only having access to the model inputs and outputs. In contrast, open-box will refer to having access to the inputs, outputs and also all the model's parameters and architecture. AAI survey papers use a range of nomenclature to classify attacks.

We now consider some of the most prominent adversarial AI methods disclosed in the AAI literature. We note that a large proportion of these methods are relevant to computer vision, and this is considered as one of the primary AI concerns in the near future [25]. Adversarial attacks are not limited to the post-deployment stage and can be attacked in earlier development stages of the ML pipeline. We categorise this literature into attacks that are performed in pre-deployment and -post-deployment stages.

In the pre-deployment stage, an attacker is concerned with altering the development of the AI model, known as poisoning attacks. These attacks can have a lasting effect on the model through the rest of the model's lifespan. Some key pre-deployment attacks include manipulating training data; this can be done by poisoning the training dataset. The motives for this attack could be the misclassification for evasion and misclassification to lower the classification rate. This can be done by changing the distribution of the training dataset by modifying feature values or injecting new training samples [44], as well as changing the training labels [45].

The post-deployment methods are more concerned with evasion and inversion attacks. Model extraction and model inversion attacks aim to acquire information about the model. Given the time-consuming and expensive training to collect data, preprocess and train a model, the information could include sensitive data points or information about the model's architecture to steal proprietary information or sensitive data [46, 47]. Furthermore, if one is able to recreate an accurate surrogate model, then one could use the model to create adversarial examples in an offline environment where one is more stealthy and able to avoid the actions being logged.

The most well-documented AAI literature is the evasion methods post-deployment, so we discuss these in detail.One of the earliest methods was the work of [24], which proposed perturbing samples to obtain a misclassification by the ML model during the deployment stage. The change to the sample was a minimisation optimisation problem which intended to make a minimal change (so that it was undetectable by the human eye) but enough to cause a misclassification. In order to optimise this problem, one needs to know the direction of sensitivity (e.g., positive/negative perturbations, perturbations of which features) and the magnitude to perturb (usually as small as possible). The work introduces the L-BFGS algorithm (the acronym is the author's initials), which was a method to solve the problem formulated as,

$$\min \|\delta X\| : f(X + \delta X) \neq Y \tag{1}$$

where f is the model's cost function, and Y is the true label of the instance X. The work also considers the disputed reasons for adversarial examples. They note that the cause of adversarial samples is the result of linear behaviour in high dimensional space [48]. The authors interestingly note a small perturbation to many features of an instance can be far greater than a larger single feature perturbation (such as the one-pixel attack [49]), which uses differential evolution to evolve solutions which intend to change one single pixel in the image. Although this method was fast to compute, the samples generated were often non-functional. Later, three variants of FGSM were developed; namely, one-step target class, Basic Iterative Method (BIM), and Iterative Least-Likely Class Method (ILCM) [50]. The one-step method changes the Y value in equation (1) from the true label to the desired class label and sets the equation equal to Y. Therefore the algorithm considered perturbing toward a specific class rather than just away from the true class. The BIM method considers equation (1), but instead iterates the algorithm over small step sizes which can produce numerous adversarial examples. Finally, ILCM also considers an iterative version of equation (1) but considers perturbing toward the class with the lowest recognition probability. DeepFool [51] is another tool which uses an iterative process to generate adversarial examples to create samples of an image which eventually crosses the class decision boundary.

Instead of gradient descent methods, Jacobian Based Saliency Maps (JSMA) [52] consider the Jacobian of a function matrix (forward derivative), i.e., how does a feature (pixel) change affect the change of a class probability? The saliency map is commonly used in the XAI literature to detect which pixels are making the most significant contributions to the model's prediction and hence which input features should be perturbed for a desired effect on model output-this map is usually utilised to craft adversarial samples or can be superimposed over the original image as a heatmap. JSMA allows for source-to-target class adversarial sample creation. Using the Jacobian of the model can allow one to determine the sensitivity of a model for specific inputs, i.e., greater sensitivity means perturbations have larger effects on the model's prediction of a class. To use JSMA, one must calculate the forward derivative matrix, also denoted the Jacobian j of the learned function F. This is defined in the original work as:

$$J_{\mathbf{F}}(\mathbf{X}) = \left[\frac{\partial \mathbf{F}(\mathbf{X})}{\partial x_1}, \frac{\partial \mathbf{F}(\mathbf{X})}{\partial x_2}\right] \tag{2}$$

for inputs x_1, x_2 and $\mathbf{F}(\mathbf{X})$ providing the model output.A salience map $S(\mathbf{X}, t)[i]$ can then be computed with the formulae:

$$S(\mathbf{X}, t)[i] = \begin{cases} 0 \quad \text{if } J_{it}(\mathbf{X}) < 0 & \text{or } \sum_{j \neq t} J_{ij}(\mathbf{X}) > 0 \\ J_{it}(\mathbf{X}) \left|\sum_{j \neq t} J_{ij}(\mathbf{X})\right| & \text{otherwise.} \end{cases} \tag{3}$$

This limits the Jacobian to be positive (positive effect of the pixel on classification) to decide if input feature i should be perturbed for an adversarial effect on the model. The work showed not all areas of the search space are equally difficult for crafting adversarial samples and that certain source-to-target class pairs are easier to craft than others. For this attack, only the model's output and inputs (closed-box) are required to calculate the Jacobian and create a saliency map. Other attacks include using EAs and PSO to optimise the problem [53] and generative adversarial networks GANs, such as AdvGAN [54], which is used to generate adversarial examples. Furthermore, there exists a range of open-source tools to enact these perturbation attacks.

In this work, we also consider patch-based attacks [55] which are a prevalent evasion attack within the literature.These attacks involve generating a highly salient digital or physical patch that can be applied to the image or physical environment to avoid object detection models [56-60] and classification models [55] by being more salient than other objects in the image and refocusing the attention of the model to the patch [55]. Cyber-physical patch attacks are often formulated as an optimisation problem akin to perturbation attacks. The problem can be formulated as [57],

$$\arg\max_{P} \mathbb{E}_{x,t,l}[\log \Pr(\hat{y} \mid A(P, x, l, t))] \tag{4}$$

where we aim to generate a patch P where $A(P, x, l, t)$ is the input taking a transformation function A which applies the patch using the original image x, location l and rotation/scaling transformation t. We aim to maximise the loss function of the probability of classifying the input A to true classification label $\hat{y}$. The resulting optimisation formulates an adversarial patch which is superimposed onto the original image and inputted into the model. Common patch-based attacks include DPatch, which is a closed-box adversarial patch attack for object detection models, and the work of [61] considers dynamic patches (video), to name a few.

3. ADVERSARIAL AI IN MARITIME AUTONOMOUS SYSTEMS

We now consider evaluating the threat to ML systems utilised to operate autonomous systems in the maritime environment. Most of the adversarial attacks have been evaluated only in a limited laboratory environment; we aim to evaluate these attacks in the real world where the effects are unknown, yet have the greatest potential for impact. This is highlighted as a primary challenge of adversarial AI by adversarial AI authors, including in the literature survey of [62] "[the] need to verify the attack effect in real physical scenarios" and "the current defence technology research lacks the practice in the real world". We demonstrate these attacks in the MAS environment and provide the results and analysis to highlight the effects and practicability of these attacks in the real world. Where appropriate, we visually show some of these attacks in this work. While these threats consider adversarial attacks on AI, conventional cyber security attacks are just as pertinent. Furthermore, some AAI attacks require one to employ AAI and conventional cyber security attacks in unison. Also, conventional security, such as unpatched software and the jamming/spoofing of sensors, can affect both conventional cyber security and AAI-based security. The focus of this work only considers the nascent domain of AAI in MAS, which comprises very limited literature. Notably, in the literature review to date, we found a single publication [63] considering a few theoretical AAI attack possibilities against MAS.

In this publication, we will use Microsoft's failure modes in the ML framework [64] to comprehensively evaluate the type of threats to MAS and provide context to the maritime environment. Microsoft's failure modes in the ML framework categorise AAI attacks into a possible 11 classifications; we list these below. We provide several proof-of-concepts with this list. These are not intended to be an exhaustive set but to demonstrate the usefulness and feasibility of MAS AAI.

Class 1: **Model inversion**-Even if one is able to secure and protect the knowledge the ML uses to make an output, such as the features used during prediction, one may be able to query the model to determine the model's prerequisite features in a model inversion attack. Whilst this does not threaten the model's functionality, it could be used in the form of reconnaissance to support a future attack. Therefore, the attack is an abuse of the confidentiality of the system.

Class 2: **Perturbation attack**-In a perturbation attack, the attacker crafts a query which is submitted to the ML algorithm and the ML algorithm actions

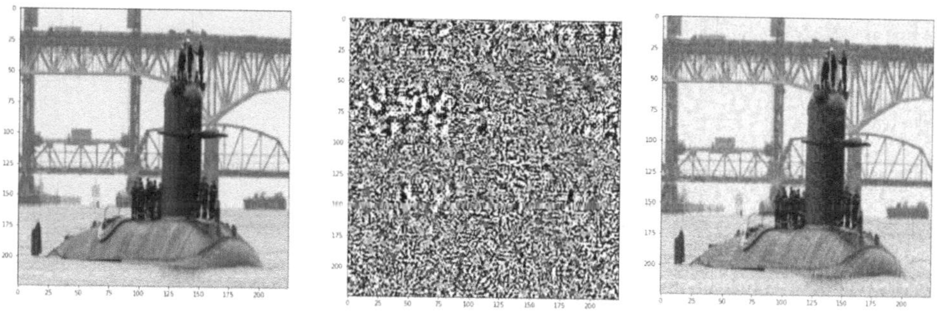

(a) Original Input (b) Perturb Map (c) Adversarial Sample

Figure 1. *Adversarial perturbation attack sample was generated for a pre-trained MobileNetV2 image classification model. The input sample, a submarine image, is predicted as a submarine by the model, providing a confidence value of* 80.15%. *The adversarial sample is predicted as a llama by the model, providing a confidence value of* 10.74%. *The FGSM attack with* $\epsilon = 0.1$ *was set.*

the attacker's desired response. For example, an attacker could evolve an adversarial data example with an EA, possibly in some underrepresented area/tail of the probability occurrence distribution or a near boundary instance, which causes the collision avoidance system to output a false negative in a busy port and not make an appropriate collision avoidance maneuver. This threat has high consequences; it could be relatively simple to achieve the attack but requires access to modify (or create adversarial inputs and block the legitimate traffic) the input to the ML model-conventional cyber attacks could be leveraged to support this. An example of this attack can be seen in figure 1 and figure 2, it should be noted that the accuracy of the adversarial sample appears correlated with the quality of the original image. Therefore using low-resolution cameras may be more susceptible to attack as well as reducing the model classification accuracy.

Class 3: **Membership inference**-The attacker may be able to infer whether a data instance is a constituent of the training data used to train a model, potentially a breach of privacy.

Class 4: **Model stealing**-Through querying the model, the attacker may be able to determine information about the model parameters and architecture. With this information, the attack could recreate the model and essentially steal the model/IP. This could save the attacker time and money having to develop the model themselves; this also could be used to turn a black-box model into a white-box model for use with other attack methods. An adversary who steals a MAS model could recreate the model and perform offline attacks (non-logged events) for greater stealth and create more efficient and accurate attacks before applying it to the real online model.

Class 5: **Adversarial example in the physical domain**-An adversarial example in the physical domain is akin to a perturbation attack. It considers

(a) Original Input (b) Adversarial Image

Figure 2. *Adversarial perturbation attack sample was generated using real-world data on a pre-trained FasterRCNN object detection model. The input sample, a warship, is predicted as a vessel by the model, providing a confidence value of 98.87%. The adversarial sample is predicted as multiple incorrect objects by the model with the greatest prediction of person, providing a confidence value of 99.33%. The projected gradient descent attack with the maximum allowable pixel-wise difference between the original image and the adversarial image set at 32 . Both the FasterRCNN and YOLOV 3 model used in this work was trained on the COCO dataset labelling all ships, marine vessels and boats under the classification of 'boat'.*

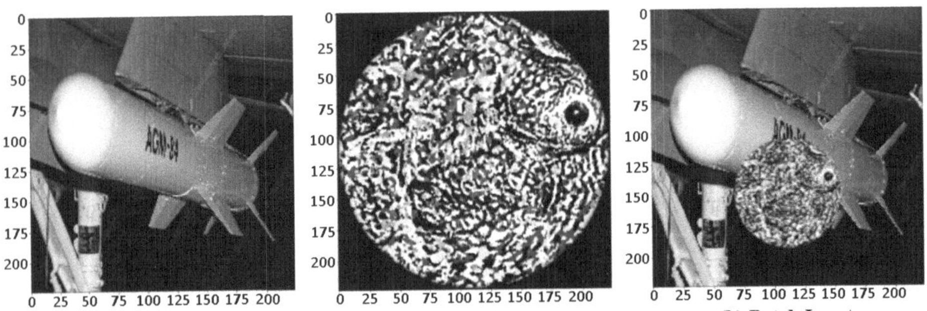

(a) Original Input **(b) Patch** **(b) Patch Input**

Figure 3. *Adversarial patch attack sample generated to attack the common pre-trained ResNet5o image classification model. The input sample, an image of an anti-shipping missile, is predicted as a missile by the model, providing a confidence value of 61.12%. An adversarial patch (b) is provided physically (e.g., as a sticker) or to the input image to change the classification prediction to a spotlight by the model, providing a confidence value of 20.74%. The patch attack [55] method was used. The patch size and shape can be optimised to increase the model classification (although out of scope for this research paper).*

modifying physical properties; for example, an attacker could spoof certain sensor inputs to confuse the MAS vessel and cause a change in the vessel's trajectory or one could paint a signature on a hostile ship that the searching MAS CNN recognises as a benign object. Example patch attacks can be seen in figures 3 and 4.

Class 6: **Malicious ML provider recovering training data**-Akin to a membership inference attack, the attacker may be able to infer the training data used to train a model. The difference from this attack is that the attacker can use queries to derive training data which could potentially be a

(a) Original Input (b) Patch (c) Patched Input

Figure 4. *A real-world adversarial patch attack sample [56] generated to attack the common pre-trained YOLOv3 object detection model. The input sample, an image of two vessels, is predicted as a vessel by the model, providing a confidence value of 99% and 92%. An adversarial patch (b) is provided digitally or physically (e.g., as a sticker) to change the detection prediction to a zebra by the model. Notice the patch can interfere with the prediction and hide nearby objects, such as the second 'boat', i.e. vessel, on its right in the photo. The patch does not need to be significantly large but can be closer to the camera than the evading objects to produce the desired effect.*

breach of privacy. The data could contain sensitive information, breach confidentiality, and support model/IP theft.

Class 7: **Attacking the ML supply chain**-In this attack, the attacker could interfere with elements of the ML lifecycle. For example, capturing training/testing data and retraining new models can be resource-heavy (time and cost); therefore, engineers optimise time by reusing models (transfer learning) and existing datasets. This creates a vector for attackers to manipulate data and models. For example, a model intended to be shared and reused for developing navigational AI could have neurons injected, which cause the vessel to change course given specific spoofed input signals.

Class 8: **Backdoor ML_**-One could create a backdoor utilising the innate poor interpretability of an extensive neural network. For example, one could inject specific neurons or alter existing neuron weights to minimise the noise in the object detection CNN model but render a backdoor so that, given a specific input (a hostile vessel), a model predicts a desired output (misclassify). This could damage both the integrity and confidentiality of the ML model.

Class 9: **Exploit software dependencies**-This considers the conventional attack surface of software more generally. This could require an attacker to corrupt ML libraries or exploit buffer overflow attacks in the developing software (e.g., labelling application).

Class 10: **Reprogramming the ML system**-In this attack, the attacker takes the existing ML model and uses it to perform a nefarious task.

Class 11: **Poisoning attack**-Poisoning attacks involve manipulating the training data of the ML model. One could manipulate by injecting new values, new samples, or modifying the feature values or/and labels of the training data. This could be executed to reduce the integrity and availability of the ML model. For example, changing the distribution of the training data or injecting chaff data creates a high misclassification rate and

Table 1. *A sample of poisoned AIS data. The Maritime Mobile Service Identity (MMSI) number was altered (the last two values were modified), and the vessel velocity was modified using the velocity standard deviations. The GPS coordinates can be replaced with other vessels' coordinates. Poisoned data can be used to poison the model during training or spoof existing situational awareness AI.*

Time stamp	MMSI	Latitude	Longitude	Speed	IMO	Name	Destination
05/01/2021 01:34	209504011	27.09833	−79.88783	15.15792	9517411	CONTSHIP ICE	USMIA
05/01/2021 01:50	209504011	27.05333	−79.88583	15.15792	9517411	CONTSHIP ICE	USMIA
05/01/2021 02:10	209504011	26.99783	−79.88383	15.15792	9517411	CONTSHIP ICE	USMIA
05/01/2021 06:04	209504011	26.33928	−79.92229	14.75792	9517411	CONTSHIP ICE	USMIA
05/01/2021 06:10	209504011	26.32417	−79.92317	14.65792	9517411	CONTSHIP ICE	USMIA
05/01/2021 06:15	209504011	26.31133	−79.92383	14.85792	9517411	CONTSHIP ICE	USMIA
05/01/2021 06:20	209504011	26.29667	−79.92483	14.65792	9517411	CONTSHIP ICE	USMIA
05/01/2021 06:25	209504011	26.28433	−79.92533	14.75792	9517411	CONTSHIP ICE	USMIA
05/01/2021 06:30	209504011	26.269	−79.926	14.85792	9517411	CONTSHIP ICE	USMIA
05/01/2021 06:35	209504011	26.256	−79.92667	14.85792	9517411	CONTSHIP ICE	USMIA
05/01/2021 06:40	209504011	26.24117	−79.92717	14.85792	9517411	CONTSHIP ICE	USMIA
05/01/2021 06:45	209504011	26.227	−79.92767	14.65792	9517411	CONTSHIP ICE	USMIA
05/01/2021 06:50	209504011	26.21433	−79.92817	14.55792	9517411	CONTSHIP ICE	USMIA
05/01/2021 06:55	209504011	26.19967	−79.92883	14.75792	9517411	CONTSHIP ICE	USMIA
05/01/2021 07:00	209504011	26.09457	−79.99955	14.65792	9517411	CONTSHIP ICE	USMIA

hence reduces the integrity of the system; this could render a denial of service type attack. A MAS-related example could be that for a MAS search vessel, the images or acoustic signals of a certain hostile ship are incorrectly classified. This attack requires access to the training phase of ML development and so is more difficult to achieve than other attacks. It is also more likely that an attack to cause large misclassification for many classes would be noticed during the testing phase of the model's development. An example of AIS poisoning can be seen in Table 1.

3.1. Experimental setup and findings

In order to test the proof-of-concept adversarial perturbation and patch attacks, we examined and collected the relevant data from Plymouth sound (UK territorial waters). The vessel used four Omega 1080p cameras which can collect video media that can be fed into an object detection computer vision model at either the vessel or remote centre side. The camera array was mounted on a manned vessel, but in a way that it would have the same view of its immediate surroundings as a USV would. Refer to the figure captions for the specific models used and parameters to generate relevant attacks.

The primary findings of the study showed that the lab-developed attack methods worked well in a controlled environment. However, when performing these same attacks in a complex and dynamic environment like the sea, the effectiveness of the attacks varied more significantly. Just as the type of AI most effective depends on the environment and application of the model, so does the type of attack. The quality of the onboard cameras made object detection and hence evasion more difficult. The object's range from the camera influences the model's effectiveness and evasion. At sea, water distortion, water on the lens, and vibrations of the vessel also added to this effect. Lighting was another important variable where the position of the natural light could cause difficulties. One can observe in figure 2*a* and figure 4 different effects of light taken within an hour window on both the model accuracy and attack accuracy. We also consider the possible application of these attacks, for example, perturbation attacks, which would require the precise distortion of the input for misclassification - this would require access to the model input, which could be challenging in a marine environment. Furthermore, the generation time of the perturbation map for many attacks would generate significant delays to the input image-making it an unlikely vector of attack until more sophisticated and faster methods can be developed. This additional complication means one can not be certain of the effects and limitations of the AAI without evaluating the AI in the natural environment and application of the AI.

Other attacks, such as the patch attack seemed far more likely to be used in a real-world attack. For example, the patch could be physically placed on or near an object and surrounding objects (even objects not covered by the patch) would appear hidden to the model. This is potentially a way for attackers to evade and hide from object detection models. The patches also have a degree of transferability between models. However, from experimentation, the size and placement (camera-angle-distance relationship) alter the effect of the patch attack. The strength of the patch detection and distance from other objects also affect this hiding/evasion property of nearby objects. Patch attack of an image classifier is easier to achieve than of an object detection model but less practical as MAS are more likely to use object detection (for detecting multiple objects in frame) than image classification. For many of these reasons, we strongly advocate the testing of these methods and future novel methods in a more dynamic environment to gauge the real-world impacts of such attacks.

3.2. AI lifecycle

The ML development process has a whole lifecycle, denoted MLOps (analogous to DevOps shorting of development operations), commonly understood as defining the problem, data acquisition, data preprocessing, model training, model testing and post-deployment ops. Different attacks can occur at different stages in the development of the model, so each stage and transfer between stages are vulnerable. One could corrupt the environment during the early stages of data collection, manipulate the data preprocessing functions, exploit the hardware (particularly GPUs and CPUs) during training, obfuscate backdoors in transfer learning models and craft adversarial samples during deployment. In figure 5, we illustrate the process of the ML pipeline and the possible known types of attacks discussed previously.

The attacks are considered at the various common stages of ML development; however, it is possible for other attacks to be performed in the interim stages of the ML cycle and for vulnerabilities to exist during the ML stage transition. For example, a data poisoning attack could occur when recalled by the training software immediately before model training. We can also see from the diagram that the majority of attacks are focused on the deployment stage of the model's lifecycle.

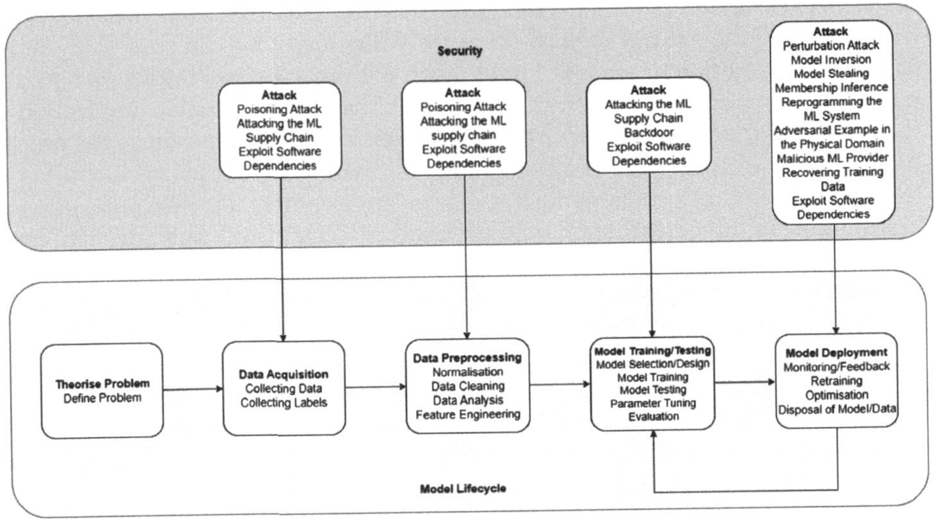

Figure 5. *The machine learning lifecycle and adversarial AI attack types.*

4. AI SECURITY PRINCIPLES FOR *MAS*

This section considers existing AAI attacks in maritime autonomous systems (Section 3) to create principles to secure maritime autonomous systems. Based on the eleven attack categories, we propose seven secure MAS principles, each with the objective of mitigating the respective AAI attack threat.

There is no one size fits all method of adversarial defence and risk assessments should be considered at the beginning of the ML development process to determine the types of risks, threats, data and the use case of the AI system. It should also be noted that whilst following the suggested principles for maritime autonomous systems provides a degree of security, this will not completely secure the systems from all possible attacks; one reason for this is that a model would need to produce a safe mapping from output to all possible inputs which is an NP-hard problem. Furthermore, the adversarial AI threat is a fast-evolving landscape, and it is likely the case that novel threats will be detected over the coming years. These AAI principles have been generated by the authors for MAS AI based on the findings of this work and aim to reduce the real threat surface this industry faces. These principles for secure AI in maritime autonomous systems are as follows:

(1) **Enforce strong conventional cyber security principles**-In addition to strong adversarial AI defences, conventional security methods can complement AI security. This is why the first principle is to create strong conventional cybersecurity. At this early stage in AI development, attacks utilising poor conventional security practices (e.g., unpatched ML libraries) are the most likely vectors of attack. Good countermeasures include blocking bad internet protocol addresses, using CAPTCHA before inputs can be made and throttling/limiting queries to the model. Log inputs and events. Ensure ML libraries and systems are patched and up to date. Limit user access to the model and implement least privilege authority practices. Secure acquisition/storage/transport/decommissioning of model and data can prevent transfer learning/data/model poisoning-based attacks. These strong conventional security measures can protect against Reprogramming ML, attacking the supply chain, and exploiting software dependencies threats.

(2) **Develop risk assessment/security assessment before starting ML projects**-Whilst this is a property of enforcing strong conventional cyber security, it was too important not to have its own principle. Consider the application of the ML, the types of ML and their associated vulnerabilities and the generation process to develop a risk assessment. Consider who, why and how an attack might benefit from attacking the application. Mission-critical and security-sensitive AI would likely require a more secure approach to AI development. Furthermore, some applications may further increase risk by utilising processes such as continual learning, which prove additional attack vectors for attackers [27]. A real-world example of this attack was the Microsoft Tay Twitter-trained bot in the first few hours of deployment [65]. Moreover, whilst convenient, the reuse of models (transfer learning/AI repurposing) and data increases the opportunities for exploitation.

(3) **Maximise the model's robustness**-Maximising model robustness reduces the attacking space available to prevent perturbation and adversarial examples. The exploitation of model robustness is one of the simplest of attacks to implement given the scalability of the attack (many models are not robust against all possible adversarial space). It is also possible to use the attack to cause a failure (sometimes not so obvious) of mission-critical systems. Maximising the model's robustness should also provide the additional benefits of protecting against accidental adversarial attacks/errors. Furthermore, the work of [27] also considers forcing model architecture and capacity proportionate to training data to improve robustness and reduce unnecessary feature space whilst covering the distribution of training data.

(4) **Maximise explainability and insight for trusted developers and minimise for untrusted users**-Explainability should play a significant role in supporting the development of adversarial AI defences in the coming years. Having greater explainability provides many benefits, but in the context of security, having a better understanding of the ML system's decision processes can support; locating poisoned models, the system limitations, transferability, robustness, and trustworthiness to enhance the security of the system. However, this knowledge could also be used to find and exploit weaknesses, such as locating adversarial space. Therefore one should limit the explainability outputted by the model to untrusted users as well as the technical details of the model, e.g., parameter values and model architecture.

(5) **Regulate the input and output of the model**-This principle ties in with revealing too much information about the model, which could be used for nefarious activities against the model, e.g., one could avoid revealing exact probabilities of detection for a classification model to prevent some gradient-based attacks. Regulating the input of the model can prevent adversarial queries from successfully triggering backdoors or exploiting the model [27].

(6) **Recognise the exploitation of the model and understand the risks of exploitation**-Having indicators of compromise for the model will not stop adversarial attacks from happening but could allow one to identify and isolate threats. Understanding the effects of a compromised system will allow one to understand the risks and develop effective tailored security approaches in depth.

(7) **Sensor redundancy/harmonisation and data correlation**-Utilising multiple fused sensor inputs can be used to bring assurance to situational awareness modules. For example, relying exclusively on a single sensor to deduce the presence or absence of objects is likely to increase the ease of attack. However, sensor fusion of the CV, LiDAR, forward-looking sonar, AIS, and RADAR, all feeding into a navigational AI system, would increase the overall robustness of the system. The requirement of fooling multiple sensors would be exponentially more challenging to mount a successful attack than that of a single sensor, and anomalous behaviour becomes more apparent if not all sensors are fooled, which could lead to a lower confidence weighting provided to data which significantly differs / appears adversarial from other sensors' results.

4.1. Countermeasures

This section details possible countermeasures which could be used to implement the six principles proposed above to protect MAS AI against AAI; these are summarised in Table 2. We first consider adversarial training. Adversarial training requires synthesising adversarial samples from a model and using those adversarial samples as training data to train the model to produce an element of model robustness against adversarial attacks [24, 48]. The samples can be iteratively generated by retraining the model and regenerating adversarial training data [66]. By creating new samples, we increase the distribution of the dataset for the robustness and accuracy of the model. The first instance of adversarial training was [24], which used FSGM to create and then inject samples into the training data. Many variations and improvements of adversarial training exist, such as using GANs [67-69]. DNN verification tools can be used to locate adversarial samples [70]. It is worth noting to search all the sample space is an NP-hard problem. The three-step null label method blocks the transferability of attacks between models. The method adds a new null label to a one-hot encoding, and the network is trained with some adversarial samples which are labelled as null, therefore if the model input is classified highly as null, this indicates an adversarial input [71] and reduces adversarial attacks happening between models.

Further countermeasures include regularisation. Regularisation is used in ML to prevent overfitting of a model during training (adding a penalty, i.e. regular term, to a cost function); this can reduce the possible adversarial attack space by making the model more robust to small perturbations. Methods of regularisation include feature pruning to prune activations and neurons from a network [72]. Neuron dropout can be used during model training to stochastically remove neurons which can prevent overfitting on small datasets and potentially remove a backdoor [73]. Other methods include adding a layer of random noise to the model after the input layer so that during forward propagation, the noise creates slightly different outcomes to make the model more robust against small permutations [74].

Ensemble methods is a term used to represent the combination of multiple ML models constituting an overall model. Common methods include using gradient boosting such as XGBoost [75]. This method can reduce the likelihood of training data poisoning as the individual models are trained on different datasets; therefore, when combining the models, the good models can reduce the effect of the poisoned models [76]. It is also possible that adversarial samples are fewer with a greater distribution of training data.

Input validation (or sanitation) and manipulation/preprocessing, which can control the data going into the model can help prevent attacks. Input reconstruction can be used to remove the adversarial effect from input data analogous to input sanitisations to prevent Structured Query Language (SQL) attacks. Input reconstruction was suggested in the work of [77], which proposed transforms applied to input images before making model predictions (clipping, JPEG compression, rescaling depth, etc.). Feature compression/data compression, as in ComCNN [78], can be used to reduce the feature depth of the input, e.g., reduce the colour depth of pixels and increase robustness at the cost of reduced input accuracy. Inputs could also be filtered, smoothed, and have random noise applied on input to sanitise the data. Regression analysis can be used to locate data outliers during input [79]. Image preprocessing (such as random image padding) can be used to prevent a backdoor attack from being triggered. Input denoising works by attempting to remove noise from the input. Tools include (high-level representation guided denoiser HGD [80]). GANs can be used to clean data by recreating a similar image to the input. MagNet [81] and defence-GAN [82] are tools that can also be used to recreate input images with reduced adversarial noise on a more similar manifold to the benign data.

A countermeasure that would be useful against the perturbation attacks shown in this paper, and others, is gradient masking. Gradient masking can reduce the likelihood of an attacker acquiring the model's gradient and hence reduce against gradient-based adversarial attacks [83]. There is no reduction in the adversarial sample size. However, this attack aims to make it more difficult for open-box probing, by masking the useful gradient, at finding these samples. The effect works by creating less smooth (sharper) boundaries for classification. Gradient masking methods include model distillation and dropout.

Model distillation requires training a smaller, less complex model based on the original complex model. This reduction/compression in model complexity, whilst

Table 2. *The associated defensive countermeasures to prevent exploitation from each adversarial attack. There is much overlap between the defences resulting in the effect of the sum of multiple defensive measures being greater than its individual constituents.*

AAI Attack	Defences
Perturbation Attack	Adversarial training, Regularisation, Ensemble methods, Input validation and manipulation/preprocessing, Gradient masking, Model distillation, Adversarial sample detection, Explainability
Poisoning Attack	Regularisation, Ensemble methods, Input validation and manipulation/preprocessing, Explainability
Model Inversion	Input validation and manipulation/preprocessing, Adversarial sample detection, Explainability, Preventing information loss
Membership Inference	Input validation and manipulation/preprocessing, Adversarial sample detection, Explainability, Preventing information loss
Model stealing	Preventing information loss
Reprogramming the ML	Regularisation, Ensemble methods, Gradient masking, Model distillation, Explainability, Preventing information loss
Adversarial Example in Physical Domain	Adversarial training, Regularisation, Ensemble methods, Input validation and manipulation/preprocessing, Gradient masking, Model distillation, Adversarial sample detection, Explainability
ML Provider Recovering Training Data	Regularisation, Ensemble methods, Input validation and manipulation/preprocessing, Gradient masking, Model distillation, Explainability, Preventing information loss
Attacking the ML Supply Chain	Strong conventional cybersecurity practices
Backdoor ML	Regularisation, Ensemble methods, Input validation and manipulation/preprocessing, Model distillation, Explainability, Preventing
Exploit Software Dependencies	Strong conventional cybersecurity practices

maintaining similar model accuracy, can help prevent adversarial attacks by creating a model with smoother loss and hence less sensitive to small perturbations [84]. The original output is used as a soft label, and the original label is used as a hard label.

Adversarial sample detection can help protect MAS AI against adversarial AI via input monitoring. Instead of sterilising the input, one could attempt to detect if the input is an adversarial sample before being accepted or rejected by the model.

The input can then be determined as adversarial or benign before being decided whether to be entered into the model. For example, these detection models can be created as a binary classifier and determine if the input follows a similar distribution to the training data [85].

Explainability covers a number of terms, such as trustworthiness, causality, transferability, and informativeness which all support the understanding and hence the security of the ML models. Explainability is a hot topic, and many methods exist to support model explainability. Improving AI explainability is critical as this sector develops AI for mission-critical operations.

Preventing information loss considers the threat of model stealing in a few ways. To protect against data stealing, one could use PATE [86], which splits the training data into subsets and trains multiple models on the subsets before the models are combined, and the systems vote on the predicted outcome. Watermarking can also be used to place a unique watermark in the model, which can be evaluated to determine if it was stolen [87].

5. CONCLUSION

This work has provided an evaluation of AI security in maritime autonomous systems. A literature review revealed the potential vulnerabilities in MAS AI that could be exposed through a set of adversarial AI test cases strategically designed to test AI used in MAS operations. However, this study of the current state-of-the-art MAS security has also highlighted the inherent vulnerabilities of only testing adversarial AI in limited laboratory environments. Given the extreme differences in marine environments based on location, weather, and time of day, it is also clear that any AAI dataset must also be evaluated in a real-world environment to be truly useful and cyber-resilient maritime AI. After the evaluation of these results in situ, we developed a series of secure AI in MAS principles which can be used to mitigate these threats across the AI's lifecycle.

In further work, we recognise the limited preparation and understating of AAI in MAS technologies by developers, security professionals, and marine regulators. Therefore, knowledge could be disseminated by the secure AI principles and AAI employee training. We would also consider the evaluation of other attacks and their associated defences in the maritime environment (in a range of conditions); we would then consider the effects of underwater distortion etc. Further, we aim to evaluate a range of real-world AI (existing commercial and military AI systems) against AAI-this will allow one to gauge the secondary effects of the attack too, e.g., if one interferes with the CV object detection, how would that impact the collision avoidance module of a vessel? As well as evaluate the effectiveness of AAI and defence in a complex and dynamic environment. Furthermore, we would like to consider the probability of each attack in a maritime autonomous environment with some attacks more effective and likely than others in the real-world environment. As we see greater accessibility of AI, we are also likely to see an increase in the misuse of AI (e.g., AI to support clandestine / smuggling operations) as well an increase in the exploitation of AI systems. The importance of the security of AI is increasing with its use in mission-critical systems (e.g., we are seeing increasing use of maritime autonomous systems by militaries [88, 89]). Whilst many of these AAI attacks have not been utilised in the real world, as the potential financial gain of these attacks increase and the increased use of AI in mission-critical systems, adversaries will look to exploit these methods and the requirement to prepare for the fast-evolving AAI threat landscape is today.

Acknowledgments

This work was supported by the Turing's Defence and Security programme through a partnership with the UK government in accordance with the framework agreement between GCHQ & The Alan Turing Institute. The authors would also like to thank the University of Plymouth for their use of their autonomous fleet in order to collect real world data.

REFERENCES

1. Felski A, Zwolak K. The ocean-going autonomous ship_challenges and threats. J Mar Sci Eng. 2020; 8(1) : 41
2. Kretschmann L, Burmeister H-C, Jahn C. Analyzing the economic benefit of unmanned autonomous ships: an exploratory cost-comparison between an autonomous and a conventional bulk carrier. Res Transp Bus Manag. 2017;25: 76-86.
3. Morris D. Worlds first autonomous ship to launch in 2018. Fortune [Internet]. 2017 [cited 2017 Jul 22]. Available from https://fortune.com/2017/07/22/first-autonomous-ship-yara-birkeland/.
4. Munim ZH. Autonomous ships: a review, innovative applications and future maritime business models. Supply Chain Forum: Int J. 2019;20: 266-279.
5. Porathe T, Prison J, Man Y. Situation awareness in remote control centres for unmanned ships. In: Proceedings of Human Factors in Ship Design &o Operation, 26-27 February 2014, London, UK. Buckinghamshire, UK: CORE; 2014. 93 p.
6. Tsvetkova A, Hellström M. Creating value through autonomous shipping: an ecosystem perspective. Marit Econ Logist. 2022;24: 255-277.
7. Ziajka-Poznańska E, Montewka J. Costs and benefits of autonomous shipping - a literature review. Appl Sci. 2021;11(10):4553.
8. Royce R. Remote and autonomous ships. In: AAWA position paper. Oslo, Norway: DNV; 2016.
9. Anderson M. Bon voyage for the autonomous ship mayflower. IEEE Spectr. 2019;57(1):36-39.
10. Caruana R, Lou Y, Gehrke J, Koch P, Sturm M, Elhadad N. Intelligible models for healthcare: predicting pneumonia risk and hospital 30-day readmission. In: Proceedings of the 21th ACM SIGKDD International Conference on Knowledge Discovery and Data Mining. New York, USA: ACM; 2015. p. 1721-1730.
11. Kong Z, Xue J, Wang Y, Huang L, Niu Z, Li F. A survey on adversarial attack in the age of artificial intelligence. Wirel Commun Mob Comput. 2021;2021: 4907754.
12. Qiu S, Liu Q, Zhou S, Wu C. Review of artificial intelligence adversarial attack and defense technologies. Appl Sci. 2019;9(5):909.

13. Kaluza P, Kölzsch A, Gastner MT, Blasius B. The complex network of global cargo ship movements. J R Soc Interface. 2010;7(48):1093-1103.
14. Askari HR, Hossain MN. Towards utilising autonomous ships: a viable advance in industry 4.0. J Int Marit Saf Environ Aff Shipp. 2022;6(1):39-49.
15. Fan C, Wróbel K, Montewka J, Gil M, Wan C, Zhang D. A framework to identify factors influencing navigational risk for maritime autonomous surface ships. Ocean Eng. 2020;202: 107188.
16. Thombre S, Zhao Z, Ramm-Schmidt H, García JMV, Malkamäki T, Nikolskiy S, et al. Sensors and AI techniques for situational awareness in autonomous ships: a review. In: IEEE Transactions on Intelligent Transportation Systems. Piscataway, NJ: IEEE; 2020.
17. Noel A, Shreyanka K, Gowtham K, Satya K. Autonomous ship navigation methods: a review. In: International Conference on Marine Engineering and Technology Oman 2019 (ICMET Oman) [Internet]; 2019 Nov 5-7; Muscat, Oman. Military Technological College Oman; 2019. Available from https://doi.org/10.24868/icmet.oman.2019.028.
18. Bentes C, Frost A, Velotto D, Tings B. Ship-iceberg discrimination with convolutional neural networks in high resolution SAR images. In: Proceedings of EUSAR 2016: 11th European Conference on Synthetic Aperture Radar. Hamburg, Germany: VDE; 2016. p. 1-4.
19. Wang J, Xiao Y, Li T, Chen CLP. A survey of technologies for unmanned merchant ships. IEEE Access. 2020;8: 224461-224486.
20. Kim H, Kim S-H, Jeon M, Kim JH, Song S, Paik K-J. A study on path optimisation method of an unmanned surface vehicle under environmental loads using genetic algorithm. Ocean Eng. 2017;142: 616-624.
21. Song CH. Global path planning method for USV system based on improved ant colony algorithm. In: Applied mechanics and materials. vol. 568, Switzerland: Trans Tech Publications; 2014. p. 785-788.
22. Zhang Y, Gong D-w, Zhang J-h. Robot path planning in uncertain environment using multi-objective particle swarm optimisation. Neurocomputing. 2013;103: 172-185.
23. Arrieta AB, Díaz-Rodríguez N, Del Ser J, Bennetot A, Tabik S, Barbado A, et al. Explainable artificial intelligence (XAI): concepts, taxonomies, opportunities and challenges toward responsible AI. Inf Fusion . 2020;58: 82-115.
24. Szegedy C, Zaremba W, Sutskever I, Bruna J, Erhan D, Goodfellow I, et al. Intriguing properties of neural networks [Internet]. arXiv; 2013. Available from: https://arxiv.org/abs/1312.6199.
25. Mirsky Y, Demontis A, Kotak J, Shankar R, Gelei D, Yang L, et al. The threat of offensive AI to organizations. Comput Secur. 2022;124: 103006.
26. Caroline B, Christian B, Stephan B, Luis B, Giuseppe D, Damiani E, et al. Securing machine learning algorithms. Athens, Greece: ENISA; 2021.
27. Kate S. Introducing our new machine learning security principles. Boca Raton, FL: CRC Press; 2022.
28. Seymour J, Tully P. Weaponizing data science for social engineering: automated E2E spear phishing on twitter. Black Hat USA. 2016;37: 1-39.
29. Dang H, Liu F, Stehouwer J, Liu X, Jain AK. On the detection of digital face manipulation. In: Proceedings of the IEEE/CVF Conference on Computer Vision and Pattern Recognition. Seattle, WA, USA: IEEE; 2020. p. 5781 – 5790.

30. Akhtar N, Mian A. Threat of adversarial attacks on deep learning in computer vision: a survey. IEEE Access. 2018;6: 14410-14430.

31. Elsayed G, Shankar S, Cheung B, Papernot N, Kurakin A, Goodfellow I, et al. Adversarial examples that fool both computer vision and time-limited humans. In: 32nd Conference on Neural Information Processing Systems (NeurIPS 2018), Montréal, Canada. Red Hook, NY: Curran Associates Inc.; 2018. 31 p.

32. Wang Z, She Q Ward TE. Generative adversarial networks in computer vision: a survey and taxonomy. ACM Comput Surv (CSUR). 2021;54(2):1-38.

33. Al-Dujaili A, Huang A, Hemberg E, OReilly U-M. Adversarial deep learning for robust detection of binary encoded malware. In: 2018 IEEE Security and Privacy Workshops (SPW). San Francisco, USA: IEEE; 2018. p. 76 – 82.

34. Kolosnjaji B, Demontis A, Biggio B, Maiorca D, Giacinto G, Eckert C, et al. Adversarial malware binaries: evading deep learning for malware detection in executables. In: 2018 26th European Signal Processing Conference (EUSIPCO). San Francisco, USA: IEEE; 2018. p. 533-537.

35. Li D, Li Q Ye Y, Xu S. Arms race in adversarial malware detection: a survey. ACM Comput Surv (CSUR). 2021;55(1):1-35.

36. Maiorca D, Biggio B, Giacinto G. Towards adversarial malware detection: lessons learned from pdf-based attacks. ACM Comput Surv (CSUR). 2019;52(4):1-36.

37. Morris JX, Lifland E, Yoo JY, Grigsby J, Jin D, Qi Y. Textattack: a framework for adversarial attacks, data augmentation, and adversarial training in NLP [Internet]. arXiv; 2020. Available from: https://arxiv.org/abs/2005.05909.

38. Wallace E, Feng S, Kandpal N, Gardner M, Singh S. Universal adversarial triggers for attacking and analyzing NLP [Internet]. arXiv; 2019. Available from: https://arxiv.org/abs/1908.07125.

39. Juuti M, Szyller S, Marchal S, Asokan N. PRADA: protecting against DNN model stealing attacks. In: 2019 IEEE European Symposium on Security and Privacy (EuroS&P). Stockholm, Sweden: IEEE; 2019. p. 512-527.

40. Wang B, Gong NZ. Stealing hyperparameters in machine learning. In: 2018 IEEE Symposium on Security and Privacy (SP). San Francisco, USA: IEEE; 2018. p. 36-52.

41. Kessler J. Data protection in the wake of the GDPR: California's solution for protecting "the world's most valuable resource". South Calif Law Rev. 2019;93: 99.

42. Sewak M, Sahay SK, Rathore H. Adversarialuscator: an adversarial-DRL based obfuscator and metamorphic malware swarm generator. In: 2021 International Joint Conference on Neural Networks (IJCNN). Piscataway, NJ: IEEE; 2021. p. 1-9.

43. Gu T, Dolan-Gavitt B, Garg S. Badnets: identifying vulnerabilities in the machine learning model supply chain [Internet]. arXiv; 2017. Available from: https://arxiv.org/abs/1708.06733.

44. Barreno M, Nelson B, Sears R, Joseph AD, Tygar JD. Can machine learning be secure? In: Proceedings of the 2006 ACM Symposium on Information, Computer and Communications Security. New York, USA: ACM; 2006. p. 16 – 25.

45. Biggio B, Nelson B, Laskov P. Support vector machines under adversarial label noise. In: Asian Conference on Machine Learning. Cambridge, MA: PMLR; 2011. p. 97-112.

46. Fredrikson M, Jha S, Ristenpart T. Model inversion attacks that exploit confidence information and basic countermeasures. In: Proceedings of the 22nd ACM SIGSAC

Conference on Computer and Communications Security. New York: ACM; 2015. p. 1322-1333.

47. Orekondy T, Schiele B, Fritz M. Knockoff nets: stealing functionality of black-box models. In: Proceedings of the IEEE/CVF Conference on Computer Vision and Pattern Recognition. Piscataway, NJ: IEEE; 2019. p. 4954 – 4963.
48. Goodfellow IJ, Shlens J, Szegedy C. Explaining and harnessing adversarial examples [Internet]. arXiv; 2014. Available from: https://arxiv.org/abs/1412.6572.
49. Su J, Vargas DV, Sakurai K. One pixel attack for fooling deep neural networks. IEEE Trans Evol Comput. 2019;23(5):828-841.
50. Kurakin A, Goodfellow I, Bengio S. Adversarial machine learning at scale [Internet]. arXiv; 2016. Available from: https://arxiv.org/abs/1611.01236.
51. Moosavi-Dezfooli S-M, Fawzi A, Frossard P. Deepfool: a simple and accurate method to fool deep neural networks. In: Proceedings of the IEEE Conference on Computer Vision and Pattern Recognition. Las Vegas, NV, USA: IEEE; 2016. p. 2574-2582.
52. Papernot N, McDaniel P, Jha S, Fredrikson M, Celik ZB, Swami A. The limitations of deep learning in adversarial settings. In: IEEE European Symposium on Security and Privacy (EuroS@P). Piscataway, NJ: IEEE; 2016. p. 372-387.
53. Chen J, Su M, Shen S, Xiong H, Zheng H. Poba-ga: perturbation optimized black-box adversarial attacks via genetic algorithm. Comput Secur. 2019;85: 89-106.
54. Xiao C, Li B, Zhu J-Y, He W, Liu M, Song D. Generating adversarial examples with adversarial networks [Internet]. arXiv; 2018. Available from: https://arxiv.org/abs/1801.02610.
55. Brown TB, Mané D, Roy A, Abadi M, Gilmer J. Adversarial patch [Internet]. arXiv; 2017. Available from: https://arxiv.org/abs/1712.09665.
56. Lee M, Kolter Z. On physical adversarial patches for object detection [Internet]. arXiv; 2019. Available from: https://arxiv.org/abs/1906.11897.
57. Liu X, Yang H, Liu Z, Song L, Li H, Chen Y. Dpatch: an adversarial patch attack on object detectors [Internet]. arXiv; 2018. Available from: https://arxiv.org/abs/1806.02299.
58. Song D, Eykholt K, Evtimov I, Fernandes E, Li B, Rahmati A, et al. Physical adversarial examples for object detectors. In: 12th USENIX Workshop on Offensive Technologies (WOOT 18). Berkeley, CA, USA: USENIX; 2018.
59. Wu H, Yunas S, Rowlands S, Ruan W, Wahlstrom J. Adversarial detection: attacking object detection in real time [Internet]. arXiv; 2022. Available from: https://arxiv.org/abs/2209.01962.
60. Yang C, Kortylewski A, Xie C, Cao Y, Yuille A. Patchattack: a black-box texture-based attack with reinforcement learning. In: Computer Vision-ECCV 2020: 16th European Conference, Glasgow, UK, August 23-28, 2020, Proceedings, Part XXVI. Berlin: Springer; 2020. p. 681-698.
61. Hoory S, Shapira T, Shabtai A, Elovici Y. Dynamic adversarial patch for evading object detection models [Internet]. arXiv; 2020. Available from: https://arxiv.org/abs/2010.13070.
62. Liang H, He E, Zhao Y, Jia Z, Li H. Adversarial attack and defense: a survey. Electronics. 2022;11(8):1283.

63. Yoo J-W, Jo Y-H, Cha Y-K. Artificial intelligence for autonomous ship: potential cyber threats and security. J Korea Inst Inf Secur Cryptol. 2022;32(2):447-463.

64. Kumar RSS, Brien DO, Albert K, Viljöen S, Snover J. Failure modes in machine learning systems [Internet]. arXiv; 2019. Available from: https://arxiv.org/abs/1911.11034.

65. Wolf MJ, Miller K, Grodzinsky FS. Why we should have seen that coming: comments on microsoft's tay "experiment", and wider implications. ACM SIGCAS Comput Soc. 2017;47(3):54-64.

66. Kim S, Kim H. Zero-centered fixed-point quantization with iterative retraining for deep convolutional neural network-based object detectors. IEEE Access. 2021;9: 20828-20839.

67. Kannan H, Kurakin A, Goodfellow I. Adversarial logit pairing [Internet]. arXiv; 2018. Available from: https://arxiv.org/abs/1803.06373.

68. Lee H, Han S, Lee J. Generative adversarial trainer: defense to adversarial perturbations with gan [Internet]. arXiv; 2017. Available from: https://arxiv.org/abs/1705.03387.

69. Madry A, Makelov A, Schmidt L, Tsipras D, Vladu A. Towards deep learning models resistant to adversarial attacks [Internet]. arXiv; 2017. Available from: https://arxiv.org/abs/1706.06083.

70. Qian YG, Zhang XM, Wang B, Li W, Chen JH, Zhou WJ, et al. Towards robust DNNs: a Taylor expansion-based method for generating powerful adversarial examples [Internet]. arXiv; 2020. Available from:. https://arxiv.org/abs/2001.08389.

71. Hosseini H, Chen Y, Kannan S, Zhang B, Poovendran R. Blocking transferability of adversarial examples in black-box learning systems [Internet]. arXiv; 2017. Available from: https://arxiv.org/abs/1703.04318.

72. Dhillon GS, Azizzadenesheli K, Lipton ZC, Bernstein J, Kossaifi J, Khanna A, et al. Stochastic activation pruning for robust adversarial defense [Internet]. arXiv; 2018. Available from: https://arxiv.org/abs/1803.01442.

73. Liu K, Dolan-Gavitt B, Garg S. Fine-pruning: defending against backdooring attacks on deep neural networks. In: Research in Attacks, Intrusions, and Defenses: 21st International Symposium, RAID 2018, Proceedings 21; 2018 Sep 10-12; Heraklion, Crete, Greece. Cham: Springer; 2018. p. 273-294.

74. Liu X, Cheng M, Zhang H, Hsieh C-J. Towards robust neural networks via random self-ensemble. In: Proceedings of the European Conference on Computer Vision (ECCV). Cham: Springer; 2018. p. 369-385.

75. Chen T, He T, Benesty M, Khotilovich V, Tang Y, Cho H. 2015. Xgboost: extreme gradient boosting, R package version 0.4-2. 1-4.

76. Li D, Li Q. Adversarial deep ensemble: evasion attacks and defenses for malware detection. IEEE Trans Inf Forensics Secur. 2020;15: 3886-3900.

77. Guo C, Rana M, Cisse M, Van Der Maaten L. Countering adversarial images using input transformations [Internet]. arXiv; 2017. Available from: https://arxiv.org/abs/1711.00117.

78. Jia X, Wei X, Cao X, Foroosh H. ComDefend: an efficient image compression model to defend adversarial examples. In: Proceedings of the IEEE/CVF Conference on Computer Vision and Pattern Recognition. Long Beach, CA, USA: IEEE; 2019. p. 6084-6092.

79. Jagielski M, Oprea A, Biggio B, Liu C, Nita-Rotaru C, Li B. Manipulating machine learning: poisoning attacks and countermeasures for regression learning. In: 2018 IEEE Symposium on Security and Privacy (SP). Piscataway, NJ: IEEE; 2018. p. 19-35.

80. Liao F, Liang M, Dong Y, Pang T, Hu X, Zhu J. Defense against adversarial attacks using high-level representation guided denoiser. In: Proceedings of the IEEE Conference on Computer Vision and Pattern Recognition. Bellingham, WA: SPIE; 2018. p. 1778-1787.

81. Meng D, Chen H. Magnet: a two-pronged defense against adversarial examples. In: Proceedings of the 2017 ACM SIGSAC Conference on Computer and Communications Security. New York, USA: ACM; 2017. p. 135 – 147.

82. Samangouei P, Kabkab M, Chellappa R. Defense-gan: protecting classifiers against adversarial attacks using generative models [Internet]. arXiv; 2018. Available from: https://arxiv.org/abs/1805.06605.

83. Folz J, Palacio S, Hees J, Dengel A. Adversarial defense based on structure-to-signal autoencoders. In: 2020 IEEE Winter Conference on Applications of Computer Vision (WACV). Piscataway, NJ: IEEE; 2020. p. 3568-3577.

84. Papernot N, McDaniel P, Wu X, Jha S, Swami A. Distillation as a defense to adversarial perturbations against deep neural networks. In: 2016 IEEE Symposium on Security and Privacy (SP). Piscataway, NJ: IEEE; 2016. p. 582 – 597.

85. Tanay T, Griffin L. A boundary tilting persepective on the phenomenon of adversarial examples [Internet]. arXiv; 2016. Available from: https://arxiv.org/abs/1608.07690.

86. Papernot N, Abadi M, Erlingsson U, Goodfellow I, Talwar K. Semi-supervised knowledge transfer for deep learning from private training data [Internet]. arXiv; 2016. Available from: https://arxiv.org/abs/1610.05755.

87. Adi Y, Baum C, Cisse M, Pinkas B, Keshet J. Turning your weakness into a strength: watermarking deep neural networks by backdooring. In: 27th USENIX Security Symposium (USENIX Security 18). Berkeley, CA, USA: USENIX; 2018. p. 1615-1631.

88. Hall A. Autonomous minehunter to trial uncrewed operations in the gulf. Navy News [Internet]; 2023 [cited 2023 Feb 13]. Available from: https://www.royalnavy.mod.uk/news-and-latest-activity/news/2023/february/13/20230213autonomous-minehunter-to-trial-uncrewed-operations-in-the-gulf.

89. Hall A. Dstl and DASA research underpins royal navy maritime autonomy. Navy News [Internet]; 2023 [cited 2023 Jan 26]. Available from: https://www.gov.uk/government/news/dstl-and-dasa-researchunderpins-royal-navy-maritime-autonomy.

CHAPTER 5

IoT OTH Maritime Surveillance Service over Satellite Network in Equatorial Environment: Analysis, Design and Deployment

Ranko Petrovic[1,2], Dejan Simic[1], Zoran Cica[3], Dejan Drajic[3], Marko Nerandzic[2,3] and Dejan Nikolic[4]

[1] *Faculty of Organizational Sciences, University of Belgrade, 11000 Belgrade, Serbia*
[2] *Vlatacom Institute of High Technologies, 11070 Belgrade, Serbia*
[3] *School of Electrical Engineering, University of Belgrade, 11120 Belgrade, Serbia*
[4] *HENSOLDT Sensors GmbH, 88090 Immenstaad, Germany*

ABSTRACT

This paper explores the challenges and constraints when over the horizon (OTH) maritime surveillance service utilizes an Internet of Things (IoT) as its backbone. The service is based on High Frequency Surface Wave Radars (HFSWRs) and relies on a satellite communication network as its communication infrastructure in harsh environments. The complete IoT OTH maritime surveillance network is currently deployed in the Gulf of Guinea, which due to its tropical climate represents an unfavorable environment for sensors and communications. In this paper, we have examined the service performance under various meteorological conditions specific to the Gulf of Guinea. To the best of our knowledge, this is the first analysis of IoT OTH maritime surveillance service in equatorial environment. Our analysis aims to mathematically describe the impact of harsh weather conditions on the performance of the service in order to mitigate it with careful overall system design and provide constant quality of the service. Analyses presented in the paper show that average service latency is about 90 s*, but it can rise to about* 120 s*, which is used as a key information during the sensor data fusion algorithm design. Validity of the analyses is demonstrated through high quality of service with an outage probability of just* 0.1% *in the driest months* *up to the 0.7% in the rainiest months. The work presented here can be used as a guideline for deployment of maritime surveillance service solutions in other equatorial regions. Moreover, the gained experience presented in this paper will significantly facilitate future expansions of the existing maritime surveillance network with more HFSWRs. This will be done in such a way that it will not affect the quality of service of the entire system on a large scale.*

Keywords: *Africa; communication networks; harsh environment; Internet of Things; maritime safety; satellite communication*

1. INTRODUCTION

In recent decades, piracy represents a growing trend, especially in the less regulated international waters, such as South-East Asia, East and West Africa and Caribbean [1]. In the East Africa, during the last few years, piracy has decreased due to the intervention of the International organizations [2,3]. Unfortunately, this decrease of pirate attacks in the East Africa does not mean that all other regions in the World experienced the same trend. On the contrary, some regions experienced a tremendous increase, such as the Gulf of Guinea [4-6]. In situations such as this, nearly all maritime nations are forced either to increase safety of their waters and Exclusive Economic Zones (EEZs) or to suffer significant losses both in human lives and goods rightfully belonging to them. Unfortunately, EEZs are huge bodies of water stretching over hundreds of thousands square miles, which makes their complete surveillance quite challenging endeavor.

Although primarily developed for a remote measurement of the ocean surfaces and collection of oceanographic data such as wave spectra or sea currents [7-10], High Fre- quency Surface Wave Radars (HFSWRs) nowadays find an increasing usage in the vessel detection and tracking at over the horizon (OTH) distances [11-16]. The major advantage of the HFSWRs compared to the sensors usually used for maritime surveillance (such as microwave radars, electro-optical systems) is its coverage area of about 200 nautical miles, which corresponds to the EEZ. Moreover, HFSWRs provide that capability at a price tag far lesser than the price of microwave sensors and moving platforms (such as vessels and aircrafts) needed to overcome the microwave sensors disadvantages and thus cover complete EEZ. Furthermore, since HFSWRs do not require moving platforms, they can monitor the sea surface and track vessels continuously during the whole year and under all weather conditions.

Despite the fact that the term Internet of Things (IoT) is usually related to sensor networks consisting of small size sensors which have limited coverage area [17], IoT can also refer to physically large sensors which can individually cover an area greater than $100,000\ km^2$. The example of such application is development of a maritime surveillance network based on HFSWRs, named IoT OTH maritime surveillance sensor network. The HFSWR used as the primary IoT over the horizon sensor in the Gulf of Guinea (and thus in this paper) has been produced by the Vlatacom Institute [18] and thoroughly described in [19].

However, in spite of its great coverage area, a single HFSWR is not enough to provide uninterrupted surveillance over the complete EEZs of some countries. So, the network of HFSWRs needs to be deployed in order to cover the area of interest. Thus, one problemthe complete EEZ coverage is solved, yet another issue is opened-communication between sensors deployed at different remote sites along huge areas and Command and Control Centers (C2). Unfortunately, in developing countries the communication infrastructure is quite often underdeveloped [20,21], especially in remote areas. To make the already bad situation worse, the Gulf of Guinea is located in the African Equatorial belt thus adding quite harsh environmental factors that influence the signal transmission and vessel detection [22, 23].

In our previous work [24] we have demonstrated the Laboratory for IoT communication infrastructure environment for remote maritime surveillance, which simulates effects of the satellite link as a part of IoT communication layer. In this paper, we will demonstrate the complete IoT environment built around HFSWRs as primary sensors in equatorial area, which uses satellite communication as a feasible communication solution.

The performance analyses consider the complete IoT OTH maritime surveillance service, not only the transmission via the satellite link. Under the IoT OTH maritime surveillance service we consider the raw data gathering from the open sea, data processing by HFSWRs at remote sites and, at the end, the data transfer to the server on a Command and Control Center (C2) location, via the satellite network. Validity of the proposed approach will be demonstrated with the data gathered from the IoT OTH maritime surveillance service deployed in the Gulf of Guinea.

It is shown that the satellite link can be a valid solution for IoT OTH maritime surveillance service, when the whole IoT OTH sensor network and service are designed with environmental communication constraints in mind.

The main contribution of the paper lies in the description and analysis of the IoT OTH maritime surveillance service, based on HFSWRs, that uses satellite communications as the communication backbone in the equatorial region (the Gulf of Guinea). Research presented in this paper quantifies and proves that communication via satellite link is still highly reliable without affecting service availability and accuracy, despite satellite link performance degrading in the critical weather conditions. The highlight of this paper is a mathematical model for IoT OTH maritime surveillance service that provides constraints for the IoT OTH service design that encompass even the worst weather conditions and thus presents a basis for data fusion algorithm design. In the end, the experience gained from the deployed operating maritime surveillance network and analysis provided in this paper can be used to design and deploy similar systems in other equatorial regions of the world.

The rest of the paper is organized as follows: various types of communication infrastructure are analyzed in Section 2. Section 3 presents a description of the whole IoT maritime surveillance sensor network in the Gulf of Guinea, whose IoT OTH maritime surveillance service is an integral part. The Section 4 presents an analysis of typical environmental (weather) conditions which can be found in

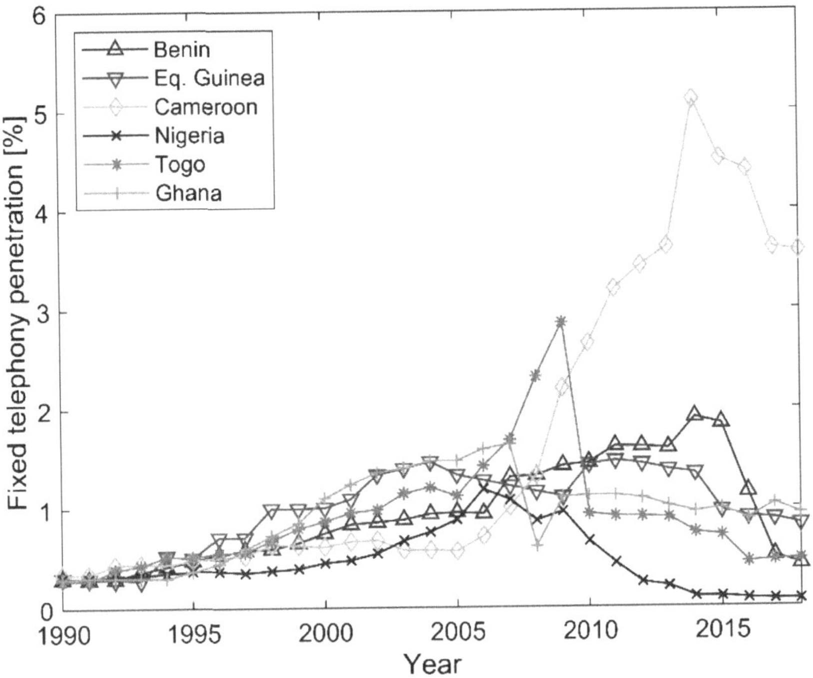

Figure 1. *Fixed telephony penetration in the Gulf of Guinea (based on data available from [25]).*

the area and have an impact on IoT OTH maritime surveillance service. Statistical analyses of the IoT OTH maritime surveillance service behavior over the year are presented in Section 5. The conclusions and future work are summarized in Section 6.

2. ANALYSIS OF AVAILABLE COMMUNICATION INFRASTRUCTURE IN THE GULF OF GUINEA

Connection between remote sites and C2 centers may be established in the following ways (the most common are listed):

- Fixed line telephony
- Fixed broadband technologies
- Mobile telephony
- Microwave links and
- Satellite links

According to the World Bank report [25], fixed telephone penetration in the Gulf of Guinea peaked in the early 2010s and since then it is in constant decline. As it can be seen from Figure 1, even when it was at the highest level, the penetration reached barely 5 percent and only in one country (Cameron), while the penetration in majority of the countries in the region was less than 2%.

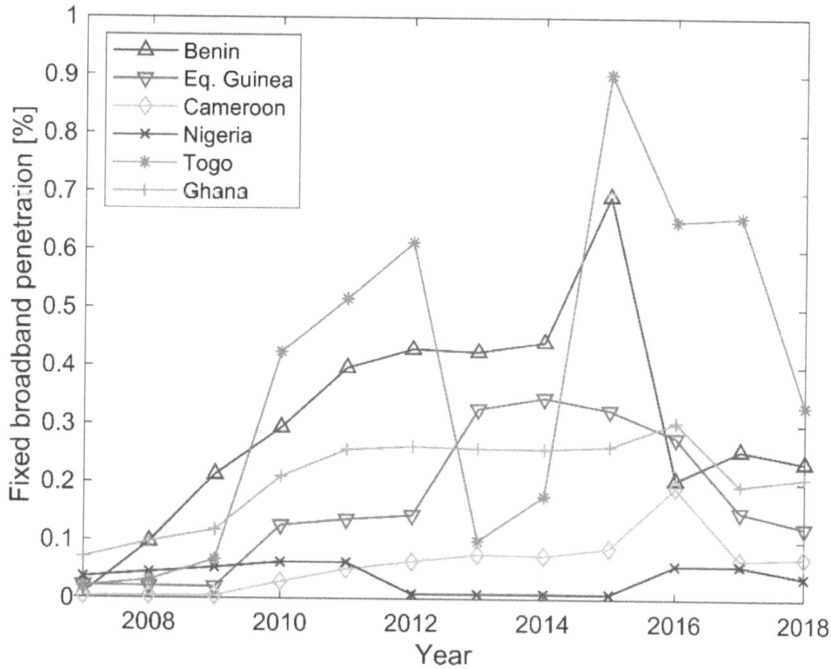

Figure 2. *Fixed Broadband penetration in the Gulf of Guinea (based on data available from [26]).*

From the aforementioned it is clear that fixed telephony cannot be chosen as a primary communication network for maritime sensor network in this part of the World, especially when it comes to uninhabited and inaccessible areas.

According to the same source [26], fixed broadband technologies situation is even more devastating, since penetration in all countries in the region is less than 1% (see Figure 2). One may argue that fixed broadband in the region is not generally available to the public, and that only large companies (mostly oil) possess access to fixed broadband of sufficient capacity. Although in some cases this is completely true, there are multiple reasons why company owned lines cannot be used as primary communication backbone for the maritime sensor network. The first and foremost is that in that manner the complete national maritime sensor network would be dependent on the privately owned foreign company, if indeed the company wants to grant or cede its capacities to the country. Secondly, those companies developed the network for their own purposes in the areas of interest, i.e., from the major population centers to their operations. Taking into account that HFSWRs are located away from both, it is clear why these networks cannot be used as a primary communication network for the maritime sensor network.

Mobile telephony might represent an optimal solution, since no additional infrastructure is needed which makes deployment costs literally negligible. Moreover, since a single HFSWR does not require high data rates (approximately 128 kb/s is quite enough) and total data-transfer is around 5 GB on a monthly

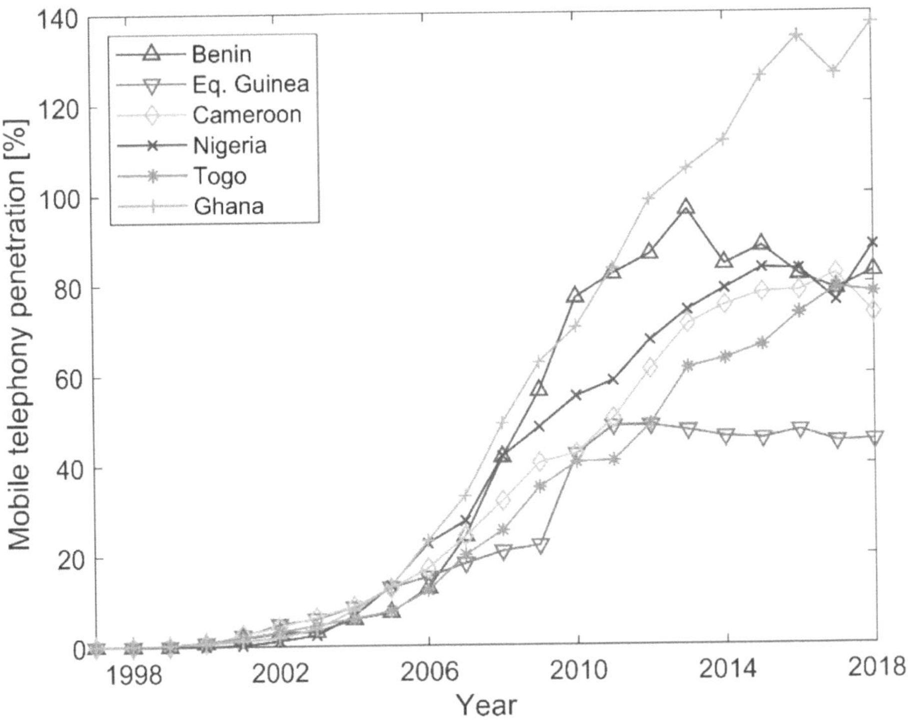

Figure 3. *Mobile telephony penetration in the Gulf of Guinea (based on data available from [27]).*

basis, any currently available mobile network would be suitable as a primary communication network. Furthermore, in many countries there are state owned or controlled mobile operators, so data security should be at a very good level as well. Despite very good penetration [27], which is constantly rising (see Figure 3), mobile telephony is not available in the entire countries, especially in scarcely inhabited areas.

In the present days, a mobile telephony network still cannot be selected as a primary communication network for the maritime sensor network in this part of the World. However, in the future, as it penetrates scarcely populated areas in the region, it may become the best option.

Communication network built with microwave (MW) links as a means of connectivity between HFSWR sites and C2s should be the most reliable and safest way. Yet, on the other hand, the costs of deployment are very high, which limits its usefulness. Taking into account that majority of the countries in the region are not in a position to further raise the maritime sensor network deployment costs, it is clear why it is not suitable here.

As the last means, there is a satellite link. The main advantages of the satellite links are increased coverage area, easy deployment and cost-effective communication solution. Although this solution may look appealing, it also has few downsides. First and foremost is the security issue, since the whole maritime sensor network is dependent on the third party (link provider). The second issue is the availability of data. HFSWR as a sensor is not very affected by meteorological factors and virtually unaffected by rain and clouds compared to the satellite links which might be very susceptible to the weather conditions. However, the IoT OTH maritime surveillance service can be designed in a way to minimize these shortcomings. The first step is measuring dependencies on meteorological factors and designing the service accordingly. The most important is to design sensor integration algorithms so the missing sensor data can be tolerated, as shown in [15]. It could be concluded that the satellite link as a means of data transfer is not a preferable solution. However, currently, in the area of the world where the system is deployed, it is the only feasible solution. This is due to the lack of other networks or lack of funds to develop a MW links network.

3. DESCRIPTION OF THE MARITIME SENSOR NETWORK IN THE GULF OF GUINEA

Full description of the observed maritime sensor network where IoT OTH maritime surveillance service is deployed can be found in [15]. In this section, only a brief description is given in order to introduce a reader with the details needed for the IoT OTH maritime surveillance service analysis, which is the paper's main topic.

The HFSWRs are deployed at two remote locations on the coast of the Gulf of Guinea that are 100 km apart. Both locations are more than 500 km away from C2 center (central location) where the data fusion is performed. All three locations are connected via the satellite links, as shown in Figure 4, representing a generic communication architecture. The data flow for the IoT OTH maritime surveillance service starts with collecting of the raw data by the sensors at two remote locations. The initial processing of the collected data during which the detection files are created is performed on local servers, situated at remote locations. Then the implemented script on local servers starts sending the detection files to central server at Command and Control Center via satellite link using TCP protocol. The algorithm for data fusion, implemented on the central server, uses successfully transferred detection files to produce the final result in the form of accurately detected targets that are being tracked. Time, and thus transmission via the mentioned satellite link, is a key factor in successful pairing of detection files from different remote locations that should participate in the fusion process.

It can be seen in Figure 4 that there are other sensors in the maritime sensor network as well as HFSWRs, such as microwave radars or electro-optical devices. However, the emphasis of this paper is on the IoT OTH maritime surveillance service based on the OTH part of the maritime sensor network and HFSWRs. This is because in the given system architecture, shown in Figure 4, only HFSWRs are capable of detecting uncooperative vessels beyond the line of horizon. The detection process starts with the data collection during the so called the Coherent Integration Time (CIT). After the data are collected, they need to be processed in order to extract vessel detections. Both processes are time consuming since the first step

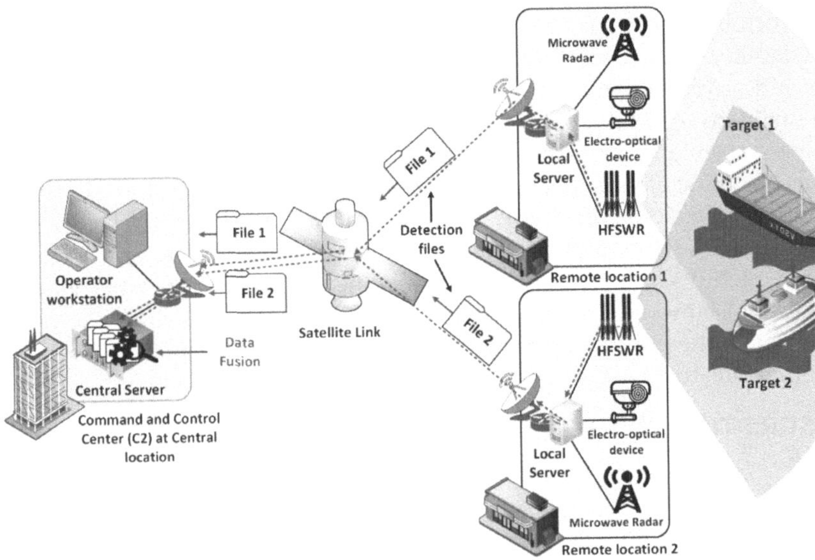

Figure 4. *Complete maritime surveillance solution architecture.*

```
  File: '2020123214538_020.cfar_reg'
  Size: 33763            Blocks: 72
Device: 10302h/66306d    Inode: 6086195
Access: (0644/-rw-r--r--)  Uid: ( 1001/
Access: 2020-05-02 22:47:05.346971594 +0100
Modify: 2020-05-02 22:47:04.583002813 +0100
Change: 2020-05-02 22:47:04.583002813 +0100
```

Figure 5. *Log file showing the times relevant for cfar_reg files.*

(data collection) lasts 33 s. The time needed for the completion of the second step (collected data processing) depends on the amount of data collected during the first step. The end result is a so called cfar_reg file (detection file) which contains possible detections of the vessels in the HFSWR coverage area. Detection files are having a time stamp in their name, written in the form of yyyydddhhmmss (four digits for year, three for day of the year and two for hour, minute and second, respectively), which is actually the time when the first step (data collection and integration) started. Complete time needed for creation of the file may be seen in Figure 5.

The created cfar_reg file is accessed by a script for sending data. The file is sent via a satellite link as shown in Figure 4. From both remote locations, new files are sent to the central location in equal periodical intervals in order to make the vessel tracking process smooth. The client side, located at a remote location, initiates the sending of files to the server side at the central location. The server side receives the files and writes them on the disk of the central server where implemented algorithms perform target tracking and multi-sensor data fusion [14-16]. As it can be seen from Figure 5, data collection started at 21:45:38, while the cfar_reg file was created (modify time) at 22:47:04. One-hour difference is due to the difference between local time and UTC time used in the creation of the file name and should be neglected. The important information lies in difference between minutes and seconds. In this case it is 86.58 s, which is a time needed to completely collect, process and transmit the data at the HFSWR level. Of course, this is just an example and actual times can vary, but it provides an explanation regarding time stamps which are discussed in the following sections.

The C2 center performs fusion on HFSWR data received from both HFSWR locations. Bandwidth assigned for the HFSWR data transmission is 128 kb/s, where 64 kb/s is assigned for control commands, and the other 64 kb/s is assigned for the transmission of detection files. Naturally, the file size impacts the file transfer time. Thus, detailed distribution of the file size analyzed over the period of one year is shown in a form of histogram in Figure 6. Please note that file size depends on various factors such as density of traffic in the HFSWR coverage area, and number of false detections, which depends on weather conditions, and ionospheric interference.

Given the distribution of the detection file sizes shown in Figure 6, the total transfer of detection file can take up to several seconds regarding that the assigned link capacity for detection file transfer is 64 kb/s. It should be noted that transfer duration is not as critical as reliability of transfer, since the lag of several seconds means nothing to the overall data fusion process at the C2 center (please see [15]). However, data missing or long lag due to the link failure or bad weather conditions may put the data fusion process in a difficult position. For this reason, TCP protocol is used for the detection file transfer, and TCP flow and error control can affect the file transfer time. In case of bad satellite link conditions frequent packet retransmissions can occur, which lead to an increase of the file transfer time. Henceforth, this paper will examine performance of the IoT OTH maritime surveillance service in various meteorological conditions. In Figure 6, it can be seen that the majority (over 80%) of the detection files are between 5 and 35kB in size. For this reason, in the remainder of the paper we have provided the analysis for the two most frequent detection file size ranges: 15 – 25kB and 25-35 kB. The conclusions derived for these two ranges apply to other detection file size ranges as well.

4. EFFECTS OF WEATHER CONDITIONS IN THE GULF OF GUINEA ON THE IOT QUALITY OF SERVICE

This section analyzes the IoT OTH maritime surveillance quality of service in meteorological conditions specific to the Gulf of Guinea. These weather conditions are:

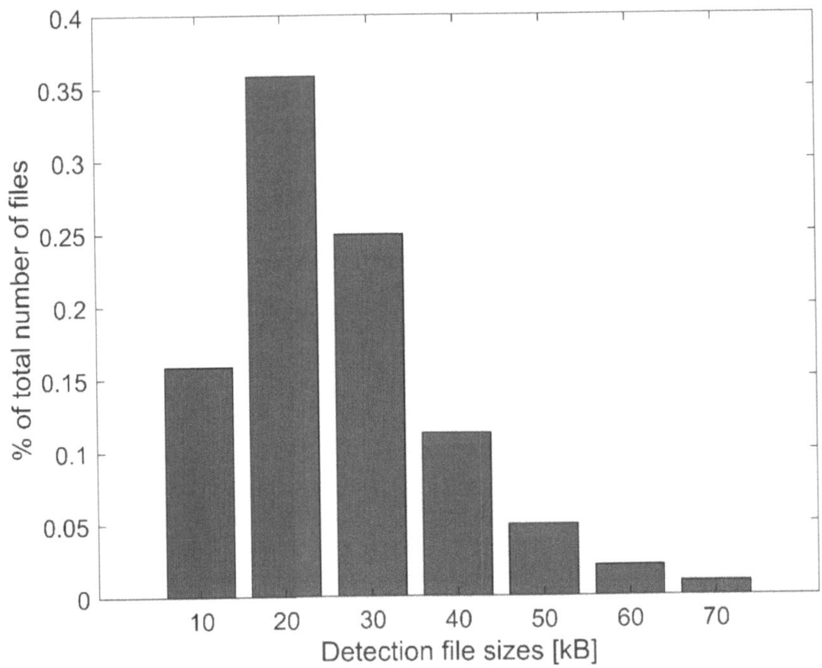

Figure 6. *Distribution of detection (cfar_reg) file sizes.*

- Clear weather, with or without light wind, both on HFSWR sites and at open sea, so called "calm sea",
- Clear weather at HFSWR sites but with strong wind at open sea,
- Storms (heavy rain and wind) at HFSWR sites, which partially cover open sea area.

Please note: Since the IoT OTH maritime surveillance service covers an area of over 100,000 square kilometers, and with HFSWR sites nearly 100 km apart (over the airline), it is almost impossible to have identical meteorological conditions throughout that space. In practice, it often happens that a combination of the aforementioned meteorological conditions occurs.

In the zone of interest there is a clear distinction between dry and rainy periods in the year. Dry season, which has its peak from December to February, is considered as a good weather period from the satellite link point of view, since rainfall is minimal and storms are quite rare. On the other hand, the period from May to July (peak of rain season) is considered as a bad weather period, since during that period rain occurs daily and violent storms are quite frequent. Figure 7 gives an overview of rain precipitation during a year in the zone of interest based on the data from the World Meteorological Organization [28]. Figure 7 shows that the period May-July is the rainiest while the period December-February is the driest.

In the observed case, a satellite connection that operates in X band has been chosen as a backbone of the overall system, since it shows better resilience to

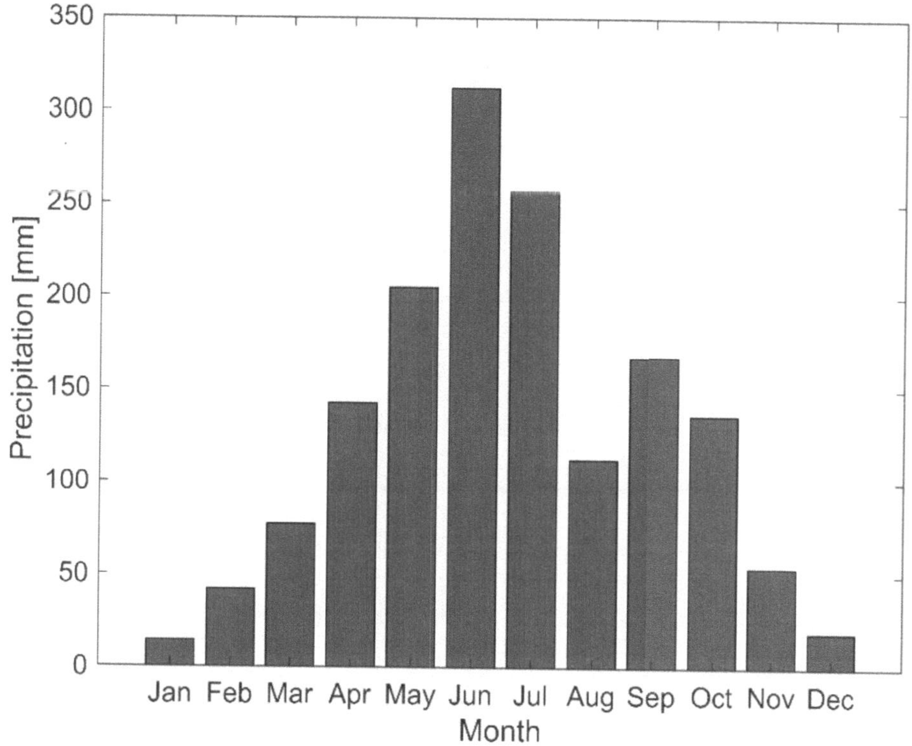

Figure 7. *Average precipitation per months.*

weather conditions in general [29] than Ku or Ka bands [30]. However, since the impact of climate in the zone on the satellite link communication can be significant, this paper will examine its influence on the overall IoT OTH maritime surveillance service as well.

In the further text, three different, most common weather scenarios (atmospheric conditions) in the Gulf of Guinea that influence performances of the developed IoT OTH maritime surveillance sensor network are analyzed in detail.

4.1. Analysis during the "Calm Sea" and Calm Weather on the HFSWR Sites

Such atmospheric conditions are characterized by clear and sunny weather with low intensity wind. The effects of such weather on the operation of HFSWRs are practically negligible. The main reason is the fact that the light wind, which characterizes such weather, has very little effect on the surface of the sea by producing waves of low altitude. Under these conditions, the sea state does not exceed 3 according to the Douglas scale [31], and according to [32], the propagation of EM waves across the sea surface is the best possible. As far as the vessel detection is concerned, it is possible even beyond the nominal HFSWR range. One such example is presented in Figure 8.

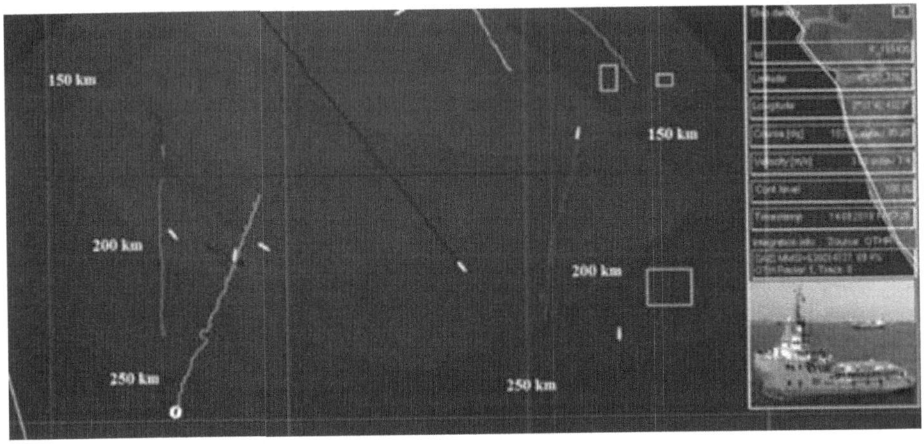

Figure 8. *Surveillance during favorable weather conditions.*

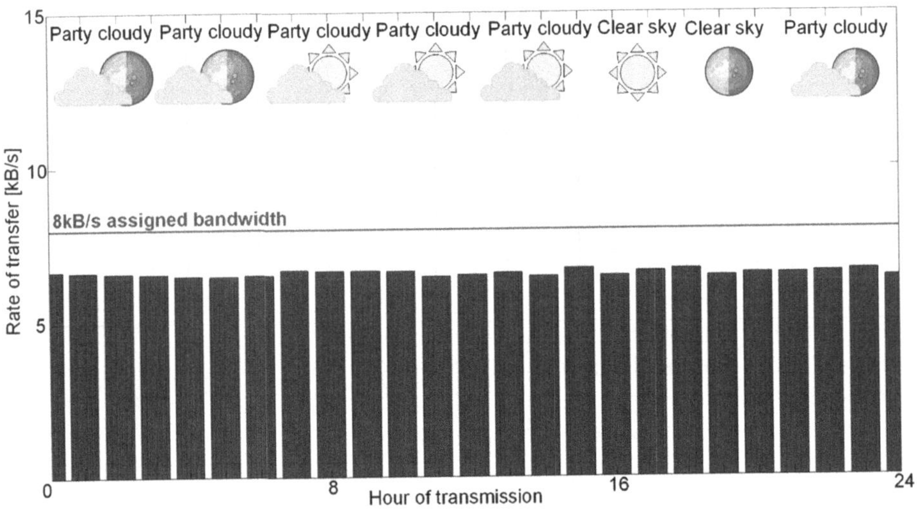

Figure 9. *Transfer rate during favorable weather conditions.*

In Figure 8, the vessel detection and tracking at a distance of 270 km is shown, which is well beyond the nominal HFSWR range. Furthermore, the clear weather does not introduce any difficulties in the operation of the satellite links, and there are no problems with communication between the HFSWRs and the command centers. This means that data will arrive in a timely manner and without significant delays. In clear weather there are no data delays, so it can be said that the IoT OTH maritime surveillance service operates in optimal conditions. Figure 9 shows the satellite link operation in one day by hour during a period in a day in the conditions of the "calm sea" (at the top of the Figure, we have shown the weather conditions for that day in parallel).

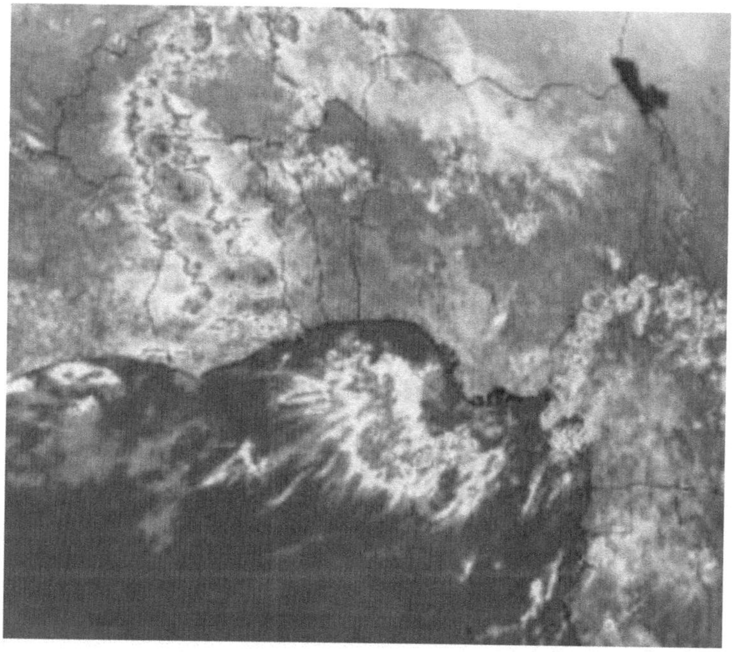

Figure 10. *Bad weather conditions over part of the HFSWR surveillance area.*

4.2. Analysis When Strong Winds Are Present in the HFSWRs Surveillance Area, but Weather on the HFSWR Site Is Calm

In Figure 10 a satellite image of the bad weather over one part of the IoT OTH maritime surveillance area is presented.

As it can be seen from Figure 10, the storm does not cover the coast itself, but "only" a large part of the HFSWR coverage area. Under these weather conditions, the range of HFSWR decreases due to the harsh surface wave propagation conditions [32]. This directly causes decline in the HFSWR vessel detection capabilities. In Figure 11 the detection capabilities of one HFSWR under the conditions described above are shown.

As it can be seen in Figure 11, the range of HFSWR is practically limited to barely 150 km. On the other hand, since the weather conditions at the HFSWR sites are favorable for satellite communications, there are no significant differences in data delays in comparison to the Case A. In other words, from the satellite communication perspective this case is very similar if not identical to the previous one. Henceforth, the analysis done in the Case A is valid here as well.

4.3. Storms (Heavy Rain and Wind) at HFSWR Sites Which Partially Cover the Observation Area

In Figure 12 a satellite image of the bad weather conditions over the HFSWR sites which partially extended to the surveillance area is shown.

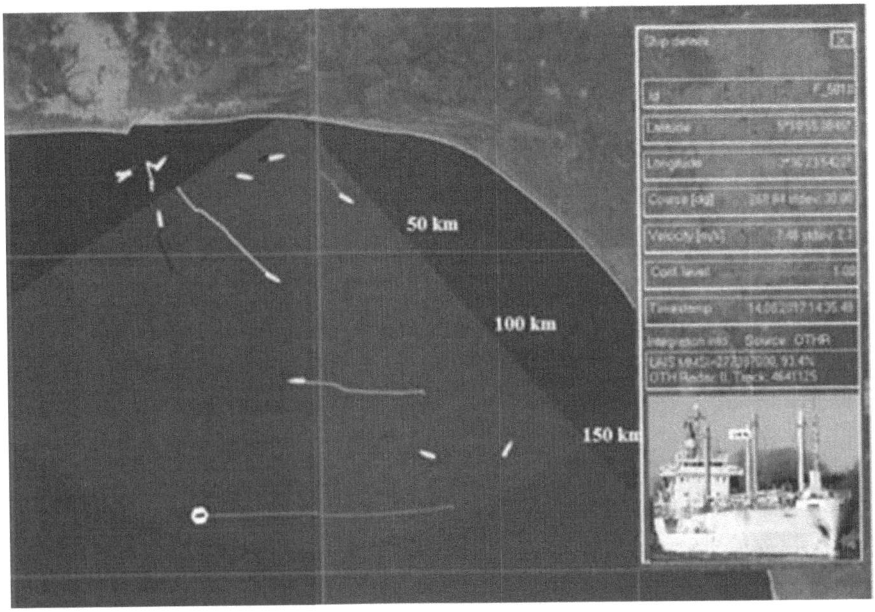

Figure 11. *HFSWR detection capabilities during bad weather conditions over part of the HFSWR surveillance.*

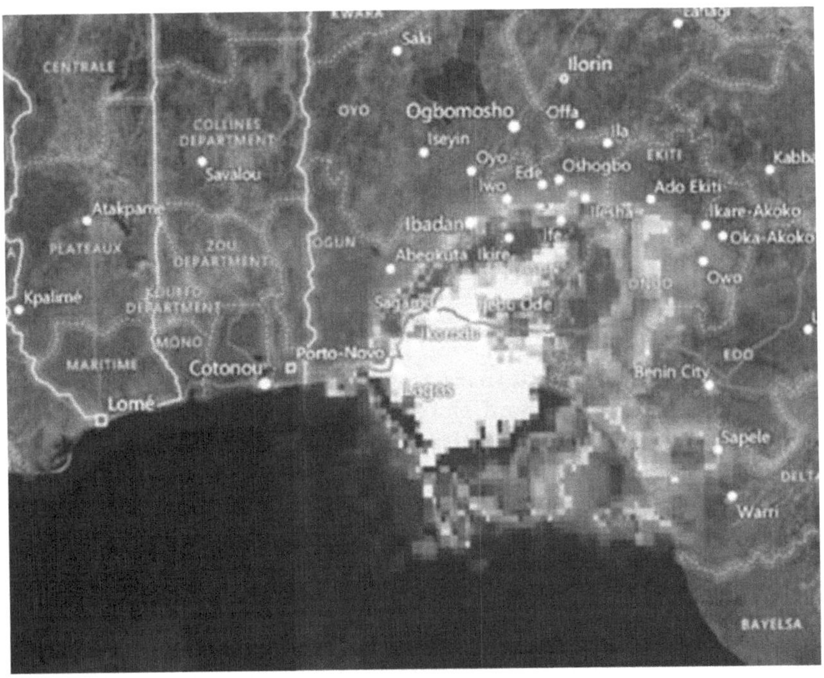

Figure 12. *Bad weather conditions over the HFSWR sites.*

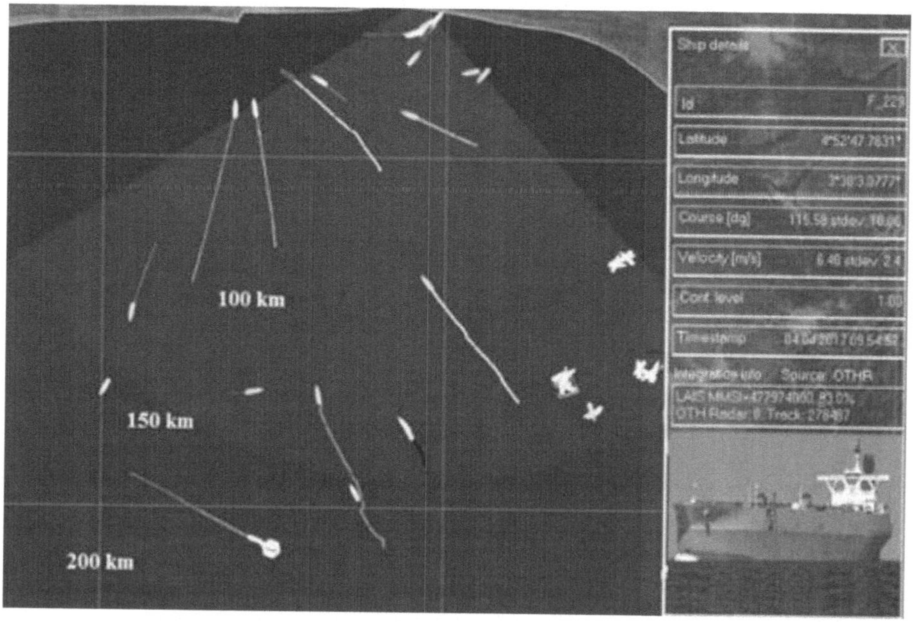

Figure 13. *HFSWR detection capabilities during bad weather conditions over part of the HFSWR surveillance area and HFSWR sites.*

From Figure 12 it can be seen that the storm has hit the coast where HFSWRs were installed, but the majority of the HFSWR surveillance area has not been affected. Even in these weather conditions, the range of HFSWRs will decrease due to harsh surface wave propagation conditions. Yet, this decrease is not as drastic as that in the previous example, since the furthest target is detected at the distance of about 180 km (Figure 13).

It should be emphasized here that detection files from HFSWRs are not received at regular intervals, as the satellite communication is very difficult, due to heavy rain. In the event of severe weather conditions over the HFSWR sites, there are occasional interruptions in the satellite communication.

Therefore, data delays occur and algorithms for multi-sensor data fusion must perform HFSWR tracks predictions and maintain tracks based on them. Despite all this, in Figure 13 it can be seen that the algorithms manage to keep the traces stable, thus providing the full information to the end user. However, this will not be possible without thorough statistical analyses of the IoT OTH maritime surveillance service, which is presented in the following section.

Figure 14 shows the satellite link operation in one day by hour during a period in a day in the conditions of the storms (at the top of the Figure, we have shown the weather conditions for that day in parallel).

Drop in the transfer rate is obvious in comparison to Figure 9 during favorable weather conditions of about 40 – 50%.

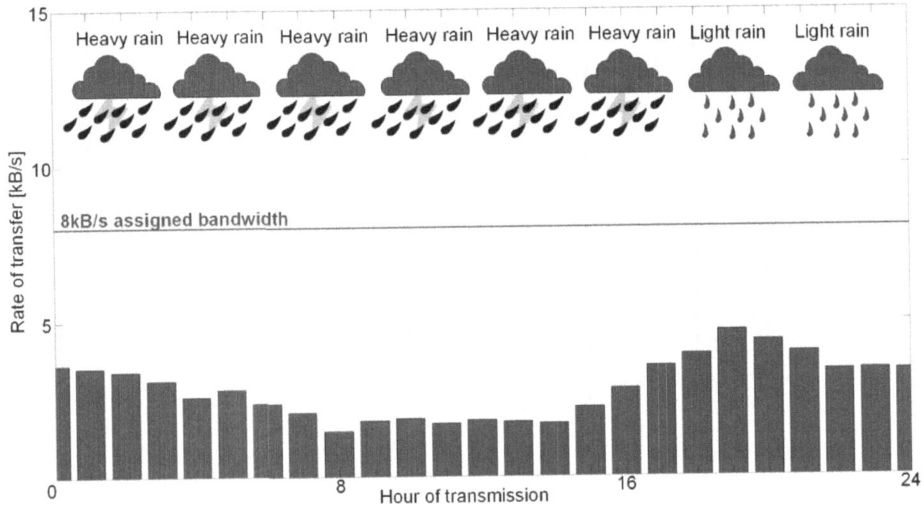

Figure 14. *Transfer rate during bad weather conditions.*

The analysis given in this section shows that weather conditions over the surveilled maritime area and over the HFSWR sites have impact on the IoT OTH maritime surveillance service, but the impact type differs. Weather conditions over the surveilled maritime area impact on the range that HFSWRs cover. Weather conditions also impact the quality of the content in the detection files which then impacts the fusion algorithms effort to provide the high-quality results to the end users. On the other hand, weather conditions over the HFSWR sites impact on the transmission of the detection files to C2 center. In case of severe weather conditions, even temporary outages of the satellite link may occur, which can lead to problems in providing the relevant information about the surveilled maritime area to the end users.

5. STATISTICAL ANALYSIS OF QUALITY OF IOT SERVICE DURING THE YEAR

In this section a statistical analysis of the quality of the IoT OTH maritime surveillance service over a good weather (dry season which has its peak from December to February) and bad weather period (from May to July, peak of rain season) will be presented. Note that the major impact on the IoT OTH maritime surveillance service is the satellite communication that represents the communication backbone of the overall maritime surveillance sensor network depicted in Figure 4. In Section 3 we have explained in detail how detection files are generated and distribution of detection file sizes is presented (Figure 6).

In Figures 15 and 16, we have shown the cumulative distribution function (CDF) of the time frame that is measured from the data collection start at the HFSWR location to the end of the detection file transfer to central location (referred as total detection file transmission time in the remainder of this section).

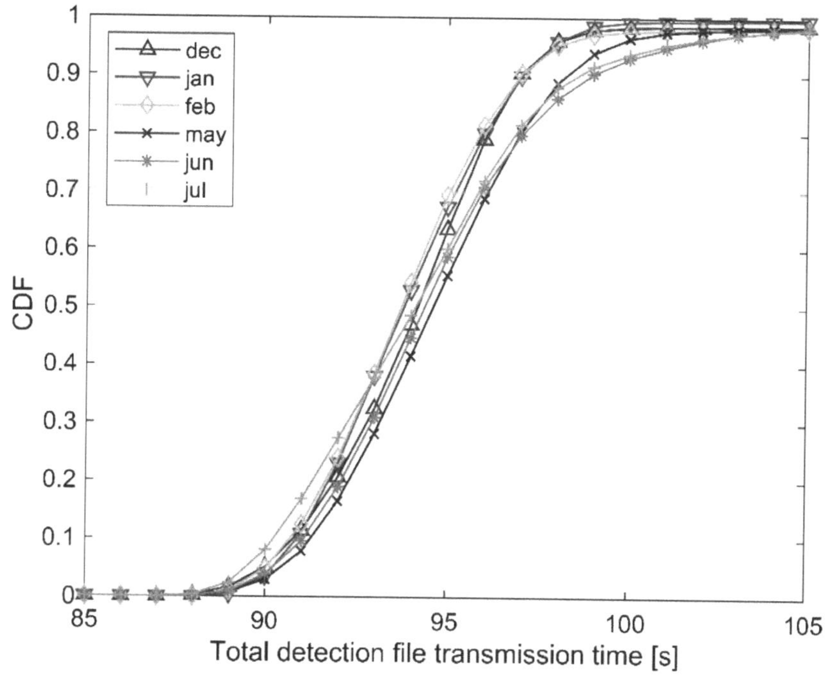

Figure 15. *CDF for detection file size range* 15 − 25kB.

Given the distribution of the detection file sizes shown in Figure 6, we have shown the cumulative distribution of total detection file transmission time for the two most common file size ranges.

Figure 15 shows the cumulative distribution for the 15-25 kB range, while Figure 16 shows the cumulative distribution for the 25-35 kB range. Figures 15 and 16 show the three rainiest months (May-July) and the three driest months (December-February), with the idea to observe whether there are some differences in the total detection file transmission times. The CDF of the rainiest months reaches the value 1 later than the CDF of the driest months. This means that, on average, it takes more time to transfer files to the central location when the weather conditions have more negative impact on the satellite link. However, given that the overall system is robust and can tolerate differences up to several seconds, the given differences in total detection file transmission times between the worst and the best weather condition months are acceptable.

Figures 17 and 18 show the comparison of total detection file transmission time CDF for the two most common file size ranges-15-25 kB and 25-35 kB for the driest and rainiest months, respectively. As expected, the file size does have an impact on the total detection file transmission time as the CDF for range 25-35 kB reaches value 1 later than the CDF for range 15-25 kB. However, the time differences are not significant and do not affect the overall IoT OTH maritime surveillance service performance.

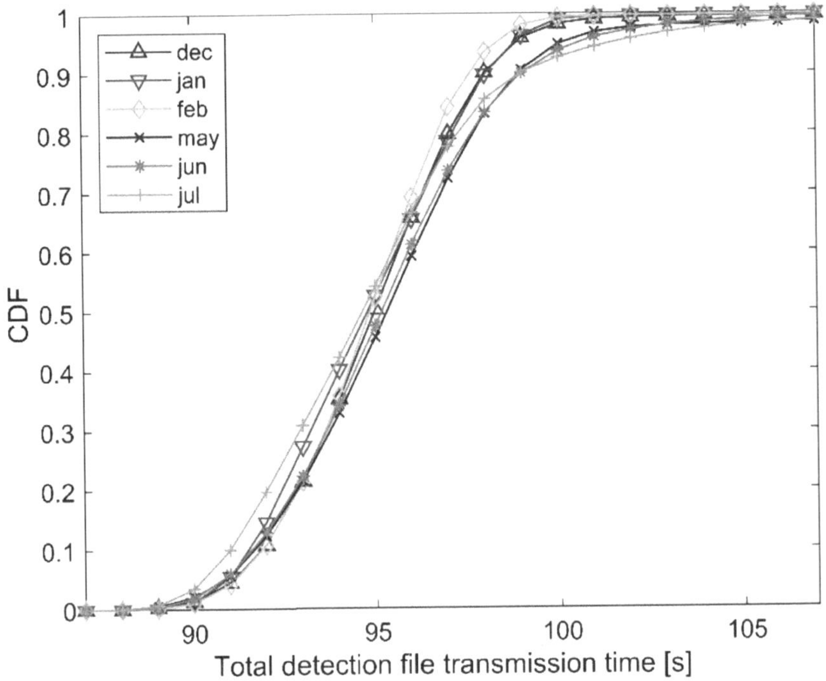

Figure 16. *CDF for detection file size range 25-35 kB.*

In order to properly estimate the service performance parameters, it is needed to analyze service behavior on a monthly basis and to derive a mathematical model of that behavior. It is important to emphasize that neither of the derived distributions describes empirical data perfectly. However, similarity of the derived distributions and empirical data is very high so it might be said that the derived distributions are properly describing the empirical data. Since those analyses may prove to be too lengthy, only the best and the worst matching will be presented and discussed further. The conclusions are valid for the whole observed time frame.

Firstly, the best matching case (June 2018) is presented (matching with 20 different probability density functions (PDF) is tested). After analysis in Matlab software package [33] is done, it has been found that the best mathematical description for the data collected during June 2018 can be achieved with PDF called "generalized extreme value" [34]. Equation (1) is a mathematical formulation of the aforementioned PDF, while the comparison of modeled PDF and empirical data is shown in Figure 19.

$$f(x \mid k, \mu, \sigma) = \left(\frac{1}{\sigma}\right) \exp\left(-\left(1 + k\frac{x-\mu}{\sigma}\right)^{\frac{-1}{k}}\right) \left(1 + k\frac{x-\mu}{\sigma}\right)^{-1-\frac{1}{k}} \tag{1}$$

Please note that:

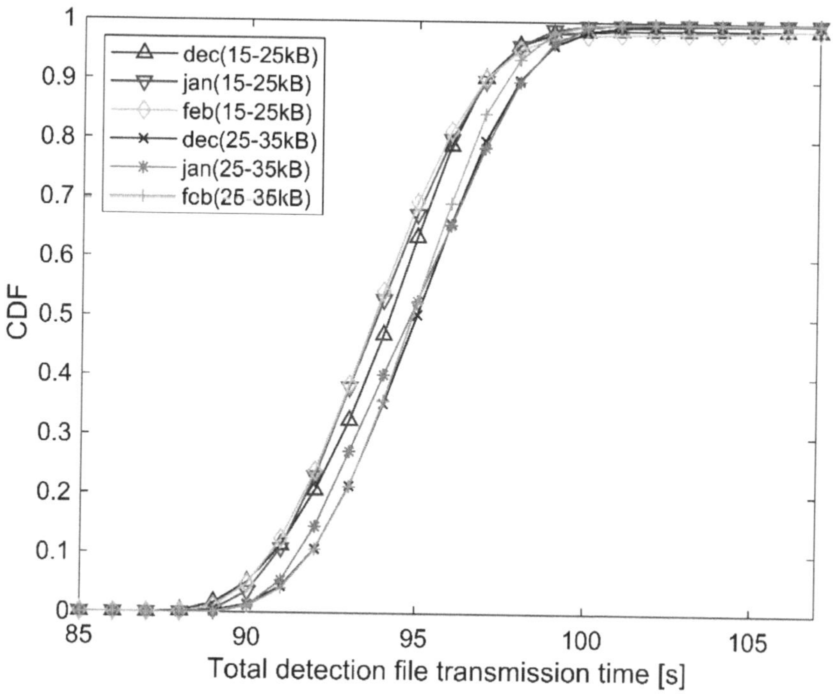

Figure 17. *CDF of* 15 – 25kB *and* 25 – 35kB *for Dec-Feb.*

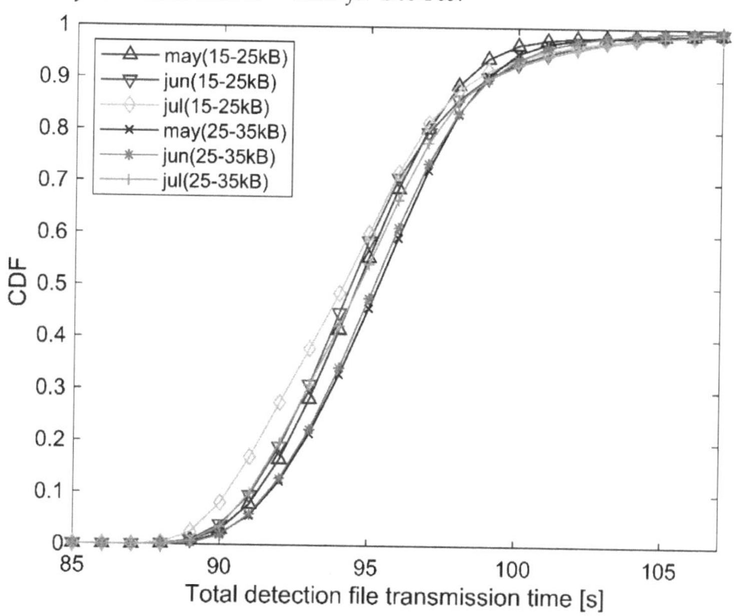

Figure 18. *CDF of* 15 – 25kB *and* 25 – 35kB *for May-July.*

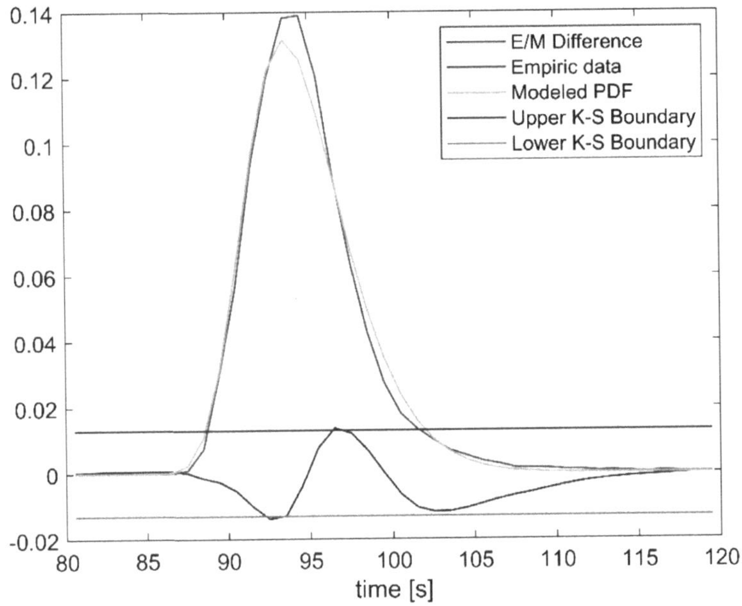

Figure 19. *Statistical modeling for June 2018.*

$$1 + k\frac{(x - \mu)}{\sigma} > 0 \tag{2}$$

Parameters needed to achieve a desired PDF are:

- $k = -0.095368$,
- $\sigma = 2.8064$,
- $\mu = 13.3195$.

As it can be seen from Figure 19 there are some differences between the empirical data and modeled PDF, which are presented with red and orange lines, respectively. These differences cause that difference in CDFs of empirical and modeled data (blue line, denoted with Empiric/Modeled (E/M) Difference) goes over the boundaries set by the KolmogorovSmirnoff (K-S) test (purple and green lines, denoted with the Upper K-S Boundary and Lower K-S Boundary). However, although some irregularities exist, it may still be said that the modeled PDF can be used to describe behavior of the empirical data with a good accuracy, because K-S boundary is violated for only $\sim$ 5% of samples with the maximum violation of $\sim$ 6%.

Next, the worst matching case (July 2018) is presented. Matlab software package is once again used for the analysis and it is concluded that generalized extreme value PDF again gives the best results. Naturally, the parameters are different and they are:

- $k = -0.13687$,
- $\sigma = 3.0326$,

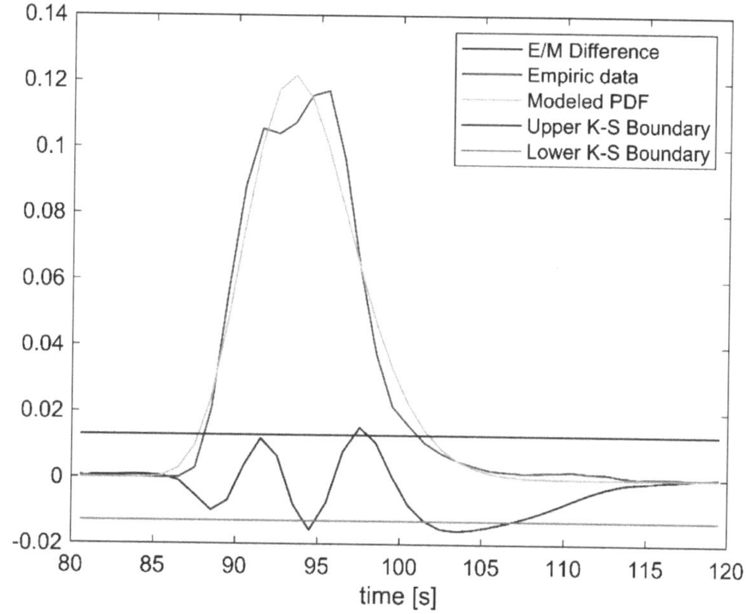

Figure 20. *Statistical modeling for July 2018.*

- $\mu = 12.9111$.

Comparison of the modeled and empirical data is shown in Figure 20.

From Figure 20 it can be seen that the differences between the modeled and empirical data are greater than in the previous case. This is easily observed either by comparing PDFs or the differences between CDFs, which consequently goes over the KolmogorovSmirnoff test boundaries on multiple occasions. However, of all the tested PDFs the one presented here gives the best results (K-S boundary is violated for $\sim 15\%$ of samples with the maximum violation of 20%) and it is taken as a mathematical model of the empirical data.

Based on the presented analysis it may be said that modeled PDFs can be reliably used to describe and predict the IoT OTH maritime surveillance service behavior. This can be further used in order to increase the data fusion process stability and thus reliability of the IoT OTH maritime surveillance service as a whole.

During the harsh weather conditions, the satellite link can perform poorly, and numerous packet retransmissions might occur and TCP connection might be disrupted. In case when the detection file is not transmitted timely to the central location this can lead to dropping the detection file at the radar location. This is due to the detection file no longer being relevant and thus not being eligible for use in the data fusion process at the C2 center. For that reason, we have observed the outage probability using the following definition: Po = Nt/Ntot, where Po is the outage probability, Nt is the number of unsuccessfully (not on time) delivered detection files to C2 center, and Ntot is the total number of generated detection files at the radar location.

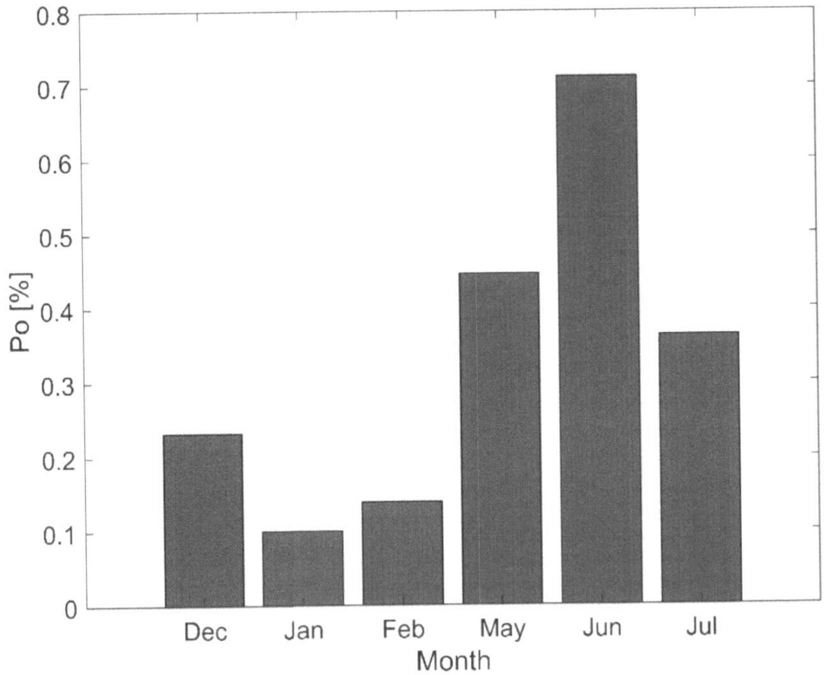

Figure 21. *Outage probability comparison between rainiest and driest months.*

In Figure 21 the outage probabilities are shown for the three rainiest and the three driest months as we expect that the rainier months should have a greater outage probability. Figure 21 shows that the rainier months indeed have a greater outage probability which comes from worse weather conditions on average (Po: 0.1 – 0.25% in good weather conditions, 0.35 – 0.7% in bad weather conditions).

In this section, we have derived the IoT OTH maritime surveillance service performance dependency on weather conditions, as a function of data latency. From the presented analyses it is clear that the average data latency is about 90 s, while the outliers can be as high as 120 s. This was a key information during design of data fusion algorithms presented in [14-16], where a critical latency for the whole end to end process, starting from the raw data collection and ending with the data display to the end user, is set to 250 s. The analyses shown in Figures 15-20 show that this requirement is always satisfied, and that the IoT OTH maritime surveillance service operates stably in all weather conditions specific for the Gulf of Guinea. The other key information (from data fusion point of view) these analyses provide is the outage probability, presented in Figure 21. It can be seen that a high variance of outage occurrence can be expected, which requires additional mechanisms in data fusion algorithms in order to maintain stable tracking. Presented analyses undoubtedly show that the transmission via the satellite link is the optimal solution for the IoT OTH maritime surveillance service in the observed environment.

6. CONCLUSIONS

In this paper the challenges and constraints when over the horizon (OTH) maritime surveillance service is built on an Internet of Things (IoT) as its backbone are examined. The service utilizes High Frequency Surface Wave Radars (HFSWRs) as primary sensors and relies on a satellite communication network as its communication infrastructure in a harsh environment. The whole IoT OTH maritime surveillance network is deployed in the Gulf of Guinea and it entered operational service. The performance analysis encloses multiple steps starting with the raw data gathering from the open sea. It continues with the data processing by HFSWRs at the remote sites and ends with the data transferring to the server on a C2 location, via the satellite network.

Thorough statistical analysis has been conducted during the extreme periods (the driest and rainiest months) in order to mathematically describe the IoT service behavior as precisely as possible. The mathematical model of the IoT OTH maritime surveillance service behavior is verified with the Kolmogorov-Smirnoff test. It should be noted that during this analysis each empirical data set is tested with nearly 20 different PDFs. Analyses presented in the paper show that average service latency is about 90 s, but it can rise to about 120 s, which is used as a key information during the sensor data fusion algorithm design. Validity of the analyses is demonstrated through high quality of service with an outage probability of just 0.1% in the driest months up to the 0.7% in the rainiest months. According to the analysis presented in this paper it is shown that the IoT OTH maritime surveillance service based on HFSWRs which relies on the satellite communication infrastructure can be a viable and reliable solution in the areas defined by harsh meteorological conditions. Moreover, the IoT service behavior can be predicted very well by the statistical models presented in this paper. At the end, the work presented here may be used as a guideline for deployment of other maritime surveillance service solutions in other equatorial regions.

For the future work we intend to expand the sensor network and gather new data in order to further improve the statistical description of the IoT OTH maritime surveillance service behavior.

Author Contributions

Conceptualization, R.P. and D.N.; methodology, D.S.; software, M.N.; validation, D.S. and Z.C.; formal analysis, R.P. and D.D.; investigation, R.P.; resources, R.P.; data curation, M.N.; writing—original draft preparation, R.P. and D.S.; writing-review and editing, Z.C., D.D. and D.N.; visualization, Z.C.; supervision, D.S.; project administration, R.P.; funding acquisition, R.P. All authors have read and agreed to the published version of the manuscript.

Acknowledgments

Authors of the paper would like to thank to Ivan Gluvacevic for his help.

REFERENCES

1. Global Report on Maritime Piracy; United Nations Institute for Training and Research (UNITAR): Geneva, Switzerland, 2014. Available online: https://www.-unitar.org/learning-solutions/publications/unosat-global-report-maritime-piracy-geospatialanalysis-1995-2013 (accessed on 10 February 2021).
2. UN News. UN Security Council Urges "Comprehensive Response" to Piracy off Somali Coast. 7 November 2017. Available online: https://news.un.org/en/story/2017/11/570172-un-security-council-urges-comprehensive-response-piracy-somalicoast (accessed on 10 February 2021).
3. European Union. European Union Naval Force ATALANTA, EU. Available online: https://eunavfor.eu/mission/ (accessed on 14 February 2021).
4. Starr, S. Maritime Piracy on the Rise in West Africa. CTC Sentin. 2014, 7, 22-26.
5. World Economic Forum. Available online: https://www.weforum.org/agenda/2019/06/west-africa-is-becoming-the-world-spiracy-capital-here-s-how-to-tackle-the-problem/ (accessed on 14 February 2021).
6. The United Nations Office on Drugs and Crime. May 2019. Available online: https://www.unodc.org/nigeria/en/press/westafrica-loses-2-3-billion-to-maritime-crime-in-three-years-as-nigeria–unodc-rally-multi-national-efforts-to-thwart-piracy-inthe-gulf-of-guinea.html (accessed on 14 February 2021).
7. Helzel, T.; Kniephoff, M.; Petersen, L. Oceanography radar system WERA: Features, accuracy, reliability and limitations. Turk. J. Elec. Eng. Comp. Sci. 2010, 18, 389-397.
8. Shay, L.K.; Martinez-Pedraja, J.; Cook, T.M.; Haus, B.K.; Weisberg, R.H. High-Frequency Radar Mapping of Surface Currents Using WERA. J. Atmos. Ocean. Technol. 2007, 24, 484-503.
9. Liu, Y.; Weisberg, R.H.; Merz, C. Assessment of CODAR SeaSonde and WERA HF Radars in Mapping Surface Currents on the West Florida Shelf. J. Atmos. Ocean. Technol. 2014, 31, 1363-1382.
10. Lipa, B.; Barrick, D.; Alonso-Martirena, A.; Fernandes, M.; Ferrer, M.I.; Nyden, B. Brahan Project High Frequency Radar Ocean Measurements: Currents, Winds, Waves and Their Interactions. Remote Sens. 2014, 6, 12094-12117.
11. Sun, W.; Ji, M.; Huang, W.; Ji, Y.; Dai, Y. Vessel Tracking Using Bistatic Compact HFSWR. Remote Sens. 2020, 12, 1266.
12. Sevgi, L.; Ponsford, A.; Chan, H.C. An integrated maritime surveillance system based on high-frequency surface-wave radars Part 1. Theoretical background and numerical simulations. IEEE Antennas Propag. Mag. 2001, 43, 28-43.
13. Ponsford, A.; Sevgi, L.; Chan, H.C. An integrated maritime surveillance system based on high-frequency surface-wave radars. Part 2. Operational status and system performance. IEEE Antennas Propag. Mag. 2001, 43, 52-63.
14. Stojkovic, N.; Nikolic, D.; Puzovic, S. Density Based Clustering Data Association Procedure for Real-Time HFSWRs Tracking at OTH Distances. IEEE Access 2020, 8, 39907-39919.
15. Nikolic, D.; Stojkovic, N.; Popovic, Z.; Tosic, N.; Lekic, N.; Stankovic, Z.; Doncov, N. Maritime over the horizon sensor inte-gration: HFSWR data fusion algorithm. Remote Sens. 2019, 11, 852.

16. Nikolic, D.; Stojkovic, N.; Lekic, N. Maritime over the Horizon Sensor Integration: High Frequency Surface-Wave-Radar and Automatic Identification System Data Integration Algorithm. Sensors 2018, 18, 1147.
17. Chiani, M.; Giorgetti, A.; Paolini, E. Sensor Radar for Object Tracking. Proc. IEEE 2018, 106, 1022-1041.
18. vOTHR. Available online: https://www.vlatacominstitute.com/over-the-horizon-radar (accessed on 15 April 2021).
19. Nikolic, D.; Stojkovic, N.; Petrovic, P.; Tosic, N.; Lekic, N.; Stankovic, Z.; Doncov, N. The high frequency surface wave radar solution for vessel tracking beyond the horizon. Facta Univ. Ser. Electron. Energetics 2020, 33, 37-59.
20. ITU-D: ICT Data and Statistics (IDS). Available online: https://www.itu.int/ITU-D/ict/statistics/ict/ (accessed on 15 April 2021).
21. World Bank Open Data. Available online: https://data.worldbank.org/ (accessed on 17 April 2021).
22. Peel, M.C.; Finlayson, B.L.; McMahon, T.A. Updated world map of the Köppen-Geiger climate classification. Hydrol. Earth Syst. Sci. 2007, 11, 1633-1644.
23. Christian, H.J.; Blakeslee, R.; Boccippio, D.J.; Boeck, W.L.; Buechler, D.E.; Driscoll, K.T.; Goodman, S.J.; Hall, J.M.; Koshak, W.J.; Mach, D.M.; et al. Global frequency and distribution of lightning as observed from space by the Optical Transient Detector. J. Geophys. Res. Space Phys. 2003, 108, ACL 4-1-ACL 4-15.
24. Petrovic, R.; Simic, D.; Drajic, D.; Cica, Z.; Nikolic, D.; Peric, M. Designing Laboratory for IoT Communication Infrastructure Environment for Remote Maritime Surveillance in Equatorial Areas Based on the Gulf of Guinea Field Experiences. Sensors 2020, 20, 1349.
25. Fixed Telephone Subscriptions (per 100 People). Available online: https://data.worldbank.org/indicator/IT.MLT.MAIN.P2 ?view = map (accessed on 15 April 2021).
26. Fixed Broadband Subscriptions (per 100 People). Available online: https://data.worldbank.org/indicator/IT.NET.BBND.P2 (accessed on 15 April 2021).
27. Mobile Cellular Subscriptions (per 100 People). Available online: https://data.worldbank.org/indicator/IT.CEL.SETS.P2 (accessed on 15 April 2021).
28. World Weather Information Service. Available online: https://worldweather.wmo.int/ (accessed on 15 April 2021).
29. Scaramuzza, P.; Rubino, C.; Caruso, M.; Tiebout, M.; Bevilacqua, A.; Neviani, A. Class-J SiGe X -Band Power Amplifier Using a Ladder Filter-Based AM-PM Distortion Reduction Technique. IEEE Trans. Circuits Syst. I Regul. Pap. 2018, 65, 3780-3789.
30. Yeo, J.X.; Lee, Y.H.; Ong, J.T. Rain Attenuation Prediction Model for Satellite Communications in Tropical Regions. IEEE Trans. Antennas Propag. 2014, 62, 5775-5781.
31. NOAA. Table 3700: Sea State. Available online: https://www.nodc.noaa.gov/woce/woce_v3/wocedata_1/woce-uot/document/wmocode.htm (accessed on 15 April 2021).
32. International Telecommunication Union. Reccommendation ITU-R P.368-9 Ground-Wave Propagation Curves for Frequencies between 10KHz and 30MHz; ITU-R: Geneva, Switzerland, 2007.

33. Matlab-MathWorks. Available online: https://www.mathworks.com/products/matlab.html (accessed on 14 February 2021).

34. Papoulis, A.; Pillai, S.U. Probability, Random Variables and Stochastic Processes, 4th ed.; McGraw-Hill: New York, NY, USA, 2002.

CHAPTER 6

Altitude Optimization and Task Allocation of UAV-Assisted MEC Communication System

Shuqi Huang[1]**, Jun Zhang**[2] **and Yi Wu**[1]

[1]*Fujian Provincial Engineering Technology Research Center of Photoelectric Sensing Application, Fujian Normal University, Fuzhou 350007, China*
[2]*Jiangsu Key Laboratory of Wireless Communications, Nanjing University of Posts and Telecommunications, Nanjing 210003, China*

ABSTRACT

Unmanned aerial vehicles (UAVs) are widely used in wireless communication systems due to their flexible mobility and high maneuverability. The combination of UAVs and mobile edge computing (MEC) is regarded as a promising technology to provide high-quality computing services for latency-sensitive applications. In this paper, a novel UAV-assisted MEC uplink maritime communication system is proposed, where an MEC server is equipped on UAV to provide flexible assistance to maritime user. In particular, the task of user can be divided into two parts: one portion is offloaded to UAV and the remaining portion is offloaded to onshore base station for computing. We formulate an optimization problem to minimize the total system latency by designing the optimal flying altitude of UAV and the optimal task allocation ratio. We derive a semi closed-form expression of the optimal flying altitude of UAV and a closed-form expression of the optimal task allocation ratio. Simulation results demonstrate the precision of the theoretical analyses and show some interesting insights.

Keywords: *UAV communication; mobile edge computing; delay; UAV altitude; task allocation*

1. INTRODUCTION

With the rapid development of maritime activities such as marine tourism, oceanic mineral exploration, and maritime rescue, reliable and high-speed maritime wireless communication is receiving more and more attention [1]. In recent years, the use of intelligent applications has been rapidly popularized with the evolution of Internet of Things (IoT) [2-4]. The emergence of these latency-sensitive and compute-intensive applications, as well as the growth of maritime activities, has greatly improved the demand for maritime wireless communication. However, the satellite communication systems and high frequency / very high frequency (HF / VHF) communications used in traditional maritime communication networks have certain limitations [5, 6]. Although satellite communications have a wide coverage area, they also have disadvantages such as high cost and large transmission delay. HF / VHF communication systems have lower costs, but they are not suitable for transmitting large amounts of data services due to insufficient channel bandwidth.

To tackle the above issues, mobile edge computing (MEC) can be an effective solution to provide high-quality computing services for compute-intensive maritime terminals. MEC places computing resources at the edge of the network, which can effectively reduce latency and significantly improve quality of service [7-10]. Many research works on MEC technology have received increasing attention. In [11], the task delay among users was minimized to optimize the tasks division ratios and offloading the transmit power. The authors in [12] proposed a MEC task offloading problem in ultra-dense networks, aiming to minimize latency while saving battery life of user devices. The offloading decision, user transmit power, and computation resource allocation were jointly optimized in [13] to minimize the weighted sum of energy consumption. In [14], the authors minimized the overall cost of energy, computation, and delay for all users by jointly optimizing the offloading decisions of all users' tasks as well as the allocation of computation and communication resources. The authors in [15] minimized the sum of weighted task latency for all users by optimizing offloading decision, transmission power, and resource allocation. However, the above works in the field of MEC networks ignore a fact that the deployment of traditional MEC servers is fixed. In addition, due to the limited coverage of the ground MEC server, it is difficult for ground MEC to meet the computing requirement when the user devices are mobile.

In view of the above deficiencies, unmanned aerial vehicles (UAVs) can be regarded as a prospective solution. As a highly flexible aircraft, UAVs have been widely used in maritime wireless communication due to their advantages of high maneuverability, high scalability, and convenient deployment [16-18]. Hence, UAV has been proposed as an aerial computing platform to provide computing services for user devices, which is called UAV-assisted MEC. The use of UAV-assisted MEC can yield significant advantages over traditional ground MEC in terms of coverage, operational cost, and flexibility. Furthermore, line-of-sight (LoS) links from UAV to ground mobile users can also mitigate the impact of obstacles and shadows [19, 20]. As a result, the combination of UAV and MEC can effectively improve the performance of maritime communication networks.

For UAV-assisted MEC system, the authors in [21] formulated a joint optimization problem to maximize the sum of the maximum delay among all users at each time slot. In [22], the authors minimized the weighted-sum energy consumption of UAV and users by jointly optimizing the UAV trajectory and computation resource allocation. In [23], the authors jointly optimized the time slot size, users scheduling, computation resource allocation, and UAVs' trajectories to investigate a minimum time of task completion in a UAV-aided MEC system. However, the above research works adopt partial offloading approach to offload computational tasks to the UAV for computation. Since the computational capacity of user and UAV is limited, it is impractical to offload computational tasks to the UAV when the computational tasks are large. The work in [24] considered a UAV-enabled MEC system in which the UAV can act as an MEC server for computation or offload the tasks to base stations (BSs) for computation. It is worth noting that the LoS channel was adopted in [24], which ignored the effect of obstacle blockage in urban scenarios. Different from this paper, we use probabilistic channel that contain both LoS and non-LoS (NLoS) links, which makes our research more consistent with realistic scenarios.

The flying altitude of the UAV is an important issue, which is the focus of UAV-assisted wireless communication networks. However, the above existing works assume that the UAV flies at a fixed altitude without considering the optimization of the flying altitude. The work in [25] mainly focused on optimizing flying altitude and the number of UAVs to minimize a total power consumption. The altitude optimization problem of UAV based adhoc network was considered through minimizing the average network outage probability in [26]. For mixed radio frequency and underwater optical communication system, the authors determined an optimal altitude of the UAV by minimizing the system outage probability in [27]. However, most of the existing works in UAV-assisted MEC system have not investigated closed-form solution for the optimal flying altitude of the UAV. In addition, the above works have made significant contributions in the field of UAV-assisted MEC communication, but UAV-assisted MEC in maritime communication has not yet received wide attention.

In practice, UAV-assisted MEC maritime communication system investigated in our paper is meaningful. Firstly, the flexible mobility of UAV compensates the disadvantage of poor channel quality between ground MEC servers and users. Second, our research has certain potential to be applied in practical scenarios. For example, for maritime environments where network infrastructure is scarce and generated data are difficult to collect and monitor, using UAV as an MEC server can help maritime user process the generated compute-intensive data. Finally, we employ onshore BS and UAV to serve maritime user simultaneously, which can alleviate the computational burden of the network and further improve the computational performance of the system.

In this paper, we consider a maritime communication system in which a UAV equipped with an MEC server is used to provide computing services to maritime user. To better fit the reality, we adopt probabilistic channel between user and UAV, which is composed of both LoS and NLoS links. In addition to offloading computational tasks to UAV, the user can also offload to onshore BS with high computational capability for computation. The main contributions of this paper are summarised as follows.

- We propose a new UAV-assisted MEC framework in order to provide computing services for user with stringent latency requirement. In our proposed design, we formulate a task offloading problem for minimizing the total system delay to optimize the flying altitude of UAV and the task allocation ratio.
- To effectively solve the optimization problem, we derive a semi closed-form expression for the optimal flying altitude of the UAV and a closed-form expression for the optimal task allocation ratio, respectively.
- Simulation results demonstrate the accuracy of the analytical results and show that the optimal flying altitude depends on the horizontal distance between the UAV and the user. We also show the impacts of the transmit power, the bandwidth, and the CPU frequency on the optimal task allocation ratio.

The rest of this paper is organized as follows. We present the system model and the problem formulation for a UAV-assisted MEC uplink maritime communication system in Section 2. Then, we derive a semi closed-form expression of the optimal flying altitude of the UAV and a closed-form expression of the optimal task allocation ratio in Section 3. The simulation results are presented in Section 4 to confirm the accuracy of the theoretical results. Finally, we conclude the paper in Section 5.

2. SYSTEM MODEL AND PROBLEM STATEMENT

In this section, we present the system model and the formulation of the optimization problem. For ease of reference, the important notations in the system model are listed in Table 1.

2.1. System Model

As shown in Figure 1, we consider a UAV-assisted MEC uplink maritime communication system including a UAV, a maritime user, and an onshore BS, in which, they are all equipped with a single antenna. We use (x_U, y_U, H), $(x_G, y_G, 0)$, and (x_B, y_B, h) to denote the three dimensional position of the UAV, the user, and the BS, respectively. We express the distance between the user and the projection of the UAV on the ground as $r = \sqrt{(x_U - x_G)^2 + (y_U - y_G)^2}$. The user simultaneously offloads the tasks to the BS and the UAV for calculation based on frequency division multiple access (FDMA) scheme. The Sea-to-air (S2A) wireless communication link between user and UAV both contains LoS link and NLoS link [28]. The LoS probability of this link can be expressed as [29, 30]

$$p_{\text{LoS}} = \frac{1}{1 + a\exp[-b(\theta - a)]}, \tag{1}$$

where a and b are constants determined by the environment, $\theta = \arctan\left(\frac{H}{r}\right)$ denotes the elevation angle between UAV and the user. The probability of NLoS

Table 1. *List of notations.*

Symbol	Definition
H	Flying altitude of UAV
T	Total system delay
c	Speed of light
α	Task allocation ratio
L	Total computing tasks
a, b	Constants determined by the environment
θ	Elevation angle between UAV and the user
ω	Beamwidth of UAV's directional antenna
r	Distance between the user and the projection of the UAV on the ground
ε, φ	Path loss exponent
f_c	Carrier frequency
$\eta_{\mathrm{LoS}}, \eta_{\mathrm{NLoS}}$	Extra path loss of LoS link and NLoS link
p_{LoS}	LoS probability between the user and the UAV
d_{S2A}	Distance between the UAV and the user
d_{S2G}	Distance between the user and the BS
$\mathrm{PL}_{\mathrm{LoS}}, \mathrm{PL}_{\mathrm{NLoS}}$	Path loss of LoS link and NLoS link
$\overline{\mathrm{PL}}$	Average path loss of S2A channel
β_0	Channel gain at the reference distance of 1m
ξ_{S2G}	Small-scale fading component
h_{S2G}	Channel gain of Rayleigh fading
P	Transmit power of the user
G	Gain of the directional antenna
N_0	Spectral density of noise power
$B_{\mathrm{UAV}}, B_{\mathrm{BS}}$	Bandwidth allocated by the user to offload tasks to the UAV and the BS
$R_{\mathrm{UAV}}, R_{\mathrm{BS}}$	Achievable offloading rate from the user to the UAV and the BS
$t_{\mathrm{S2A}}, t_{\mathrm{S2G}}$	Transmission delay of the user on S2A and S2G links
$f_{\mathrm{UAV}}, f_{\mathrm{BS}}$	CPU frequency allocated for computation on the UAV and the BS
$t_{\mathrm{UAV}}, t_{\mathrm{BS}}$	Computation delay for the user processing tasks on the UAV and the BS

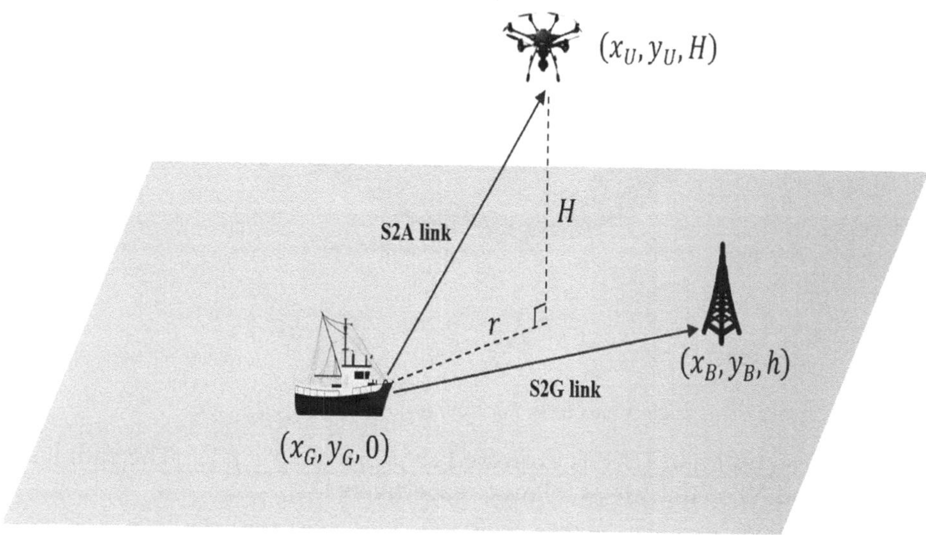

Figure 1. *A UAV-assisted MEC uplink communication system.*

Table 2. *Simulation parameters.*

Description	Value
LoS link extra path loss	$\eta_{\text{LoS}} = 1$
NLoS link extra path loss	$\eta_{\text{NLoS}} = 20$
Carrier frequency	$f_c = 2.4$ GHz
Small-scale fading component	$\xi_{S2G} = 0.7$
Noise power spectral density	$N_0 = 5 \times 10^{-15}$ W/Hz
Total computing task	$L = 5 \times 10^5$ bit
Cycle for one bit	$C = 100$ cycle/bit
Speed of light	$c = 3.0 \times 10^8$ m/s
Transmit power of the user	$P = 1$ W

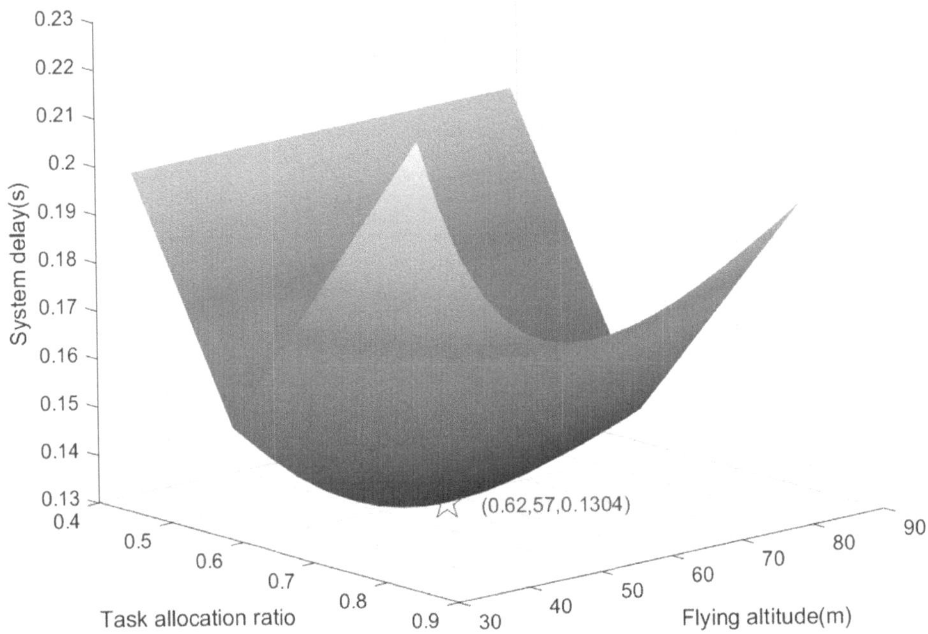

Figure 2. *Total system delay versus task allocation ratio and flying altitude.*

link between user and UAV is expressed as $p_{\text{NLOS}} = 1 - p_{\text{LoS}}$. The path loss of LoS link and NLoS link are as follows, respectively [31],

$$\text{PL}_{\text{LoS}} = \left(\frac{4\pi f_c d_{\text{S2 A}}}{c}\right)^{\varepsilon} \eta_{\text{LoS}}, \tag{2}$$

$$\text{PL}_{\text{NLoS}} = \left(\frac{4\pi f_c d_{\text{S2 A}}}{c}\right)^{\varepsilon} \eta_{\text{NLoS}}, \tag{3}$$

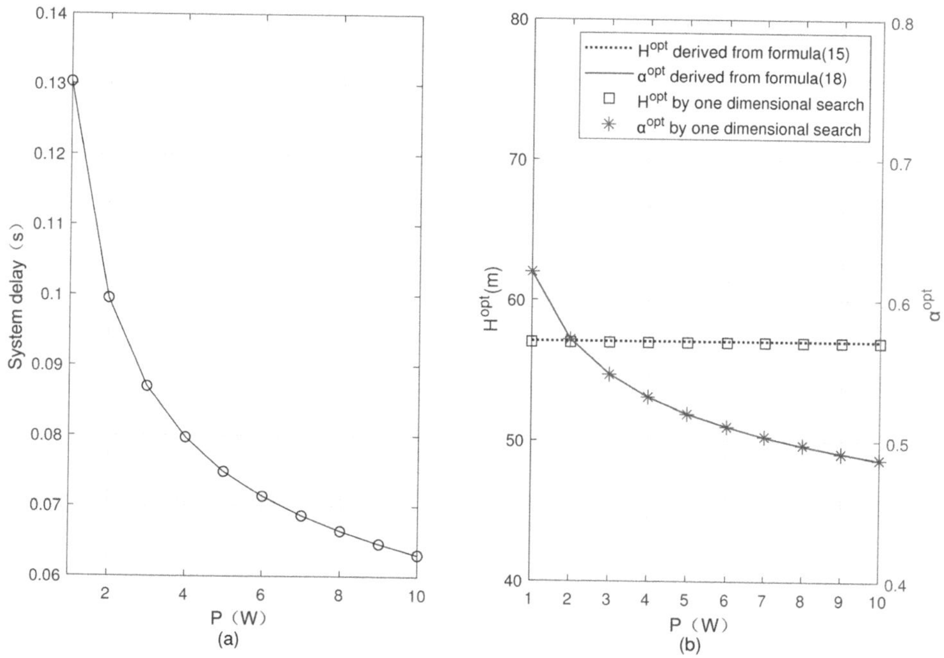

Figure 3. *Minimum system delay, optimal flying altitude, and optimal task allocation ratio versus transmit power of the user. (a) Minimum system delay, (b) optimal flying altitude and optimal task allocation ratio.*

where c denotes the speed of light, f_c denotes the carrier frequency, $d_{S2\,A} = \sqrt{H^2 + r^2}$ is the distance between the UAV and the user, ε denotes the path loss exponent, η_{LoS} and η_{NLoS} denote the extra path loss of LoS link and NLoS link, respectively, and we set $\varepsilon = 2$ [32,33]. The average path loss of S2A channel is expressed as follows

$$\overline{\mathsf{PL}} = p_{\mathsf{LoS}}\mathsf{PL}_{\mathsf{LoS}} + p_{\mathsf{NLoS}}\mathsf{PL}_{\mathsf{NLoS}}. \tag{4}$$

The Sea-to-Ground (S2G) wireless communication link between user and BS is modeled as the Rayleigh fading channel [34], and the channel gain is expressed as

$$h_{S2G} = \beta_0 \left(d_{S2G}\right)^{-\varphi} \xi_{S2G} \tag{5}$$

where β_0 is the channel gain at the reference distance of 1 m, d_{S2G} represents the distance between the user and the BS, φ denotes the path loss exponent, and we set $\varphi = 3.5, \xi_{S2G}$ denotes the small-scale fading component conforming to the Rayleigh fading model.

The achievable rate of the uplink transmission task from the user to the UAV is given as

$$R_{UAV} = B_{UAV} \log_2 \left(1 + \frac{PG}{\overline{\mathrm{PL}} N_0 B_{UAV}}\right) \tag{6}$$

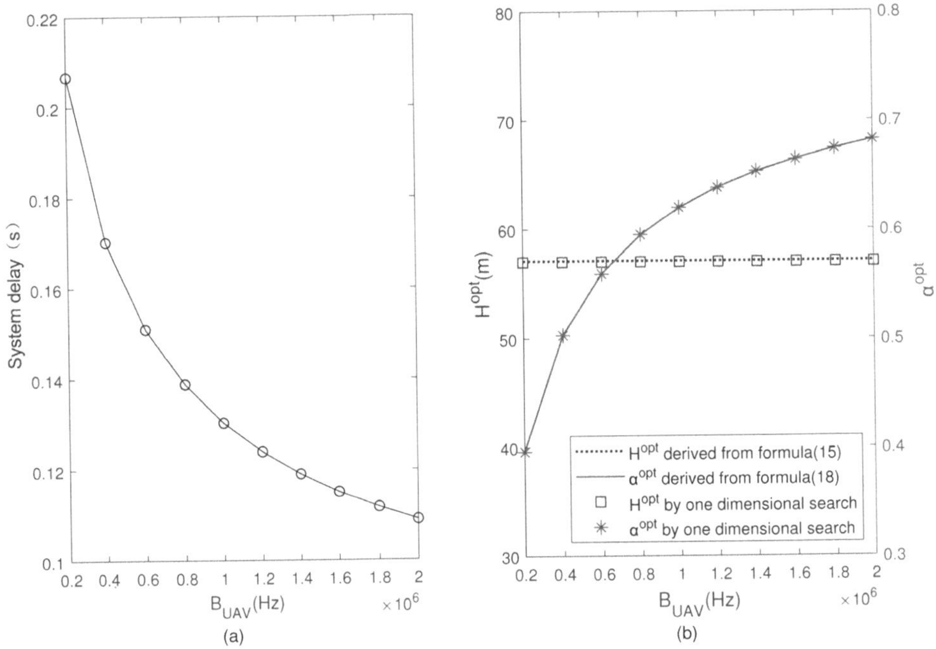

Figure 4. *Minimum system delay, optimal flying altitude, and optimal task allocation ratio versus bandwidth allocated to UAV for task offloading. (a) Minimum system delay, (b) optimal flying altitude and optimal task allocation ratio.*

where B_{UAV} represents the bandwidth allocated by the user to offload tasks to the UAV, P is the transmit power of the user, and N_0 is the spectral density of noise power. G denotes the gain of the directional antenna [35], it can be described as

$$G = \begin{cases} \frac{G_0}{\left(\frac{\omega}{2}\right)^2}, & 0 \le \phi \le \frac{\omega}{2} \\ g \approx 0, & \text{otherwise} \end{cases} \tag{7}$$

where ω denotes the beamwidth of UAV's directional antenna, $G_0 = 30,000/2^2 \times (\pi/180)^2 \approx 2.2846, \phi$ denotes the sector angle, g is the channel gain beyond the beamwidth of the directional antenna. The achievable rate of the uplink transmission task from the user to the BS can be expressed as

$$R_{\text{BS}} = B_{\text{BS}} \log_2 \left(1 + \frac{Ph_{\text{S2G}}}{N_0 B_{\text{BS}}}\right) \tag{8}$$

where B_{BS} denotes the bandwidth allocated by the user to offload the task to the BS.

Let L denotes the total computing tasks, the task allocation ratio is described as $\alpha \in [0,1]$. The task allocation ratio from the user to the UAV and BS is α and $1-\alpha$, respectively. The total delay of the system is divided into two parts, one is the delay of transmission from the user offloading task to the UAV on the S2A link

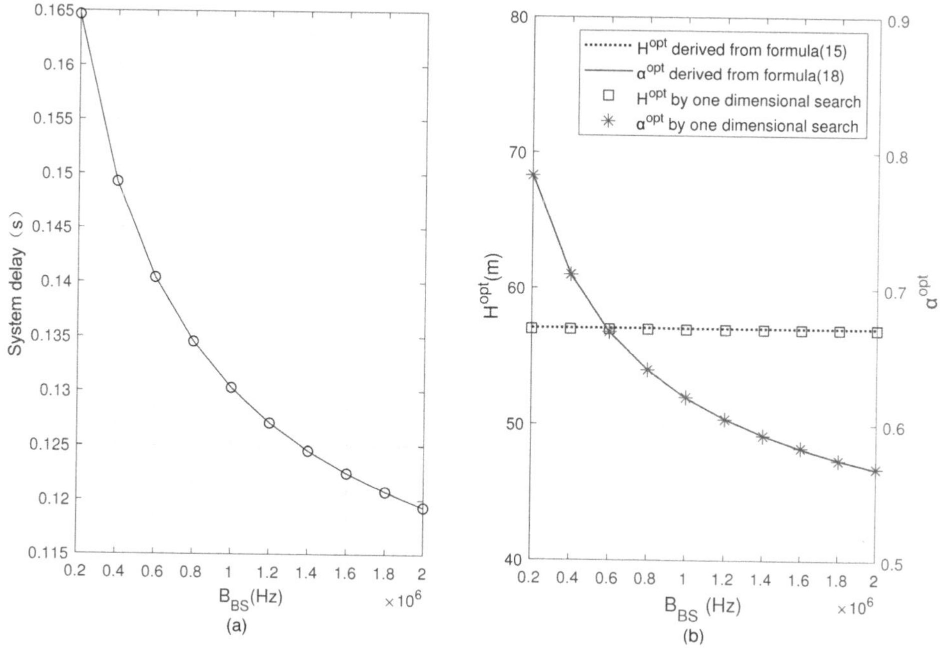

Figure 5. *Minimum system delay, optimal flying altitude, and optimal task allocation ratio versus bandwidth allocated to BS for task offloading. (a) Minimum system delay, (b) optimal flying altitude and optimal task allocation ratio.*

and the computation delay on the UAV, the other part is the transmission delay of the user offloading task to the BS on the S2G link and the computation delay on the BS.

The transmission delay of the user on S2A and S2G links can be, respectively, expressed as

$$t_{\mathrm{S2\,A}} = \frac{\alpha L}{R_{\mathrm{UAV}}}, \tag{9}$$

$$t_{\mathrm{S2G}} = \frac{(1-\alpha)L}{R_{\mathrm{BS}}}. \tag{10}$$

The computation delay for the user processing tasks on the UAV and the BS can be, respectively, expressed as

$$t_{\mathrm{UAV}} = \frac{\alpha LC}{f_{\mathrm{UAV}}}, \tag{11}$$

$$t_{\mathrm{BS}} = \frac{(1-\alpha)LC}{f_{\mathrm{BS}}}, \tag{12}$$

where C represents the number of cycles required to compute one bit task, f_{UAV} is the CPU frequency allocated for computation on the UAV, and f_{BS} is the CPU

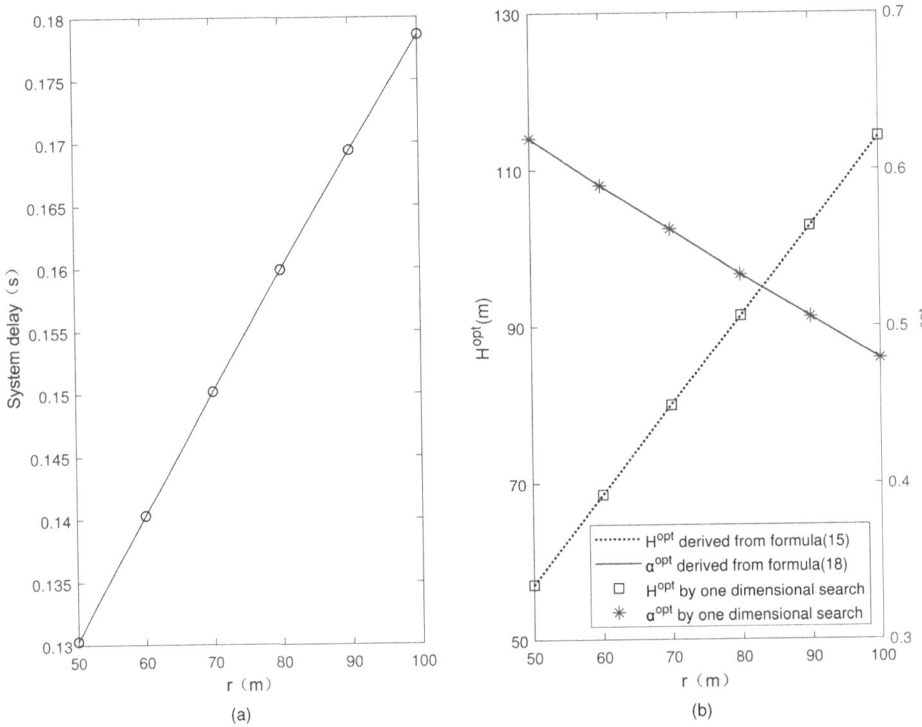

Figure 6. *Minimum system delay, optimal flying altitude, and optimal task allocation ratio versus the distance between the UAV's projection on the ground and the user. (a) Minimum system delay, (b) optimal flying altitude and optimal task allocation ratio.*

frequency allocated for computation on the BS. The total system delay can be described as

$$T = \max \{t_{S2\,A} + t_{UAV}, t_{S2G} + t_{BS}\} \tag{13}$$

2.2. Problem Statement

Based on the above discussion, we formulate a new optimization problem that minimizes the total delay of the system by optimizing the optimal flying altitude of the UAV and the optimal task allocation ratio. Specifically, the optimization problem can be expressed as follows

$$\min_{\{H,\alpha\}} T \tag{14a}$$

$$\text{s.t. } 0 \leq \alpha \leq 1, \tag{14b}$$

$$H_{\min} \leq H \leq H_{\max}, \tag{14c}$$

where inequality (14b) represents the constraint of task allocation ratio, and inequality (14c) is the flying altitude constraint of the UAV.

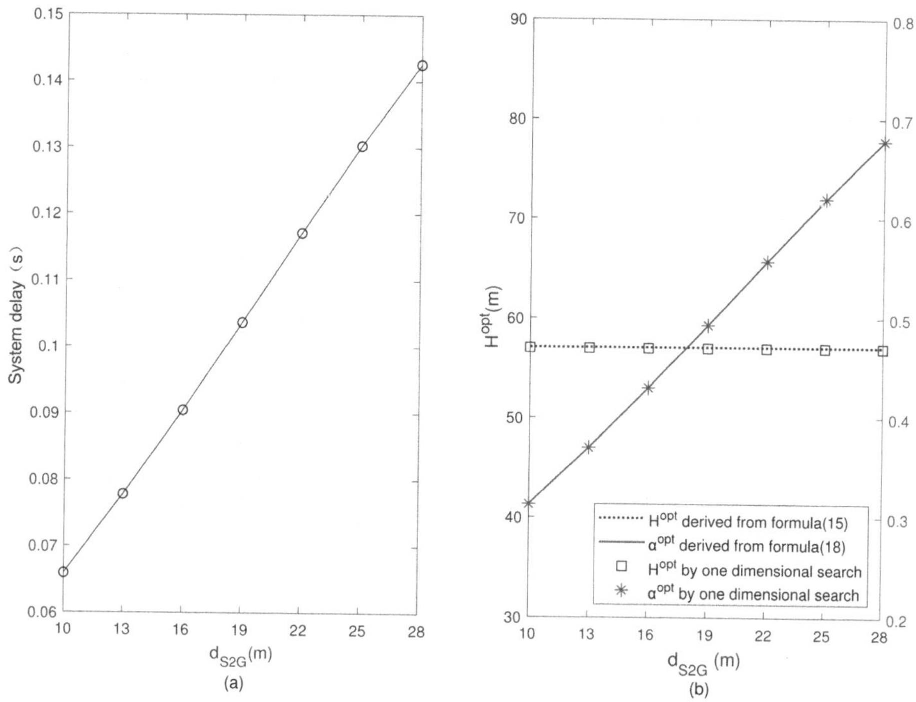

Figure 7. *Minimum system delay, optimal flying altitude, and optimal task allocation ratio versus the distance between the user and the BS. (a) Minimum system delay, (b) optimal flying altitude and optimal task allocation ratio.*

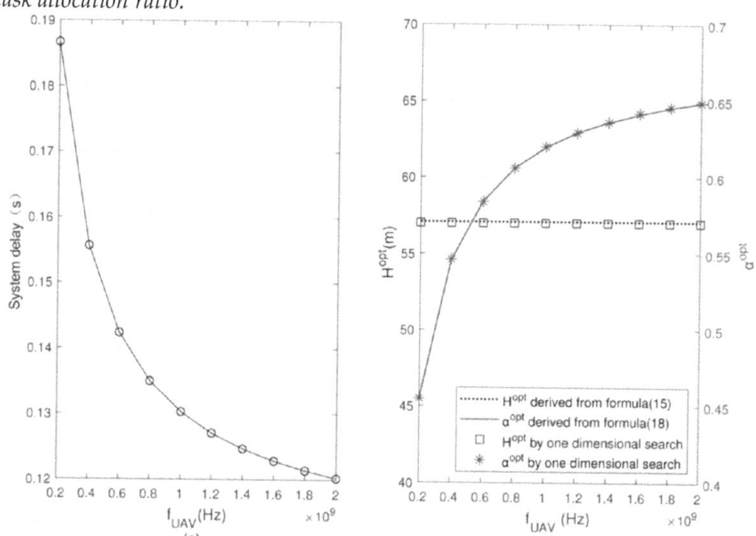

Figure 8. *Minimum system delay, optimal flying altitude, and optimal task allocation ratio versus the CPU frequency allocated for computation on the UAV. (a) Minimum system delay, (b) optimal flying altitude and optimal task allocation ratio.*

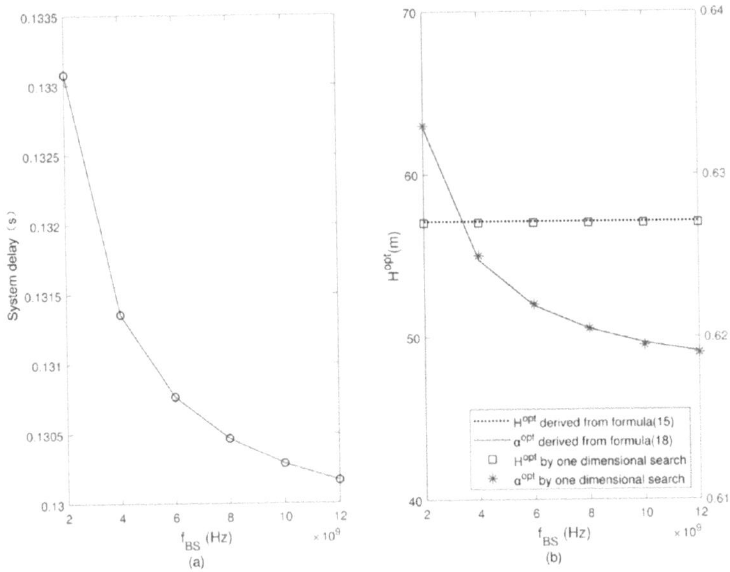

Figure 9. *Minimum system delay, optimal flying altitude, and optimal task allocation ratio versus the CPU frequency allocated for computation on the BS. (a) Minimum system delay, (b) optimal flying altitude and optimal task allocation ratio.*

3. ALTITUDE OF UAV AND TASK ALLOCATION OPTIMIZATION

In this section, the main results are presented to solve the problem of (14a). We derive a semi closed-form expression of the optimal altitude of the UAV and a closed-form expression of the optimal task allocation ratio, which are shown in the following results.

Theorem 1. *According to optimization problem (14a), the optimal flying altitude of the UAV can be expressed as*

$$H^{opt} = \frac{-At(H)}{2\eta_{\text{NLoS}}[1+at(H)]^2 + 2\left(\eta_{\text{LoS}} - \eta_{\text{NLoS}}\right)[1+at(H)]}, \tag{15}$$

where

$$A = \frac{180}{\pi}\left(\eta_{LoS} - \eta_{NLoS}\right) abr, \tag{16}$$

$$t(H) = \exp\left[-b\left(\frac{180}{\pi}\arctan\left(\frac{H}{r}\right) - a\right)\right], \tag{17}$$

the optimal task allocation ratio of the UAV is given by

$$\alpha^{opt} = \frac{\frac{1}{R_{\text{BS}}} + \frac{C}{f_{\text{BS}}}}{\frac{1}{R_{\text{UAV}}} + \frac{C}{f_{\text{UAV}}} + \frac{1}{R_{\text{BS}}} + \frac{C}{f_{\text{BS}}}}. \tag{18}$$

Proof of Theorem 1. Our aim is to minimize the total system delay, which depends on the maximum value of the transmission and computation delay of both S2A and S2G links, i.e., $\min\max\{t_{S2\,A} + t_{UAV}, t_{S2G} + t_{BS}\}$. On the one hand, in the case of low flying altitude H, due to the stronger path loss of NLoS than LoS, the average path loss increases, which makes the transmission delay of the S2A link increase, resulting in an increase of the total system delay. On the other hand, when the flying altitude H is higher, the rate of user uplink transmission tasks to the UAV decreases, which makes the transmission delay of the S2A link increase, resulting in an increase of the total system delay. Therefore, there exists an optimal H that makes $t_{S2\,A} + t_{UAV} = t_{S2G} + t_{BS}$. We denote $t_{S2\,A} + t_{UAV}$ as T_1, and denote $t_{S2G} + t_{BS}$ as T_2. Substituting (10) and (12) into T_2, we obtain $T_2 = \frac{(1-\alpha)L}{R_{BS}} + \frac{(1-\alpha)LC}{f_{BS}}$, which is independent of H. Substituting the rate of the user's uplink transmission task to UAV in (6) into (9), the transmission delay of the S2A link can be obtained, and the expression of T_1 is expressed

$$T_1 = \frac{\alpha L}{B_{UAV}\log_2\left(1 + \frac{PG}{QN_0 B_{UAV} g(H)}\right)} + \frac{\alpha LC}{f_{UAV}} \tag{19}$$

where

$$Q = \left(\frac{4\pi f_c}{c}\right)^2, \tag{20}$$

$$s(H) = H^2 + r^2 \tag{21}$$

$$g(H) = s(H)\left[\eta_{NLoS} + \frac{\eta_{LoS} - \eta_{NLoS}}{1 + at(H)}\right]. \tag{22}$$

In order to find an optimal altitude H^{opt}, then according to (19), which is equivalent to minimizing $g(H)$, the first order derivative with respect to H is obtained as

$$g'(H) = 2\eta_{NLoS}H + \frac{2\left(\eta_{LoS} - \eta_{NLoS}\right)H}{1 + at(H)} + \frac{At(H)}{[1 + at(H)]^2}. \tag{23}$$

Let $g'(H)$ equals zero, and since $1 + at(H) > 0$ holds, we have

$$2\eta_{NLoS}H[1 + at(H)]^2 + 2\left(\eta_{LoS} - \eta_{NLoS}\right)H[1 + at(H)] + At(H) = 0 \tag{24}$$

The optimal flying altitude of the UAV in (15) can be obtained by arranging the above formula.

Substituting (9) and (11) into T_1, we get $T_1 = \alpha\left(\frac{L}{R_{UAV}} + \frac{LC}{f_{UAV}}\right)$, and T_1 is monotonically increasing with respect to α, while $T_2 = (1-\alpha)\left(\frac{L}{R_{BS}} + \frac{LC}{f_{BS}}\right)$ is monotonically decreasing with respect to α. Then according to (13), T has the minimum value when satisfying $T_1 = T_2$, and $T_1 = T_2$ can be expressed as

$$\frac{\alpha L}{R_{\text{UAV}}} + \frac{\alpha LC}{f_{\text{UAV}}} = \frac{(1-\alpha)L}{R_{\text{BS}}} + \frac{(1-\alpha)LC}{f_{\text{BS}}}. \tag{25}$$

By arranging the above formula, we can get the optimal task allocation ratio in (18).

From Theorem 1, we can see that the optimal flying altitude of the UAV depends on the distance between the UAV's projection and the user. Therefore, UAV can adjust its flying altitude in time according to the change of the horizontal distance between itself and the user. We also find that the optimal task allocation ratio is proportional to the bandwidth allocated to the UAV by the user. Moreover, the transmit power of the user and the bandwidth allocated to the BS also affect the optimal task allocation ratio.

4. NUMERICAL RESULTS

In this section, numerical results are given to verify the accuracy of the theoretical results. We set the distance between the projection of the UAV on the ground and the user is $r = 50$ m, environmental parameters are $a = 9.61, b = 0.16$ [36]. The channel gain at the reference distance of 1 m is $\beta_0 = -30$ dB, the distance between the user and the BS is $d_{\text{S2G}} = 25$ m. The bandwidth allocated by the user to offload tasks to the UAV and the BS are $B_{\text{UAV}} = B_{\text{BS}} = 1$MHz, the beamwidth of UAV's directional antenna is $\omega = 100°$. The CPU frequency allocated for computation on the UAV and the BS are $f_{\text{UAV}} = 1$GHz, $f_{\text{BS}} = 10$GHz. The other simulation parameters are shown in Table 2 . In Table 1, $\eta_{\text{LoS}} = 1$ and $\eta_{\text{NLoS}} = 20$ are the parameters determined by the environment [29, 36], and we choose the frequency band of 2.4GHz as the carrier frequency due to the extensive use in UAV products and related work [37,38].

By iterating (15), the optimal flying altitude H^{opt} can be obtained as 57 meters, and then substitute the iterative H^{opt} into (18), the optimal task allocation ratio α^{opt} can be obtained as 0.62.

Figure 2 shows the total system delay versus the task allocation ratio and flying altitude of the UAV. We can see that the total system delay decreases and then increases as the task allocation ratio and flying altitude of the UAV increase, i.e., there exists an optimal flying altitude and optimal task allocation ratio that can minimize the total system delay, which is consistent with our analyses in Section 3.

Figure 3 shows the minimum system delay, H^{opt}, and α^{opt} versus P. It can be seen that as P increases, α^{opt} gradually decreases and the minimum total delay of the system also decreases. It is because the computing capability of UAV is weaker than that of BS, and when user's transmit power increases, the user offloads more tasks to BS, and then the task allocation ratio of the user offloading tasks to UAV α decreases accordingly. We also see that with the increase of P, H^{opt} remains unchanged. This is because the optimal flying altitude of the UAV is irrelevant to P. By comparing the results obtained based on (15) and (18) with the results obtained by one-dimensional search, we can observe that the two are basically identical.

Figures 4 and 5 depict the minimum system delay, H^{opt}, and α^{opt} versus B_{UAV} and B_{BS}, respectively. Comparing Figures 4 and 5, it can be observed that the minimum total delay of the system decreases with the increase of both B_{UAV} and B_{BS}, α^{opt} increases in Figure 4 but decreases in Figure 5. It is because when B_{UAV} increases, the user offloads more computing tasks to the UAV, and the task allocation ratio of offloading to UAV α increases. However, when B_{BS} increases, the user offloads more computing tasks to the BS, and the task allocation ratio of offloading to BS $1-\alpha$ increases, i.e., α decreases. In the two figures we can also see that H^{opt} remains constant, since the optimal flight altitude is independent of B_{UAV} and B_{BS}. The correctness of the analytical results can be verified by comparing the results based on Formulas (15) and (18) with the results obtained from the one-dimensional search.

In Figures 6 and 7, we plot the minimum system delay, H^{opt}, and α^{opt} versus r and d_{S2G}, respectively. From Figure 6, we can observe that with the increase of r, the minimum system delay and H^{opt} increase, but α^{opt} decreases. The reason is that as r increases, the rate of the user's uplink transmission to the UAV decreases, and the minimum total delay of the system increases accordingly. In order to minimize the total delay of the system, more computing tasks of the user are offloaded to the BS, so the task allocation ratio of offloading to the UAV α decreases. However, α^{opt} in Figure 7 increases with the increase of d_{S2G}. This is because the decrease of the transmission rate between the user and BS leads to an increase of the total system delay, and computing tasks will be more offloaded to the UAV in order to minimize the total system delay. We compare the results obtained based on Formulas (15) and (18) with the results obtained by one-dimensional search, and we can see that the results coincide basically, indicating the accuracy of our analyses in Section 3.

Figures 8 and 9 show the minimum system delay, H^{opt}, and α^{opt} versus f_{UAV} and f_{BS}, respectively. By comparing Figures 8 and 9 , we see that when f_{UAV} and f_{BS} increase, respectively, the minimum total delay of the system decreases, α^{opt} gradually increases in Figure 8 but decreases in Figure 9. The reason is that when f_{UAV} increases, more computing tasks are offloaded to the UAV, then the task allocation ratio of task offloading to the UAV α increases. However, when f_{BS} increases, the user offloads more computing tasks to the BS, and the task allocation ratio of offloading to BS $1-\alpha$ increases, i.e., α decreases. From the two figures, we also see that the optimal flying altitude H^{opt} remains unchanged. This is because the optimal flying altitude of the UAV in Formula (15) is irrelevant to f_{UAV} and f_{BS}. By comparing the results obtained based on Formulas (15) and (18) with the results obtained from the one-dimensional search, it can be seen that both are basically consistent.

5. CONCLUSIONS

In this paper, we investigated a UAV-assisted MEC uplink maritime communication system in which the UAV is employed as an MEC server to provide computing services to the user. In addition, the user can offload tasks to BS for computation. Our objective is to minimize the total delay of the system by optimizing the flying altitude of UAV and the task allocation ratio. The numerical results verified the effectiveness of the theoretical analyses and reveal that the optimal flying altitude of the UAV is determined by the horizontal distance between the UAV and the user. We also showed the relationship between the optimal task allocation ratio and system parameters such as the transmit power, the bandwidth, and the CPU frequency.

Author Contributions

Writing-original draft preparation, methodology and software, S.H.; writing-review and editing, J.Z.; project administration and funding acquisition, Y.W. All authors have read and agreed to the published version of the manuscript.

REFERENCES

1. Wang, Y.; Feng, W.; Wang, J.; Quek, T.Q.S. Hybrid Satellite-UAV-Terrestrial Networks for 6G Ubiquitous Coverage: A Maritime Communications Perspective. IEEE J. Sel. Areas Commun. 2021, 39, 3475-3490.
2. Han, R.; Wen, Y.; Bai, L.; Liu, J.; Choi, J. Rate Splitting on Mobile Edge Computing for UAV-Aided IoT Systems. IEEE Trans. Cogn Commun. Netw. 2020, 6, 1193-1203.
3. Liu, C.; Feng, W.; Chen, Y.; Wang, C.; Ge, N. Cell-Free Satellite-UAV Networks for 6G Wide-Area Internet of Things. IEEE J. Sel. Areas Commun. 2021, 39, 1116-1131.
4. Malik, R.; Vu, M. Energy-Efficient Joint Wireless Charging and Computation Offloading in MEC Systems. IEEE J. Sel. Top. Signal Process. 2021, 15, 1110-1126.
5. Lei, C.; Feng, W.; Chen, Y.; Ge, N. Joint Power and Channel Allocation for Safeguarding Cognitive Satellite-UAV Networks. In Proceedings of the 2021 IEEE Global Communications Conference (GLOBECOM), Madrid, Spain, 7-11 December 2021; pp. 1-6.
6. Li, X.; Feng, W.; Wang, J.; Chen, Y.; Ge, N.; Wang, C. Enabling 5G on the ocean: A hybrid satellite-UAV-terrestrial network solution. IEEE Wirel. Commun. 2020, 27, 116-121.
7. Wu, Y.; Zheng, J. Modeling and Analysis of the Local Delay in an MEC-Based VANET for a Suburban Area. IEEE Internet Things J 2022, 9, 7065-7079.
8. Abrar, M.; Ajmal, U.; Almohaimeed, Z.M.; Gui, X.; Akram, R.; Masroor, R. Energy Efficient UAV-Enabled Mobile Edge Computing for IoT Devices: A Review. IEEE Access 2021, 9, 127779-127798.
9. Wei, P.; Guo, K.; Li, Y.; Wang, J.; Feng, W.; Jin, S.; Ge, N.; Liang, Y. Reinforcement Learning-Empowered Mobile Edge Computing for 6G Edge Intelligence. IEEE Access 2022, 10, 65156-65192.
10. Zhang, L.; Ansari, N. Latency-Aware IoT Service Provisioning in UAV-Aided Mobile-Edge Computing Networks. IEEE Internet Things J. 2020, 7, 10573-10580.
11. Fang, F.; Xu, Y.; Ding, Z.; Shen, C.; Peng, M.; Karagiannidis, G.K. Optimal Resource Allocation for Delay Minimization in NOMA-MEC Networks. IEEE Trans. Commun. 2020, 68, 7867-7881.
12. Chen, M.; Hao, Y. Task Offloading for Mobile Edge Computing in Software Defined Ultra-Dense Network. IEEE J. Sel. Areas Commun. 2018, 36, 587-597.
13. Wang, J.; Feng, D.; Zhang, S.; Liu, A.; Xia, X. Joint Computation Offloading and Resource Allocation for MEC-Enabled IoT Systems with Imperfect CSI. IEEE Internet Things J. 2021, 8, 3462-3475.

14. Chen, M.; Liang, B.; Dong, M. Joint offloading and resource allocation for computation and communication in mobile cloud with computing access point. In Proceedings of the IEEE INFOCOM 2017-IEEE Conference on Computer Communications, Atlanta, GA, USA, 1-4 May 2017; pp. 1-9
15. Jiang, F.; Wang, K.; Dong, L.; Pan, C.; Yang, K. Stacked Autoencoder-Based Deep Reinforcement Learning for Online Resource Scheduling in Large-Scale MEC Networks. IEEE Internet Things J. 2020, 7, 9278-9290.
16. Liu, Y.; Xie, S.; Zhang, Y. Cooperative Offloading and Resource Management for UAV-Enabled Mobile Edge Computing in Power IoT System. IEEE Trans. Veh. Technol. 2020, 69, 12229-12239.
17. Mei, H.; Yang, K.; Liu, Q.; Wang, K. Joint Trajectory-Resource Optimization in UAV-Enabled Edge-Cloud System with Virtualized Mobile Clone. IEEE Internet Things J. 2020, 7, 5906-5921.
18. Zhan, C.; Hu, H.; Sui, X.; Liu, Z.; Niyato, D. Completion Time and Energy Optimization in the UAV-Enabled Mobile-Edge Computing System. IEEE Internet Things J. 2020, 7, 7808-7822.
19. Xu, Y.; Zhang, T.; Yang, D.; Liu, Y.; Tao, M. Joint Resource and Trajectory Optimization for Security in UAV-Assisted MEC Systems IEEE Trans. Commun. 2021, 69, 573-588.
20. Sun, S.; Zhang, G.; Mei, H.; Wang, K.; Yang, K. Optimizing Multi-UAV Deployment in 3-D Space to Minimize Task Completion Time in UAV-Enabled Mobile Edge Computing Systems. IEEE Commun. Lett. 2021, 25, 579-583.
21. Hu, Q.; Cai, Y.; Yu, G.; Qin, Z.; Zhao, M.; Li, G.Y. Joint Offloading and Trajectory Design for UAV-Enabled Mobile Edge Computing Systems. IEEE Internet Things J. 2019, 6, 1879-1892.
22. Ji, J.; Zhu,K.; Yi, C.; Niyato, D. Energy Consumption Minimization in UAV-Assisted Mobile-Edge Computing Systems: Joint Resource Allocation and Trajectory Design. IEEE Internet Things J. 2021, 8, 8570-8584.
23. Xu, Y.; Zhang, T.; Loo, J.; Yang, D.; Xiao, L. Completion Time Minimization for UAV-Assisted Mobile-Edge Computing Systems IEEE Trans. Veh. Technol. 2021,70, 12253-12259.
24. Yu, Z.; Gong, Y.; Gong, S.; Guo, Y. Joint Task Offloading and Resource Allocation in UAV-Enabled Mobile Edge Computing. IEEE Internet Things J. 2020, 7, 3147-3159.
25. Zhang, J.; Zhang, T.; Yang, Z.; Li, B.; Wu, Y. Altitude and number optimisation for UAV-enabled wireless communications. IET Commun. 2020, 14, 1228-1233.
26. Agarwal, A.; Mishra, D. Altitude Optimization for DF Relaying Trajectory of UAV in Cooperative FANET. In Proceedings of the GLOBECOM 2020-2020 IEEE Global Communications Conference, Taipei, Taiwan, 7-11 December 2020; pp. 1-6.
27. Yadav, S.; Vats, A.; Aggarwal, M.; Ahuja, S. Performance Analysis and Altitude Optimization of UAV-Enabled Dual-Hop Mixed RF-UWOC System. IEEE Trans. Veh. Technol. 2021, 70, 12651-12661.
28. Zhang, J.; Liang, F.; Li, B.; Yang, Z.; Wu, Y.; Zhu, H. Placement optimization of caching UAV-assisted mobile relay maritime communication. China Commun. 2020, 17, 209-219.
29. Al-Hourani, A.; Kandeepan, S.; Lardner, S. Optimal LAP Altitude for Maximum Coverage. IEEE Wirel. Commun. Lett. 2014, 3 , 569-572.

30. Ouamri, M.; Alkanhel, R.; Gueguen, C.; Alohali, M.; Ghoneim, S. Modeling and analysis of uav-assisted mobile network with imperfect beam alignment. Comput. Mater. Contin. 2023, 74, 453-467.
31. Al-Hourani, A.; Kandeepan, S.; Jamalipour, A. Modeling air-to-ground path loss for low altitude platforms in urban environments. In Proceedings of the 2014 IEEE Global Communications Conference, Austin, TX, USA, 8-12 December 2014; pp. 2898 – 2904.
32. Liu, Y.; Liu, K.; Han, J.; Zhu, L.; Xiao, Z.; Xia, X. Resource Allocation and 3-D Placement for UAV-Enabled Energy-Efficient IoT Communications. IEEE Internet Things J. 2021, 8, 1322-1333.
33. Li, J.; Zhao, H.; Wang, H.; Gu, F.; Wei, J.; Yin, H.; Ren, B. Joint Optimization on Trajectory, Altitude, Velocity, and Link Scheduling for Minimum Mission Time in UAV-Aided Data Collection. IEEE Internet Things J. 2020, 7, 1464-1475.
34. You, C.; Zhang, R. Hybrid Offline-Online Design for UAV-Enabled Data Harvesting in Probabilistic LoS Channels. IEEE Trans. Wirel. Commun. 2020, 19, 3753-3768.
35. Balanis, C.A. Antenna theory: A review. Proc. IEEE 1992, 80, 7-23.
36. Shi, W.; Li, J.; Xu, W.; Zhou, H.; Zhang, N.; Zhang, S.; Shen, X. Multiple Drone-Cell Deployment Analyses and Optimization in Drone Assisted Radio Access Networks. IEEE Access 2018, 6, 12518-12529.
37. Fernandez, M.G.; Lopez, Y.A.; Arboleya, A.; Gonzalez-Valdes, B.; Andres, L.H. Antenna Diagnostics and Characterization Using Unmanned Aerial Vehicles. IEEE Access 2017, 5, 23563-23575.
38. Zhou, Y.; Cheng, N.; Lu, N.; Shen, X.S. Multi-UAV-Aided Networks: Aerial-Ground Cooperative Vehicular Networking Architecture. IEEE Veh. Technol. Mag. 2015, 10, 36-44.

CHAPTER 7

Using Satellite AIS to Analyze Vessel Speeds Off the Coast of Washington State, U.S., as a Risk Analysis for Cetacean-Vessel Collisions

Nathan C. Greig[1,2,3], Ellen M. Hines[1,2], Samantha Cope[1,4] and XiaoHang Liu[2]

[1]*Estuary & Ocean Science Center, San Francisco State University, Tiburon, CA, United States*
[2]*Department of Geography &*
Environment, San Francisco State University, San Francisco, CA, United States
[3]*Midpeninsula Regional Open Space District, Los Altos, CA, United States*
[4]*ProtectedSeas, Anthropocene Institute, Palo Alto, CA, United States*

ABSTRACT

Most species of whales are vulnerable to vessel collisions, and the probability of lethality increases logistically with vessel speed. An Automatic Identification System (AIS) can provide valuable vessel activity data, but terrestrial-based AIS has a limited spatial range. As the need for open ocean monitoring increases, AIS broadcasts relayed over earth-orbiting satellites, satellite AIS (SAIS), provides a method for expanding the range of AIS broadcast reception. We used SAIS data from 2013 and 2014 to calculate vessel density and speed over ground around the coast of Washington state in the northwestern United States. Nearby shipping lanes connecting the Ports of Seattle, Tacoma, Portland, and in Canada, Vancouver, have the greatest density of vessel traffic arriving and departing. Knowledge of shipping activity is important in this area due to the nearby presence of NOAA designated Cetacean Density and Distribution Working Group's Biologically Important Areas (BIA) for large whale species vulnerable to vessel collisions. We quantified density and speed for each vessel type that transits through BIA's. We found that cargo and tanker vessels traveled the farthest distance at the greatest speeds. As ship-strike risk assessments have traditionally relied on terrestrial AIS, we explored issues in the application of SAIS data. Temporal gaps in SAIS data led to a resulting systematic underestimation of vessel speed in calculated speed over ground. However, SAIS can be helpful in documenting minimum vessel speeds across large geographic areas and across national boundaries, especially beyond the reach of terrestrial AIS receivers. SAIS data can also be useful in examining vessel density at broad scales and could be used to assess basin-wide open ocean routes. Future use of additional satellite platforms with AIS receivers and technological advances will help rectify this issue and improve data coverage and quality.

Keywords: *satellite automatic identification system; cetacean ship strikes; geographic information systems; olympic coast national marine sanctuary; baleen whales; biologically important areas; shipping speed*

1. INTRODUCTION

Shipping, a highly globalized industry, is an important component of international trade, and is correlated with global economic patterns (Schwehr and McGillivary, 2007; Rodrigue, 2010; Frisk, 2012). While the world's vessel fleet is growing over time (Silber et al., 2010), so does the need to effectively track vessel movements. The Automatic Identification System (AIS) is a non-proprietary tracking technology standardized by the International Telecommunications Union and required for most vessels by the International Maritime Organization's Safety of Life at Sea Convention (Tetreault, 2005; United States Coast Guard [USCG], 2008). Originally conceived to improve ships' navigational safety, terrestrial AIS broadcasts are limited roughly to line-of-sight (Calder and Schwehr, 2009), so coverage does not extend well into the open ocean (Silber et al., 2014). Satellite AIS (SAIS), since 2008, can help overcome the terrestrial line-of-sight limitation by collecting AIS broadcasts from a constellation of earth-orbiting satellites (Ball, 2013; Robards et al., 2016).

Despite the initial intent of AIS as a navigational aid, the data provide valuable insight on human use of marine areas in the context of environmental conservation. Events like the establishment of large off-shore protected areas clarify the need to extend the range of traditional AIS broadcasts to better monitor vessel activity far from shore (McCauley et al., 2016). SAIS data have been employed to monitor fishing activity and protected area regulation compliance (Natale et al., 2015; de Souza et al., 2016; Rowlands et al., 2019). Terrestrial AIS has provided valuable data to assess the risk of collision between ships and whales near established shipping lanes in coastal areas (Williams and O'Hara, 2010; Wiley et al., 2011; Redfern et al., 2013; Jensen et al., 2015), but lethal interactions may also occur outside of these highly monitored areas (Rockwood et al., 2017). For off-shore areas, SAIS has the potential to contribute vessel activity data necessary for collision risk assessments (Williams and O'Hara, 2010; van der Hoop et al., 2012).

Since 2009, the International Whaling Commission has reported more than 1,200 confirmed incidents of vessel collisions with whales (Cates et al., 2017). Vessel speed is an important component of shipping's potential impact on cetaceans (Gende et al., 2011; Conn and Silber, 2013; Currie et al., 2017). The conservation benefit to whales by reducing vessel speed is well established and is generally expressed as a simple logistic relationship between vessel speed and probability of lethality (Vanderlaan and Taggart, 2007; Gende et al., 2011; Wiley et al., 2011; Conn and Silber, 2013). There is a significant positive relationship, and the greatest rate of change generally occurs between 9 and 15 knots, corresponding roughly to an increase in probability of lethality from 20 to 80% (see Vanderlaan and Taggart, 2007). Vessel speed limits help to reduce the anthropogenic mortality risk and possibly collision probability (Gende et al., 2011; Conn and Silber, 2013). AIS provides data on vessel speed which has been combined with density to examine threats of collision (Felski et al., 2015; Rockwood et al., 2017).

Our objective was to use available SAIS data to delineate spatial locations where vessel traffic density and speed were high in known areas of cetacean concentrations off the coast of Washington state. The use of SAIS data was necessary because only a little more than half of the study area water was within the potential range of terrestrial AIS.

In Washington state waters, there were 19 of 130(15%) strandings from 1980 to 2006 that showed evidence of collisions with vessels. Numerous biases in collision detection lead to underestimates in true numbers of mortalities (Douglas et al., 2008), so the actual number is likely higher (Williams et al., 2011). There are Biologically Important Areas (BIAs) in this research area for gray (Eschrichtius robustus) and humpback whales (Megaptera novaeangliae) (Calambokidis et al., 2015; Figures 1, 2). BIAs are species, region, and time specific areas which the National Oceanic and Atmospheric Administration (NOAA) Fisheries, Cetacean Density and Distribution Working Group has identified as important for reproduction, feeding, migrating, or small and residential populations (Ferguson et al., 2015).

At the time of this writing, NOAA has declared an active Unusual Mortality Event (UME) for gray whales along the United States west coast from California to Alaska indicating above-average mortality rates: 34 strandings in Washington state waters and 10 in Canadian waters between January and September of 2019 (National Oceanic and Atmospheric Administration [NOAA], 2019). The gray whale BIAs in our study area were based on migratory corridors between annual feeding and reproductive areas, from numerous survey methods and expert opinion (Calambokidis et al., 2015; Figure 2). Assessing patterns of vessel behavior in these areas can help managers evaluate potential impacts to this vulnerable population suffering from elevated mortality events. While the United States Fish and Wildlife Service removed the species-level listing of humpback whales on the Endangered Species List, some unique migratory populations are still listed as Threatened or Endangered (United States Fish and Wildlife Service [USFWS], 2016). The humpback whale BIAs were based on surveys and opportunistic sources about highly concentrated feeding animals (Calambokidis et al., 2015; Figure 2). We used these BIAs as a proxy for areas of high cetacean concentration for each species.

SAIS can now facilitate a complete global picture of vessel activity (Skauen, 2019). The growing number of satellites in orbit capable of transmitting SAIS data suggests that temporal gaps between satellite passes over an area will be minimized (McCauley et al., 2016). If this trend continues, it may become increasingly

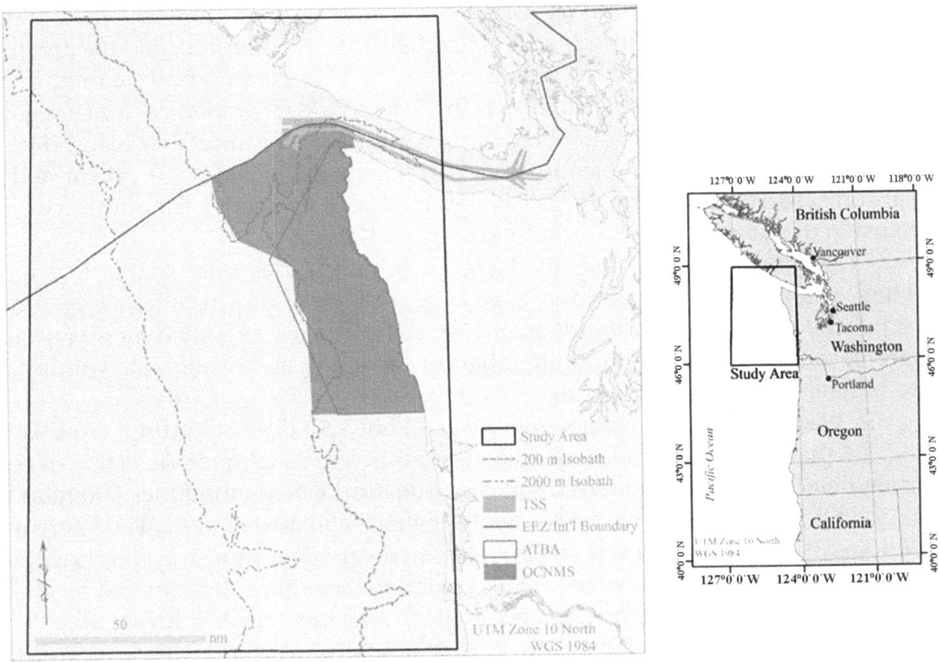

Figure 1. *Bathymetry and administrative areas in the study area along the Pacific Northwest coast including Washington state in the United States and southwest Vancouver Island in Canada. Note: Traffic Separation Scheme (TSS), Exclusive Economic Zone (EEZ), Area To Be Avoided (ATBA), Olympic Coast National Marine Sanctuary (OCNMS).*

more attractive to use SAIS data instead of terrestrial AIS data due to its spatial coverage. We took this opportunity to investigate and evaluate potential issues related to monitoring vessel density and speed using SAIS data that may need to be rectified before more pervasive use of the technology can occur. AIS is an imperfect data source as it is limited by human input, data corruption, signal noise, GPS faults, and gyrocompass or other instrument failure onboard the target vessel (Aarsæther and Moan, 2009; McGillivary et al., 2009; Silber and Bettridge, 2010; Robards et al., 2016). Further, gaps in SAIS data are also evident (Allen et al., 2018; Eriksen et al., 2018), especially in areas of high activity due to satellite congestion (Jia et al., 2019). Despite these issues, both AIS and SAIS provide insights on vessel activity otherwise unavailable.

In this research, we provide a baseline understanding of vessel activity off the Washington coast as a foundation for future risk assessments. Extensive research on risk to North Atlantic right whales on the United States east coast has shown the effectiveness of reduced speed and routing alternatives in reducing whale mortality from vessel collisions (Laist et al., 2014). However, the vessel patterns within United States waters outside the Strait of Juan de Fuca has still not been studied. Multiple species of large whales off the coast of Washington state are vulnerable to collisions (Douglas et al., 2008; Silber et al., 2010). This geographic focus area is important due to its connections to several primary ports and the presence of multiple species of slow-reproducing whales recovering from past population declines.

2. MATERIALS AND METHODS

2.1. Study Area

Our research site was offshore from the important North American west-coast ports of Vancouver, Seattle, Tacoma, and Portland (Figure 1). The specific study area was between $46 - 49°$n and $124 - 127°$W, and defined the extent of SAIS data collection. Reaching 90 to 125 nm offshore of the state of Washington in the northwestern United States, the study area extended from roughly the mouth of the Columbia River in the south to the Strait of Juan de Fuca and southern Vancouver Island in the north.

Inside the Strait of Juan de Fuca are the Ports of Tacoma and Seattle, major United States shipping ports, and Port Metro Vancouver (Figure 1). Port Metro Vancouver is the largest Canadian port, handling roughly the same amount of total tonnage as the Port of New York and more than Seattle and Tacoma combined (United States Army Corps of Engineers [USACE], 2016; Port Metro Vancouver, 2017). Tacoma and Seattle ranked 7th and 10th, respectively, in the United States for total container ship traffic in 2013 and 29th and 31st, respectively, for total tonnage in the United States in 2014 (United States Army Corps of Engineers [USACE], 2015, 2016). These ports connect internationally to East and Southeast Asia, and domestically to Alaska, Hawaii, and the West Coast of the United States. Additionally, the Port of Portland is located inland from the mouth of the Columbia River. Portland ranked 25th among United States ports for total container ship traffic in 2013 and 28th for total tonnage in 2014 (United States Army Corps of Engineers [USACE], 2015, 2016).

Within the full $75,367\ km^2$ study area, 88% is open water and 12% is land. Administrative areas in the study area include the NOAA administered Olympic Coast National Marine Sanctuary (OCNMS), an International Maritime Organization (IMO) designated Area to be Avoided (ATBA), and part of the United States Coast Guard (USCG) controlled Juan de Fuca Traffic Separation Scheme (TSS) (Figure 1). Within the study area, there are seven BIAs, six for gray whales and one for humpback whales (Table 1; and Figure 2).

2.2. SAIS Information

Satellite AIS data were collected by exactEarth Ltd. (Cambridge, ON, Canada) for the calendar years 2013 and 2014, and received from the OCNMS. Each record in the tables corresponded to an individual AIS broadcast. AIS information is comprised of static information that does not change over the course of a voyage, and dynamic information that can change as frequently as every AIS broadcast.

The original SAIS data that were received had 3,045,407 records for the year 2013 and 2,941,900 for 2014. The years 2013 and 2014 had consistent fuel sulfur regulations (10,000 ppm or 1.0%) for vessels operating within the North America Emissions Control Area, 200 nm from the coast (Environmental Protection Agency

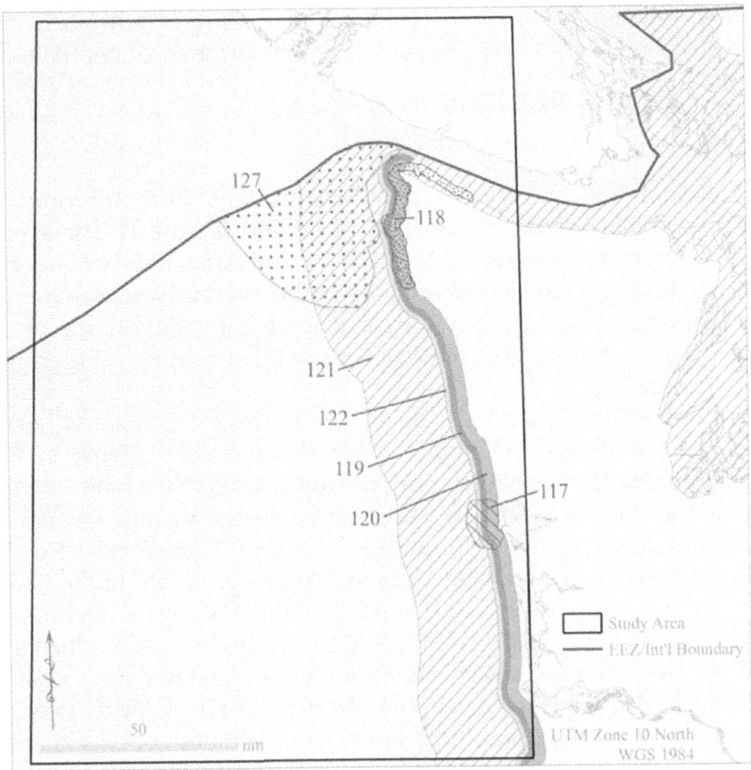

Figure 2. *Biologically Important Areas (BIA) (see Table 1 for species and type) for gray (Eschrichtius robustus) and humpback whales (Megaptera novaengliae) along the coast of the North American Pacific Northwest (Calambokidis et al., 2015). All BIAs are for gray whales with the exception of BIA 127 for humpback whales.*

[EPA], 2010). Thus, there were no temporal changes in traffic patterns based on emissions controls standards during our study period (Jensen et al., 2015). A further reduction in sulfur regulations to 0.1% occurred in 2015 (International Maritime Organization [IMO], 2019). Although evidence from California suggests that ships may alter their speed based on new regulations (Moore et al., 2018), differences in shipping patterns are seen to principally reflect longer term economic changes (Jensen et al., 2015). The results of this study are meant to inform management of past trends and provide insight on the use of SAIS data for evaluating vessel activity.

2.3. Data Preparation

We conducted data quality control, starting with removing duplicate SAIS records, defined as records having the same Maritime Mobile Service Identity (MMSI), latitude, longitude, and time. The MMSI is a unique, regulated, and coded identifier for a ship. The second quality control step was to remove all records with a missing or null MMSI. Next, we created a tabular relationship

Table 1. *Summary of Biologically Important Areas (BIA) within the North American Pacific Northwest (Calambokidis et al., 2015).*

Number	Species	Name	Type	Months	Are a within study area (nm)^2	Description
117	Gray whale	Grays Harbor	Feeding	April–November	86.8	Outside Grays Harbor
118	Gray whale	Northwest Washington	Feeding	May–November	150.1	N o rthwest tip o f Olympic Peninsula
119	Gray whale	Northbound Phase A	Migration	January–July	652.3	Within 8 km of coast
120	Gray whale	Northbound Phase B	Migration	March–July	393.8	Within 5 km of coast
121	Gray whale	Potential Presence	Migration	Jan.–July, Oct.–Dec.	4111.8	Within 47 km of coast
122	Gray whale	Southbound	Migration	October–March	828.7	Within 10 km of coast
127	Humpback whale	Northern Washington	Feeding	May–November	989.2	Northern part of United States EEZ

between dynamic SAIS information and static vessel information, using the MMSI as a primary key. We used ArcGIS (Environmental Systems Research Institute [ESRI], 1999-2018) geoprocessing and the programming language Python (Python Software Foundation, 1990-2019) running in PyScripter (Vlahos, 2005-2015) to write or modify numerous Python scripts.

The first script reprojected the point data into the Universal Transverse Mercator (UTM) Zone 10 North projection based on the Geographic Coordinate System (GCS) World Geodetic System (WGS) 1984, which allows distance to be measured in meters. AIS latitude and longitude are collected from a GPS receiver, which is based on GCS WGS 1984. The second Python script ran a spatial selection of only the SAIS points that were broadcasted from the water, and eliminated random error points located on land (Jensen et al., 2015). We added and calculated new fields for season and day/night based on the time stamp. Seasons were defined as Winter (January-March), Spring (April-June), Summer (July-September), and Autumn (October-December) (Forney and Barlow, 1998; Becker et al., 2014; Jensen et al., 2015). Day and night were defined by using published nautical twilight times from the United States Naval Observatory, Astronomical Applications Department for Forks, WA and Ocean Shores, WA for the years 2013 and 2014. The nautical twilight was defined per month by using the average time from the 15th of each month over the study area (Jensen et al., 2015).

We created ship transit line segments and evaluated transit contiguity (Jensen et al., 2015) by joining sequential SAIS data points from the same vessel to create straight line segments between these points. The time difference and distance between sequential points defined the calculated speed over ground (SOG). While reported SOG from terrestrial AIS systems has previously been used to evaluate threat of collision to cetaceans in coastal areas (van der Hoop et al., 2012; Conn and Silber, 2013), we chose to use a calculated SOG due to frequent gaps in SAIS data. Eriksen et al. (2018) used calculated SOG to identify SAIS records with large temporal or positional gaps, therefore unusable in analysis, by removing vessel transits with implausible speeds. We employed a similar approach (see below). The final part of the script evaluated transit contiguity, a single vessel on a continuous transit, based on MMSI, Trip ID, time between broadcasts, and heading difference.

We merged all 24 months' of line segments and the 22 OCNMS-specified vessel types (see Table 2) together to create complete transit lines. We analyzed the years 2013 and 2014 together to simplify calculations, despite differences in overall mean vessel speeds. In 2013, 42.8% of the data were removed due to duplicate records, whereas this was only 5.8% for 2014 . We conducted analyses at the vessel type level to minimize inherent inter-type vessel differences.

2.4. Calculated Speed Truncation

Truncation of calculated SOG values, an attempt to enumerate the highest possible legitimate speed per vessel type, was necessary because of errors in the data or processing that led to implausible SOG values and means, common in SAIS data (Eriksen et al., 2018). The truncation threshold was an estimate of the maximum plausible and attainable speed for each vessel type. The maximum calculated speed of liquefied gas carriers, cable layers, and pollution control vessels were less than one knot under the truncation threshold. Furthermore, the three vessel types in the tanker category had identical truncation thresholds. Combined, these relative measures of accuracy and precision indicate that the truncation threshold method was a workable approximation for maximum speed.

Truncation meant removing any record where the calculated speed was greater than a given threshold from the following equation:

$$TT_{\alpha} = \bar{x}_{\alpha} + (3 \times \sigma_{\alpha})$$

where TT_{α} is the truncation threshold for vessel type α, $\bar{x}_{\alpha}$ is the mean non-0 broadcast SOG value for type α, and σ_{α} is the standard deviation of non-0 broadcast SOG values for type α. This equation was derived from examining histograms and statistics of broadcast SOG values and validated by expert opinion (G. Galasso, Deputy Superintendent, Olympic Coast National Marine Sanctuary, pers. comm.). The use of nonzero broadcast SOG values was necessary because 36.7% of all broadcast SOG values were zero, skewing the mean and inflating the standard deviation.

We overlaid vessel transit lines with truncated average speed values across a range of areas of interest. This included the entire space and time of the study area, and spatial subsets that included the OCNMS, ATBA, and the BIAs during active months. We examined temporal subsets that included day versus night and the four seasons. The overall truncation rate of 1.28% allowed the vast majority of data to be retained for further analysis. However, truncation was necessitated in almost 55,000 records, due to either anomalous location or time stamp broadcasts.

Several vessel types with the highest truncation thresholds (supply ship, container ship, roll-on / roll-off (RORO) cargo ship, and public vessel) also had some of the highest rates of data truncation (Supplementary Table S1). The source of these high rates of truncation and any possible correlation within SAIS is unclear.

2.5. Hexagon Average Speed and Density

The Olympic Coast NMS has used hexagons, one square statute mile or 2.6 square kilometers in area, as a unit of measurement or observation as part of their spatial planning process (N. Wright, Marine Geographer, Olympic Coast NMS, pers. comm.). There are 29,542 homogenous hexagons for the entire study area, including the ONMS. We calculated mean SOG and number of vessel transits per month for vessel transits across each hexagon that intersected the BIAs.

Table 2. *Overall calculated and broadcast speed over ground (SOG) by vessel type for 2013 and 2014 combined.*

	Calculated speed over ground			Broadcast speed over ground		
Vessel type	**Mean (Knots)**	**SD (Knots)**	**Count**	**Mean (Knots)**	**SD (Knots)**	**Count**
Cargo						
Bulk carrier	10.5*	4.1	963279	8.5	5.4	976195
Cargo ship	11.2*	3.8	77292	9.5	5.2	78431
Container ship	16.1*	4.7	382776	13.4	7.3	392957
Refrigerated cargo	11.7*	2.6	2851	8.4	5.3	3144
RORO Cargo Ship	16.9*	5.7	45424	14.0	8.1	46857
Vehicle carrier	14.0*	4.9	194421	11.5	6.8	197485
Tanker						
Chemical carrier	11.8*	3.7	51393	9.8	5.5	52312
Liquefied gas carrier	13.5*	3.8	2627	12.8	4.5	2660
Oil tanker	12.4*	3.4	198504	10.3	5.5	201838
Tug						
Articulated tug barge	9.1*	2.7	85510	7.5	4.3	86458
Tug	6.7*	3.2	317997	5.4	3.9	322224
Fishing						
Fishing vessel	3.0*	3.6	1400609	2.5	3.4	1408449
Passenger						
Passenger ship	18.2*	4.7	50239	15.8	7.8	51086
Miscellaneous						
Cable layer	8.1	5.1	2056	8.3	5.1	2061
Dredger	3.6*	3.7	50877	2.5	3.4	51693
Drill ship	3.0	4.4	750	3.2	4.4	752
Pollution control	0.6*	1.9	55869	0.4	1.8	55941
Private vessel	7.5*	5.3	24783	6.2	5.6	25025
Public vessel	6.5*	8.1	276096	6.2	8.4	282194
Research ship	5.0*	4.4	37174	4.4	4.3	37404
Supply ship	10.4	6.6	4306	10.9*	6.7	4531
Unknown	3.8*	3.9	833	3.1	3.9	842

Vessel types are as specified by the Olympic Coast National Marine Sanctuary. The asterisk (*) indicates a statistically significantly higher mean SOG ($\alpha = 0.05$) when comparing calculated versus broadcast SOG within a vessel type.

Figure 3. *Average calculated speed over ground (SOG) and vessel transits per month in the feeding Biologically Important Areas (BIA) for gray and humpback whales for 2013 and 2014. Darker colors represent faster average speeds and more transits per month for each vessel type. The unit of observation is the square statute mile hexagon.*

3. RESULTS

3.1. Total Transits and Distance Traveled

The commercial ports of Vancouver, Seattle, Tacoma, and Portland drove much of the vessel traffic in the study area. Bulk carriers had the most cargo transits.

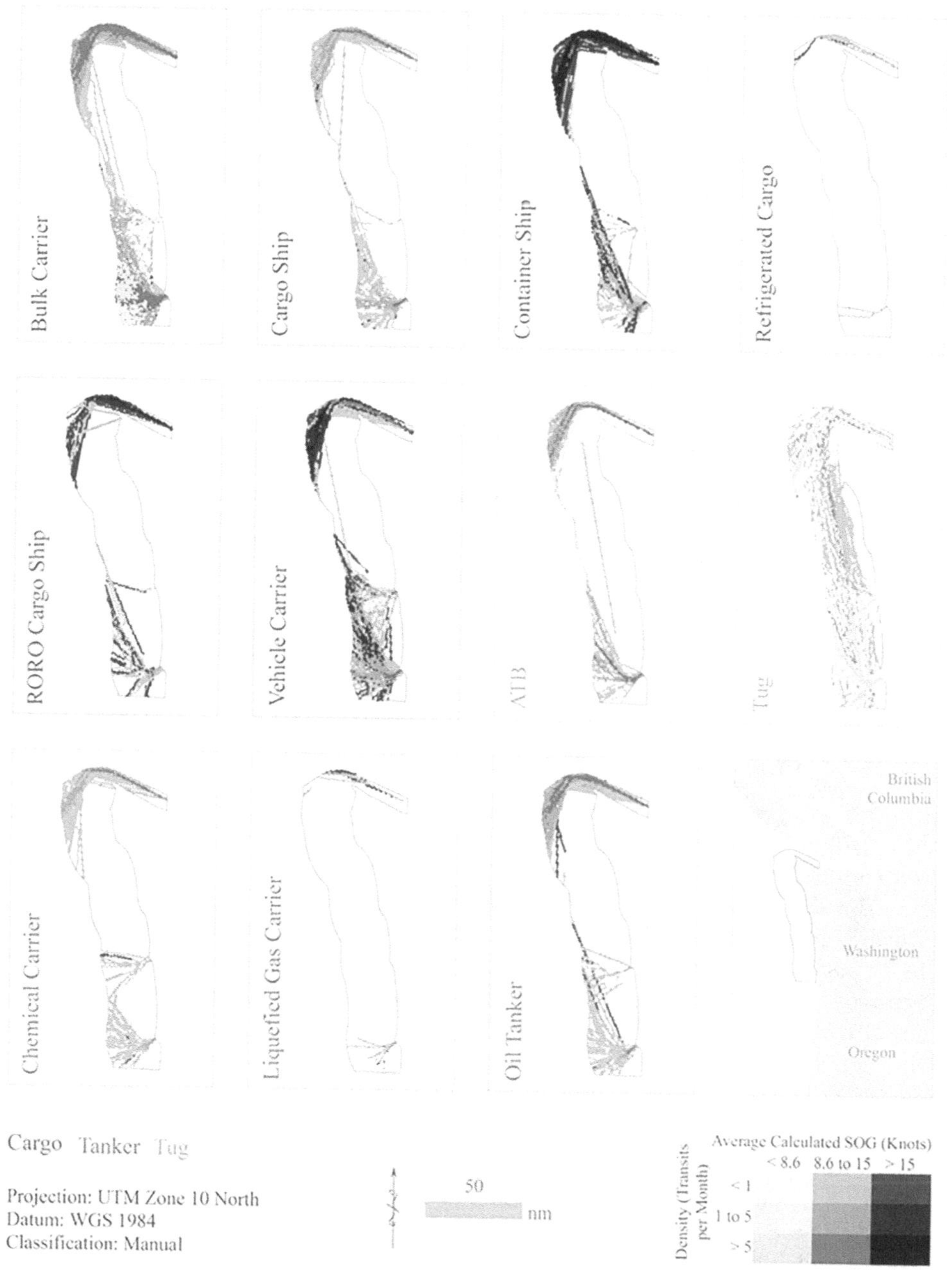

Figure 4. *Average calculated speed over ground (SOG) and vessel transits per month in the gray whale potential presence Biologically Important Area (BIA)for 2013 and 2014. Darker colors represent faster average speeds and more transits per month for each vessel type. The unit of observation is the square statute mile hexagon.*

Table 3. *Percent of calculated speed over ground (SOG) values less than 15 knots for the entire research area and within each of the Biologically Important Areas (BIA) for 2013 and 2014.*

		Biologically important areas			
Vessel type	**Overall**	**Grays Harbor**	**NW Washington**	**Potential presence**	**N Washington**
Cargo					
Bulk carrier	95.2	99.4	98.3	94.7	93.4
Cargo ship	92.2	100.0	100.0	91.3	86.2
Container ship	33.8	55.2	30.7	40.6	32.8
Refrigerated cargo	89.7		100.0	89.7	79.5
RORO cargo ship	33.7	77.8	26.9	47.4	34.8
Vehicle carrier	44.8	90.2	67.4	53.1	32.5
Tanker					
Chemical carrier	90.2	98.8	92.0	91.7	90.7
Liquefied gas carrier	61.1		0.0	82.2	75.7
Oil tanker	83.6	100.0	97.0	86.5	87.1
Tug					
Articulated tug barge	99.7		100.0	99.7	99.8
Tug	100.0	100.0	100.0	100.0	100.0
Fishing					
Fishing vessel	100.0	100.0	100.0	100.0	99.9
Passenger					
Passenger ship	17.7	100.0	34.8	25.5	17.0
Miscellaneous					
Cable layer	98.6			98.1	99.3
Dredger	100.0	100.0	100.0	99.9	100.0
Drill ship	100.0			100.0	100.0
Pollution control	99.9	100.0	99.2	99.5	100.0
Private vessel	93.2	99.1	93.7	91.2	85.4
Public vessel	84.0	93.1	58.9	68.3	83.7
Research ship	99.3	100.0	99.8	99.5	99.7
Supply ship	83.3	100.0	98.2	71.8	100.0
Unknown	100	100.0	100.0	100.0	100.0

Miscellaneous category vessels without private and public vessels accounted for a very small proportion of vessel transits and distance traveled.

The 42,629 sum total transits of all vessels in the years 2013 and 2014 covered 2,694,197 nm. Fishing vessels account for the most total vessel transits at 26.9%, followed by bulk carriers at 23.5% (Supplementary Table S2). The only other vessel type to have greater than 10% of the total transits is container ships (10.4%). Seasonally, as found in Jensen et al. (2015) outside San Francisco Bay, vessel speeds did not change overall, with the exception of passenger ships that traveled more slowly in the autumn and winter, although still at speeds greater than 15 knots (Supplementary Table S3).

The cargo category made up 41.7% of total transits, the most of any category. With the exception of public vessels and private vessels, most vessel types in the miscellaneous category registered very few transits. The remaining seven vessel types in the miscellaneous category only account for 2.2%. Vessels in the cargo and tanker categories averaged 27.7 transits per day.

Bulk carriers accounted for the most distance traveled by any one vessel type (32.0%). Only fishing vessels (16.8%) and container ships (13.2%) accounted for more than 10% of total distance traveled. Cargo category vessels traveled more than half (56.0%) of all distance traveled.

Figure 5. *Container ship calculated speed over ground (SOG) and transits per month for 2013 and 2014. The area northwest of the blue Exclusive Economic Zone (EEZ) exhibits high vessel speed and density, although it was not included in the Biologically Important Area (BIA) designations.*

3.2. Overall Average Calculated SOG

Passenger ships, including ferries, showed the greatest average calculated SOG, 18.2 knots (Table 2). These were followed by RORO cargo ships (16.9 knots), container ships (16.1 knots), and vehicle carriers (14.0 knots). The five vessel types with the greatest average calculated SOG also had the five greatest average broadcast SOG. Supply ships were the only miscellaneous category vessel type to average greater than 10 knots (10.4 knots). Public vessels had the greatest variability of speeds, with a standard deviation of 8.1 knots, followed by supply ships (6.6 knots) and RORO cargo ships (5.7 knots).

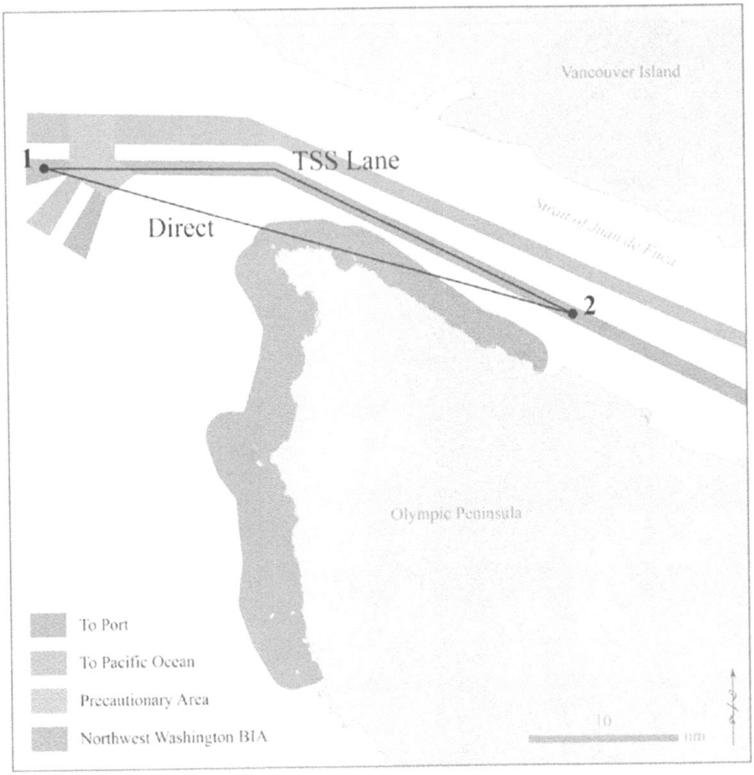

Figure 6. *Potential vessel transit lines cutting across the Northwest Washington Biologically Important Area (BIA) due to temporal gaps between sequential points in the SAIS data, (1) and (2). The vessel could actually remain inside the Traffic Separation Scheme (TSS), but could appear to cut across the BIA, leading to false transits.*

With the exceptions of three vessel types (cable layer, drill ship, and supply ship), all vessel types had a greater calculated SOG than broadcast SOG. We tested if the calculated and broadcast speeds came from the same statistical population using both the parametric Welch t-Test and the non-parametric Mann-Whitney-Wilcoxon Test. Further, with the exception of four vessel types (fishing vessels, dredgers, pollution control, and research ships), all vessel types had an equal or greater broadcast SOG standard deviation than calculated SOG standard deviation (Table 2).

3.3. Biologically Important Areas

There are three feeding BIAs in the study area, each of which was transited by most vessel types (Figure 3). The areas of the BIAs most frequented by vessels were the western and northern regions of the humpback whale feeding BIA. The south-eastern part of this BIA is inside the IMO-designated ATBA, which specifies an area that all ships greater than 400 gross tonnage should avoid for safety and environmental concerns. The shipping lanes entering and exiting the Strait of Juan de

Fuca had the highest density of vessels. Vessels in the Grays Harbor and Northwest Washington feeding BIAs for gray whales were not as common, with the exception of the Strait of Juan de Fuca. Commercial vessels infrequently transited the Grays Harbor BIA.

There were four gray whale migration BIAs within the study area. The Northbound Phase A, Northbound Phase B, and Southbound migration BIAs are located within eight-, five-, and ten-kilometer buffers of the coast, respectively. Due to this coastal proximity and the ATBA, there were relatively few vessels in any of these BIAs. However, the potential presence BIA extends 47 km from the coast and was transited by all vessel types (Figure 4). Fishing vessels, tugs, private vessels, public vessels, and research ships utilized the entire area, while most vessels in the cargo and tanker categories avoided the ATBA, which overlaps with a large portion of the central part of this BIA.

Container ships had the greatest combined average SOG and density, particularly in the northern region of the BIA. The northern portion of the BIA at the mouth of or inside the Strait of Juan de Fuca had the highest densities of most vessel types. Drill ships and unknown vessel types were very uncommon in the potential presence BIA. Tugs and ATBs showed very different movement patterns. Tugs traversed the entire BIA, but ATBs followed the pattern of commercial vessels and avoided the ATBA.

A full tabular statistical summary for calculated SOG in the four BIAs analyzed can be found in Supplementary Table S4. Most transits for most vessel types across the BIAs occurred at less than 15 knots (Table 3). Notable exceptions were the fastest vessel types (container ship, RORO cargo ship, vehicle carrier, and passenger ship). Grays Harbor had the greatest proportion of vessel speeds below 15 knots. This BIA is just offshore, so vessels were likely approaching or leaving port at slower speeds.

4. DISCUSSION

Satellite AIS data are valuable for assessing broad-scale patterns of human activity in off-shore waters (Rowlands et al., 2019) and will likely grow in value and accessibility over time (McCauley et al., 2016), as the use of SAIS-derived data allows for vessel tracking much further from the coast than is possible with terrestrial receivers. By analyzing vessel density and speed using SAIS data, our contribution to the current understanding of cetacean collision risk assessment was twofold. First, we broadly assessed vessel density and speed by vessel type and further narrowed the assessment within the active months of BIAs for local cetaceans. Second, we investigated issues related to the nature of SAIS data and their potential impact on its successful use for informing cetacean-vessel collision risk assessments in the future.

4.1. Overall Average Calculated Speed Over Ground

Most vessel types had a greater average calculated SOG than average broadcast SOG. It is important to use calculated SOG in any analysis so that the risk from potential vessel collisions is not underestimated by using broadcast SOG. Bulk carriers, cargo ships, refrigerated cargo, and chemical carriers had an average broadcast SOG less than 10 knots, but an average calculated SOG greater than 10 knots, the speed limit for North Atlantic right whale seasonal management areas (Laist et al., 2014). Similarly, container ships and RORO cargo ships crossed the 15 knot threshold when average calculated SOG was considered instead of average broadcast SOG. Exceeding these thresholds could have important management implications. More research is needed to document the difference between broadcast and calculated SOG when using SAIS data.

Container ships were one of the fastest and most common vessel types. Although passenger ships did not comprise a large proportion of transits or total distance, they were the fastest vessel type, and thus warrant special consideration in any potential future risk assessment. Fishing vessels and bulk carriers, the most common vessel types, had average calculated SOG less than or near the 10 knot speed restrictions that are commonly used in whale management areas (Laist et al., 2014). All other cargo and tanker ships had average calculated SOG values above 10 knots. Vessel categories that transited across the study area (cargo, tanker, and passenger) tended to have greater average speeds than those working within the study area (fishing, miscellaneous). These categories should warrant potentially differing policy and analysis considerations. Tugs and ATBs had characteristically different patterns, with more common tugs behaving like small vessels and less common ATBs behaving like larger cargo ships. ATBs are generally much larger than tugs, and are likely subject to ATBA restrictions.

4.2. Biologically Important Areas

The northern-most portion of the BIAs had the highest concentration of vessel traffic and the fastest average vessel speeds. The Northern Washington feeding BIA for humpback whales had the greatest number of transits among the feeding BIAs, due to its location just offshore of the TSS. While this BIA does not extend north of the United States EEZ to cover the shipping lanes between the Strait of Juan de Fuca and Alaska and Asia, it does overlap the shipping lanes toward the United States West Coast and Hawaii. The Northwest Washington feeding BIA for gray whales was infrequently transited, as it is located close to shore. The Grays Harbor feeding BIA for gray whales was transited most commonly by tugs, fishing, public, and private vessels. Commercial traffic in this area was uncommon.

The potential presence migration BIA for gray whales is spatially extensive and located from the mouth of the Strait of Juan de Fuca south along the coast, and was transited by all vessel types. The majority of large, commercial vessels abided by the ATBA restrictions. Notable vessel types inside the ATBA were tug, public, private, research, and fishing vessels. Each of these vessel types had a low average calculated SOG. The other gray whale migration BIAs are located within several kilometers of the coast (Figure 2). Exploratory analyses showed relatively few vessels transits in these areas, so they were not considered for further analysis.

The BIAs were designated solely within the United States EEZ and do not cross international boundaries (Calambokidis et al., 2015). However, the shipping routes between the Strait of Juan de Fuca and Asia and Alaska continue north and west of the United States EEZ, in an area with frequent humpback whale sightings (Calambokidis et al., 2015). The study area extended to cover Canadian waters, and vessel speed and density remain high in the shipping routes extending toward Asia and Alaska (dark red in Figure 5). We recommend that future analyses or management planning concerning whales or vessel traffic, including risk analyses, should be considered a transboundary effort.

4.3. Marine Spatial Planning

Using AIS in the risk analysis process is one potential tool in marine management (Wiley et al., 2013). Marine Spatial Planning (MSP) is designed as an adaptive spatial planning process to help manage current and future human activities in the marine environment to meet a variety of objectives and minimize useruser and user-environment interactions by engaging multiple stakeholders (Ehler and Douvere, 2007; Ehler, 2008; Foley et al., 2010; Redfern et al., 2013). Recent examples of successful MSP for cetacean protection are the shifts in the TSS outside Boston, Massachusetts, and San Francisco Bay, California (Wiley et al., 2013; United States Coast Guard [USCG], 2013). The scientific processes used stakeholder involvement throughout, created numerous alternatives, showed how challenges can help the process, and used AIS to evaluate and monitor results (Wiley et al., 2013).

Since AIS is an international standard (Tetreault, 2005), the SAIS data that were provided crossed jurisdictional boundaries along the border with Canada. Off the coast of British Columbia, risk to several species of cetaceans has been investigated (Williams and O'Hara, 2010). The findings of the present research, including calculated SOG and SAIS data, should be important considerations in any MSP processes off the coast of the state of Washington and British Columbia in Canada. The SAIS data, within its limitations and at small scale, are an effective means to delineate areas of high use for vessel traffic, even across international boundaries. Calculating vessel speed is critical to avoid underestimating vessel speed and the probability of a lethal vessel and cetacean collision.

5. LIMITATIONS

One of the main limitations of this research is the temporal resolution of the available SAIS data. The ability to precisely track vessel movement decreases with increasing time between sequential points. Vessels can potentially transit around corners, but large amounts of time between SAIS broadcasts can make transit lines appear to cut those corners. This is evident in many vessel types on the northwestern tip of the Olympic Peninsula and the near-shore BIAs (see Figure 6 as an example). This introduces the potential for a vessel to appear to cross a BIA when in fact it did not. The TSS, controlled by the USCG and used by vessels for insurance and accident coverage purposes, never intersects the Northwest Washington BIA. However, the TSS circumnavigates this BIA, and it is therefore possible that a vessel remaining in the TSS could appear to transit through the BIA. This results in calculated transits across the BIA, sometimes at high SOG, that never actually occurred. The other BIAs, the TSS, ATBA, and OCNMS are also susceptible to this limitation.

Since the shortest distance between two points is a line and the time between points remains identical despite the actual path taken, the calculated SOG, while faster than broadcast SOG, is still a systematic underestimation of true SOG, assuming random GPS error. As distance increases for a vessel to traverse around a corner and time remains the same, speed also must increase. Using the example in Figure 6, the TSS lane distance and speed are greater than the direct distance and speed between points 1 and 2. Although we may be overestimating the number of transits through some administrative areas, this would be associated with an underestimation of SOG. Most vessels are required to broadcast every few seconds while under way using engine, but the time gap between SAIS records is frequently on the order of minutes or hours. The uncertainty in vessel path is unknown, but will increase with path sinuosity.

As an emerging technology, current SAIS presents tremendous opportunities for research, but caution should be used and uncertainty addressed when using this technology for large scale applications. Temporal gaps in vessel transits add uncertainty to transit path and calculated SOG that was not quantified in this research. Duplicate records accounted for 42.8% of total SAIS records for the year 2013. Broadcast SOG values were zero in 36.7% of all records. There were numerous time, location, and missing value errors that had to be addressed prior to data analysis. Units of measurement for vessel length were also not consistent. These factors cast into question the reliability of individual values and the present quality of SAIS data as a whole, and add uncertainty to automated aggregate calculations. SAIS data in its current state should not be used in a policy enforcement context or for documenting individual presence or absence in an administrative area at large scale. However, SAIS data can be helpful to assess general or overall compliance within an area of interest. As calculated SOG is an underestimation of true vessel speed, SAIS can be helpful in documenting minimum vessel speeds across large geographic areas, especially beyond the reach of terrestrial AIS receivers. SAIS is useful in examining vessel density at broad scales, and could be used to assess basin-wide open ocean routes. Future additional satellite platforms with AIS receivers will only increase the quality of SAIS data and decrease the amount of temporal gaps. This will open potential research questions involving larger scale questions of specific areas.

Author Contributions

NG conducted the analysis and led the project. All authors listed have made a substantial, direct and intellectual contribution to the work, and approved it for publication.

Acknowledgments

We would like to thank the following people for the integral roles they played in this process. Their ideas, insights, and material support made this research possible. T. J. Moore, the scripts you wrote to create transit lines from point AIS data and calculate speed over ground were a fundamental component of this research. Nancy Wright and George Galasso, NOAA Olympic Coast National Marine

Sanctuary, for providing the data necessary for this project, and for your support and consultation. John Berge, Pacific Merchant Shipping Association and Cpt. John Veentjer, Marine Exchange of Puget Sound, for insight into the shipping industry.

REFERENCES

Aarsæther, K. G., and Moan, T. (2009). Estimating navigation patterns from AIS. J. Navig. 62, 587-607. doi: 10.1017/s0373463309990129

Allen, A. S., Yurk, H., Vagle, S., Pilkington, J., and Canessa, R. (2018). The underwater acoustic environment at Sgaan kinghlas-bowie seamount marine protected area: characterizing vessel traffic and associated noise using satellite AIS and acoustic datasets. Mar. Pollut. Bull. 128, 82-88. doi: 10.1016/j. marpolbul.2018.01.014

Ball, H. (2013). Satellite AIS for Dummies, Special Edition. Mississauga. Ontario: John Wiley & Sons Canada, Ltd.

Becker, E. A., Forney, K. A., Foley, D. G., Smith, R. C., Moore, T. J., and Barlow, J. (2014). Predicting seasonal density patterns of California cetaceans based on habitat models. Endanger. Species Res. 23, 1-22. doi: 10.3354/esr00548

Calambokidis, J., Steiger, G. H., Curtice, C., Harrison, J., Ferguson, M. C., Becker, E., et al. (2015). Biologically Important Areas for Selected Cetaceans within US Waters - West Coast Region. Aquat. Mamm. 41, 39-53. doi: 10.1578/am.41.1. 2015.39

Calder, B., and Schwehr, K. (2009). "Traffic analysis for the calibration of risk assessment methods," in Proceedings of the U.S. Hydrographic Conference, Norfolk, VA.

Cates, K., DeMaster, D. P., Brownell, R. L. Jr., Silber, G., Gende, S., et al. (2017). Strategi Plan to Mitigate the Impacts of Ship Strikes on Cetacean Populations: 2017-2020. IWC Strategic Plan to Mitigate Ship Strikes. Available online at: https://iwc.int/ship-strikes (accessed January 2, 2020).

Conn, P. B., and Silber, G. K. (2013). Vessel speed restrictions reduce risk of collision-related mortality for North Atlantic right whales. Ecosphere 4:43.

Currie, J. J., Stack, S. H., and Kaufman, G. D. (2017). Modelling whale-vessel encounters: the role of speed in mitigating collisions with humpback whales (Megaptera novaeangliae). J. Cetacean Res. Manag. 17, 57-63.

de Souza, E. N., Boerder, K., Matwin, S., and Worm, B. (2016). Improving fishing pattern detection from satellite AIS using data mining and machine learning. PLoS One 11:e0163760. doi: 10.1371/journal.pone.0163760

Douglas, A. B., Calambokidis, J., Raverty, S., Jeffries, S. J., Lambourn, D. M., and Norman, S. A. (2008). Incidence of ship strikes of large whales in Washington State. J. Mar. Biolog. Assoc. U.K. 88, 1121-1132. doi: 10.1017/ s0025315408000295

Ehler, C. (2008). Conclusions: benefits, lessons learned, and future challenges of marine spatial planning. Mar. Policy 32, 840-843. doi: 10.1016/j.marpol.2008. 03.014

Ehler, C., and Douvere, F. (2007). Visions for a Sea Change. Report of the First International workshop on Marine Spatial Planning. Intergovernmental

Oceanographic Commission and Man and the Biosphere Programme. IOC Manual and Guides, 46: ICAM Dossier, 3. Paris: UNESCO.

Environmental Protection Agency [EPA], (2010). Designation of North American Emission Control Area to Reduce Emissions from Ships. Office of Transportation and Air Quality. Regulatory Announcement EPA-420-F-10-015. Washington, DC: EPA.

Environmental Systems Research Institute [ESRI], (1999-2018). ArcGIS 10.X for Desktop. Redlands, CA: ESRI.

Eriksen, T., Greidanus, H., and Delaney, C. (2018). Metrics and provider-based results for completeness and temporal resolution of satellite-based AIS services. Mar. Policy 93, 80-92. doi: 10.1016/j.marpol.2018.03.028

Felski, A., Jaskólski, K., and Banyś, P. (2015). Comprehensive assessment of automatic Identification System (AIS) data application to anti-collision manoeuvring. J. Navig. 68, 697-717. doi: 10.1017/s0373463314000897

Ferguson, M. C., Curtice, C., Harrison, J., and Van Parijs, S. M. (2015). Biologically important areas for cetaceans within us waters - overview and rationale. Aquat. Mamm. 41, 2-16. doi: 10.1578/am.41.1.2015.2

Foley, M. M., Halpern, B. S., Micheli, F., Armsby, M. H., Caldwell, M. R., Crain, C. M., et al. (2010). Guiding ecological principles for marine spatial planning. Mar. Policy 34, 955-966. doi: 10.1002/eap.1469

Forney, K. A., and Barlow, J. (1998). Seasonal patterns in the abundance and distribution of California cetaceans, 1991-1992. Mar. Mamm. Sci. 14, 460-489. doi: 10.1111/j.1748 – 7692.1998.*tb*00737.x

Frisk, G. V. (2012). Noiseonomics: the relationship between ambient noise levels in the sea and global economic trends. Sci. Rep. 2:437. doi: 10.1038/srep00437

Gende, S. M., Hendrix, A. N., Harris, K. R., Eichenlaub, B., Nielsen, J., and Pyare, S. (2011). A Bayesian approach for understanding the role of ship speed in whale-ship encounters. Ecol. Appl. 21, 2232-2240. doi: 10.1890/10-1965.1

International Maritime Organization [IMO], (2019). Sulphur Oxides (SOx) and Particulate Matter (PM) - Regulation 14. Available at: http://www.imo.org/en/OurWork/Environment/PollutionPrevention/AirPollution/Pages/Sulphuroxides-%28SOx%29-%E2%80%93-Regulation-14.aspx (accessed November 26, 2019).

Jensen, C. M., Hines, E., Holzman, B. A., Moore, T. J., Jahncke, J., and Redfern, J. V. (2015). Spatial and temporal variability in shipping traffic off San Francisco, California. Coast. Manag. 43, 575-588. doi: 10.1080/08920753.2015.1086947

Jia, H., Smith, T., and Prakash, V. (2019). Estimating vessel payloads in bulk shipping using AIS data. IJSTL 11:25. doi: 10.1504/IJSTL.2019.09 6864

Laist, D. W., Knowlton, A. R., and Pendleton, D. (2014). Effectiveness of mandatory vessel speed limits for protecting North Atlantic right whales. Endanger. Species Res. 23, 133-147. doi: 10.3354/esr00586

McCauley, D. J., Woods, P., Sullivan, B., Bergman, B., Jablonicky, C., Roan, A., et al. (2016). Ending hide and seek at sea. Science 351, 1148-1150. doi: 10.1126/ science.aad 5686

McGillivary, P. A., Schwehr, K. D., and Fall, K. (2009). "Enhancing AIS to improve whale-ship collision avoidance and maritime security," in Proceedings of the OCEANS 2009, MTS/IEEE Biloxi - Marine Technology for Our Future: Global and Local Challenges, Biloxi, MS, 1-8.

Moore, T. J., Redfern, J. V., Carver, M., Hastings, S., Adams, J. D., and Silber, G. K. (2018). Exploring ship traffic variability off California. Ocean Coast. Manage. 163, 515-527. doi: 10.1016/j.ocecoaman.2018.03.010

Natale, F., Gibin, M., Alessandrini, A., Vespe, M., and Paulrud, A. (2015). Mapping fishing effort through AIS data. PLoS One 10:e0130746. doi: 10.1371/journal. pone. 0130746

National Oceanic and Atmospheric Administration [NOAA], (2019). 20192020 Gray Whale Unusual Mortality Event Along the West Coast. Available at: https://www.fisheries.noaa.gov/national/marine-life-distress/2019-graywhale-unusual-mortality-event-along-west-coast (accessed November 26, 2019).

Port Metro Vancouver, (2017). Statistics Overview, Decision Support Services. Available online at: https://www.portvancouver.com/wp-content/uploads/ 2018/03/2017-Stats-Overview-1.pdf (accessed January 2, 2020).

Python Software Foundation, (1990-2019). Python 2.7.3. Wilmington, DE: Python Software Foundation.

Redfern, J. V., McKenna, M. F., Moore, T. J., Calambokidis, J., DeAngelis, M. L., Becker, E. A., et al. (2013). Assessing the risk of ships striking large whales in marine spatial planning. Cons. Biol. 27, 292-302. doi: 10.1111/cobi. 12029

Robards, M. D., Gilber, G. K., Adams, J. D., Arroyo, J., Lorenzini, D., Schwehr, K., et al. (2016). Conservation science and policy applications of the marine vessel automatic identification system (AIS)-a review. Bull. Mar. Sci. 92, 75-103. doi: 10.5343/bms.2015.1034

Rockwood, R., Calambokidis, J., and Jahncke, J. (2017). High mortality of blue, humpback and fin whales from modeling of vessel collisions on the U.S. West Coast suggests population impacts and insufficient protection. PLoS One 12:183052. doi: 10.1371/journal.pone.0183052

Rodrigue, J. (2010). Maritime transportation: drivers for the shipping and port industries. Paper Commissioned for the Experts' Session on Innovation and the Future of Transport (Paris: International Transport Forum).

Rowlands, G., Brown, J., Soule, B., Boluda, P. T., and Rogers, A. D. (2019). Satellite surveillance of fishing vessel activity in the ascension island exclusive economic zone and marine protected area. Mar. Policy 101, 39-50. doi: 10.1016/j.marpol. 2018.11.006

Schwehr, K. D., and McGillivary, P. A. (2007). "Marine ship automatic identification system (ais) for enhanced coastal security capabilities: an oil spill tracking application," in Proceedings of the IEEE. OCEANS 2007, Vancouver, BC. doi: 10.1109/OCEANS.2007.4449285

Silber, G. K., Adams, J. D., and Fonnesbeck, C. J. (2014). Compliance with vessel speed restrictions to protect North Atlantic right whales. PeerJ 2:e399. doi: 10.7717/peerj.399

Silber, G. K., and Bettridge, S. (2010). Vessel Operations in Right Whale Protection Areas in 2009. NOAA Technical Memorandum NMFS-OPR-44. Silver Spring, MD: NOAA. Silber, G. K., Slutsky, J., and Bettridgea, S. (2010). Hydrodynamics of a ship/whale collision. J. Exp. Mar. Biol. Ecol. 391, 10-19. doi: 10.1016/j.jembe.2010.05.013

Skauen, A. N. (2019). Ship tracking results from state-of-the-art space-based AIS receiver systems for maritime surveillance. CEAS Space J. 11, 301-316. doi: 10.1007/s12567-019-00245-z

Tetreault, B. J. (2005). Use of automatic identification system (AIS) for maritime domain awareness (MDA). Oceans Proc. IEEE 2, 1590-1594.

United States Army Corps of Engineers [USACE], (2015). US Waterborne Container Traffic by Port/Waterway in 2013. New Orleans, LO: Waterborne Commerce Statistics Center.

United States Army Corps of Engineers [USACE], (2016). Tonnage for Selected US Ports in 2014. New Orleans, LO: Waterborne Commerce Statistics Center.

United States Coast Guard [USCG], (2008). The Nationwide Automatic Identification System Increment One (I-1) Resource Guide. Washington, DC: USCG.

United States Coast Guard [USCG], (2013). Changes to Offshore Traffic Separation Schemes off San Francisco Bay. Coast Guard District Eleven, Local Notice to Mariners 16/2013. Washington, DC: USCG.

United States Fish and Wildlife Service [USFWS], (2016). Endangered and Threatened Wildlife and Plants; Identification of 14 Distinct Population Segments of the Humpback Whale and Revision of Species-Wide Listing. 81 FR. 62259. Washington, DC: USFWS.

van der Hoop, J. M., Vanderlaan, A. S. M., and Taggart, C. T. (2012). Absolute probability estimates of lethal vessel strikes to North Atlantic right whales in Roseway Basin, Scotian Shelf. Ecol. Appl. 22, 2021-2033. doi: 10.1890/11-1841.1

Vanderlaan, A. S. M., and Taggart, C. T. (2007). Vessel collisions with whales: the probability of lethal injury based on vessel speed. Mar. Mamm. Sci. 23, 144-156. doi: 10.1111/j.1748 – 7692.2006.00098.x

Vlahos, K. (2005-2015). Python Scripter Version 2.6.0.0.

Wiley, D., Hatch, L., Thompson, M., Schwehr, K., and MacDonald, C. (2013). Marine Sanctuaries and Marine Planning: Protecting endangered marine life. Proc. Mar. Saf. Secur. Counc. 70, 10-15.

Wiley, D., Thompson, M., Pace, R. M. III, and Levenson, J. (2011). Modeling speed restrictions to mitigate lethal collisions between ships and whales in the Stellwagen Bank National Marine Sanctuary, USA. Biol. Conserv. 144, 2377-2381. doi: 10.1016/j.biocon.2011.05.007

Williams, R., Gero, S., Bejder, L., Calambokidis, J., Kraus, S. D., Lusseau, D., et al. (2011). Underestimating the damage: interpreting cetacean carcass recoveries in the context of the Deepwater Horizon/BP incident. Conserv. Lett. 4, 228-233. doi: 10.1111/ j.1755-263x.2011.00168.x

Williams, R., and O'Hara, P. (2010). Modelling ship strike risk to fin, humpback and killer whales in British Columbia, Canada. J. Cetacean Res. Manag. 11, 1-8.

CHAPTER 8

A Sliced Parabolic Equation Method to Characterize Maritime Radio Propagation

Yuzhen Wang[1,2], Ting Zhou[3,4], Tianheng Xu[1,4] and Honglin Hu[1]

[1] *Shanghai Advanced Research Institute, Chinese Academy of Sciences, Shanghai 201210, China*
[2] *School of Electronic, Electrical and Communication Engineering, University of Chinese Academy of Sciences, Beijing 100049, China*
[3] *School of Microelectronics, Shanghai University, Shanghai 200444, China*
[4] *Shanghai Frontier Innovation Research Institute, Shanghai 201100, China*

ABSTRACT

For maritime broadband communications, atmospheric ducts can enable beyond line-ofsight communications or cause severe interference. Due to the strong spatial-temporal variability of atmospheric conditions in near-shore areas, atmospheric ducts have inherent spatial heterogeneity and suddenness. This paper aims to evaluate the effect of horizontally inhomogeneous ducts on maritime radio propagation through theoretical analysis and measurement validation. To make better use of meteorological reanalysis data, we design a range-dependent atmospheric duct model. Then, a sliced parabolic equation algorithm is proposed to improve the prediction accuracy of path loss. We derive the corresponding numerical solution and analyze the feasibility of the proposed algorithm under the range-dependent duct conditions. A 3.5GHz long-distance radio propagation measurement is utilized to verify the algorithm. The spatial distribution characteristics of atmospheric ducts in the measurements are analyzed. Based on actual duct conditions, the simulation results are consistent with the measured path loss. The proposed algorithm outperforms the existing method during the multiple duct periods. We further investigate the influence of different duct horizontal characteristics on the received signal strength.

Keywords: *maritime communications; radio propagation measurement; evaporation duct; elevated duct; path loss*

1. INTRODUCTION

With the rapid development of marine activities, there has been a growing demand for high-speed and cost-efficient maritime communications [1-5]. To achieve space-airground-sea integrated communication networks for sixth-generation (6G) communications, it is essential to build a seamless maritime communication network to extend broadband coverage to the sea area [6,7]. The LTE-Maritime project proved the feasibility of high-speed wireless communication in coastal areas (up to 100 km from a shore) [8]. Unlike terrestrial scenarios, the maritime wireless propagation environment has unique meteorological characteristics, such as sea wave fluctuations [9] and the ducting effect over the sea surface [10]. Current fifth-generation (5G) cellular networks designed for terrestrial scenarios will be heavily affected by these maritime environmental factors [11]. Therefore, it is crucial to establish reliable maritime channel models for maritime wireless networks, especially for near-coast wireless networks [12].

Maritime communications face many application scenarios, such as land-to-ship, shipto-ship, and buoy-to-ship [13]. Researchers have recently launched several measurement campaigns to investigate wireless maritime channel properties [14]. In particular, recent measurements have shown that evaporation ducts and elevated ducts can greatly reduce the transmission loss [15]. The atmospheric duct is caused by a rapid decrease in atmospheric refractivity over the sea surface [16]. The trapped signal in the duct layer has an extra gain and can travel hundreds of kilometers [17]. In addition, atmospheric ducts have the characteristics of wide areas and long durations. The ducting effect usually causes unexpected remote interference and seriously degrades the performance of coastal 5G networks and LTE networks [18]. More importantly, atmospheric ducts can be utilized to improve wireless coverage and transmission efficiency, especially for land-to-ship communications [19]. It can also improve unmanned aerial vehicle (UAV)-aided maritime communication when the UAV is deployed at the proper height in atmospheric ducts [20].

Therefore, the atmospheric duct effect on maritime radio propagation and channel modeling has drawn growing attention from researchers [21]. Gunashekar et al. measured the 2GHz maritime channel and analyzed the influence of the antenna height on the ducting effect [22]. The authors in [23] measured a near-shore wireless channel at 5GHz, where the evaporation duct layer is assumed to be horizontally homogeneous. The authors in [24] conducted a 30 km maritime communication experiment and proposed a ray-tracing model to analyze large-scale fading. In [25], an 87 km maritime microwave measurement demonstrated that atmospheric ducts can enhance the received signal by more than 40 dB and revealed that the evaporation duct communication is feasible. In [26], a 107 km measurement at the C band was conducted in the Bohai Sea and results indicated that the distribution of fast fading is close to a Rayleigh distribution.

These empirical models are completed through a large number of actual measurements and useful for maritime communications. Note that the measurements of maritime radio channels require a great deal of financial and human resources. Such statistical methods are not sufficient for fast variations in dynamic maritime environments [27]. Ray-tracing and parabolic equation methods have better applicability. The authors in [28] proposed a novel ship-ship round earth loss model by incorporating the ray-tracing method and wind speed. A novel three-dimensional non-stationary UAV-to-ship channel model was proposed, which considered the multi-bounce components introduced by atmospheric ducts [29].

Recently, parabolic equation methods have been widely used for radio propagation simulations in maritime environmental conditions [30]. A parabolic equation simulation tool (PETOOL) for evaporation ducts was proposed [31,32], which can support a wide variety of evaporation duct models. Using the propagation loss obtained from PETOOL, a vertically separated MIMO ship-to-ship communication system was proposed [33]. In [34], the authors proposed a large-scale path loss model for the ducting channel. For rough sea surface conditions, an improved parabolic equation method was proposed based on a pade approximation form [35]. The fading characteristics in evaporation ducts were studied with average duct heights in the South China Sea [36]. A ship-to-ship channel model was proposed, which considered the influence of evaporation ducts on the scatterer distribution [37]. A prediction map of radio propagation was proposed with help of atmospheric duct data in the Korean coastal area [38]. The authors in [39] proposed a logdistance maritime wireless channel at 8GHz, and the meteorological data pointed out that the difference in the evaporation duct profiles at the Tx and Rx increases with increasing link distance. Machine learning has been widely used in the processing of wireless channel measurements [40]. In [41], a two-stage deep learning network was designed to characterize duct effects. In [42], a duct channel model based on a neural network was proposed to predict propagation loss. It should be noted that the training data in [41, 42] were generated in ideal evaporation ducts. To improve the maritime communication reliability, the authors in [43] proposed a hybrid method to exploit the combination of cooperative diversity and statistical evaporation duct heights.

In contrast to the homogeneous assumptions in [30-43], the duct layers are rangedependent. The strong spatial-temporal variability of the lower atmosphere in near-shore areas can cause horizontal variability in atmospheric ducts. A measurement campaign in the Indian Ocean and the South China Sea has also shown the horizontal variability of atmospheric ducts [44,45]. It is obvious that disregarding the range-dependent duct environment will lead to inaccurate prediction results [46].

To the best of the authors' knowledge, most existing models cannot reflect the effect of horizontally inhomogeneous ducts. The range-dependent characteristics should be taken into consideration. The decision methods of maritime communication with duct conditions are also lacking.

With the development of Maritime Internet of Things (M-IoT), future maritime wireless communications need to utilize meteorological information to establish sensitive maritime channels that cope with changing environments [47]. Moreover, the state-of-the-art numerical weather prediction can provide data of high spatial-temporal resolution. This also motivates us to explore a framework of meteorological information-aided maritime radio propagation characterization. Our preliminary study [48] presents a basic environmental information-aided interference simulation method and verifies the feasibility of meteorological data.

In this paper, we aim to study the effect of horizontally inhomogeneous ducts on maritime radio propagation. Considering the limitations of the existing methods, we propose a sliced parabolic equation algorithm to predict radio propagation characteristics in atmospheric ducts. The major contributions are summarized as follows.

- We investigate existing atmospheric duct models and their effects on maritime radio propagation. Subsequently, we propose a range-dependent duct model to support the actual duct environment. The spatial-temporal characteristics of atmospheric ducts are studied by using the meteorological reanalysis data.
- We propose a sliced parabolic equation algorithm to improve the prediction accuracy of path loss. Moreover, we derive the corresponding numerical solution and the approximate error expression in inhomogeneous environments.
- We validate the algorithm's effectiveness by the 3.5GHz field measurement with the actual meteorological data. Compared with the traditional method, the proposed algorithm has higher accuracy during the multiple duct periods.
- We evaluate the proposed algorithm in various range-dependent ducts, including evaporation ducts and elevated ducts. Simulation results show that the duct's horizontal characteristics have a significant impact on path loss.

The rest of the paper is organized as follows. In Section 2, maritime radio propagation scenarios and three basic duct models are introduced. Based on a range-dependent duct model, Section 3 presents a sliced parabolic equation algorithm. The numerical solution and feasibility derivation are also provided. In Section 4, the effectiveness of the proposed algorithm is verified by the 3.5GHz field measurements. Section 5 presents the simulation results in evaporation ducts and elevated ducts with different horizontal characteristics. Finally, conclusions are given in Section 4.

2. SYSTEM MODEL AND PRELIMINARIES

2.1. Maritime Radio Propagation Scenarios

A typical near-shore maritime communication network is shown in Figure 1, where various wireless communication links are adopted to provide wide coverage and a costefficient broadband data service [49]. In general, a near user group can be directly served by the coastline base station (CBS). Relay nodes or on-demand nodes such as UAVs are utilized for coverage enhancement. The CBS also serves on-demand wireless nodes. Ship-to-ship communications aim to provide high rate transmission, such as offshore device-to-device communication [50].

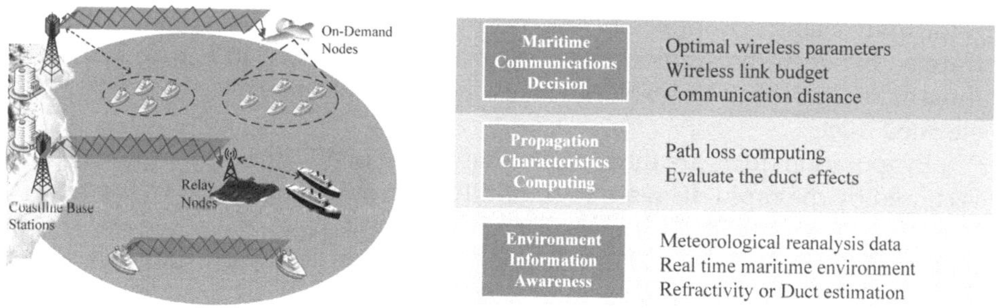

(a) (b)

Figure 1. *(a) Typical propagation scenarios in near-coast maritime communications, which can be affected by atmospheric ducts. (b) The framework of meteorological information-aided radio propagation characterization for maritime communications.*

As the near-shore maritime communication network will play increasingly important roles, the atmospheric duct is a non-negligible factor that will significantly impact the performance of maritime wireless systems, especially between 2GHz and 20GHz [15]. Atmospheric ducts usually exist above the sea surface and in coastal regions. As shown in Figure 1, the captured signal in the atmospheric ducts could propagate beyond the line of sight. Atmospheric ducts will affect these wireless links, including CBS-to-ship/relay, ship-to-ship, and CBS-to-UAV channels. The utilization of atmospheric ducts is still in the early stage.

2.2. Descriptions of Atmospheric Duct Models

The formation of atmospheric ducts depends on atmospheric refractive conditions. Without loss of generality, the refractivity N is defined as [51]

$$N = (n-1) \times 10^6 = \frac{77.6}{T}\left(P + \frac{4810e}{T}\right) \quad \text{N-units}, \tag{1}$$

where P is the atmospheric pressure (hPa), T denotes the absolute temperature (K), e is the water vapor pressure (hPa) and n is the atmospheric refractive index.

The curvature of the Earth needs to be considered in wireless channel modeling. The modified refractivity M is defined as

$$M(z) = N(z) + \left(\frac{z}{R_e}\right) \times 10^6 \quad \text{M-units}, \tag{2}$$

where z is the height above the surface level in km and R_e denotes the Earth's radius, whose average value is approximately 6,371,000 m.

Based on (2), the modified refractivity gradient is written as

$$\frac{dM}{dz} = \frac{dN}{dz} + 0.157 \tag{3}$$

The trapping condition is associated with a negative modified refractivity gradient. The duct height is given by the height where the gradient of the modified refractivity changes from negative to positive. The radio signals can be effectively trapped between the surface and ducting layer. As shown in Figure 2, there are three main types of atmospheric ducts: evaporation ducts, surface-based ducts, and elevated ducts.

Evaporation ducts are the most common type in maritime environments. This is caused by the rapid decrease in humidity with height and high water evaporation. The probability of occurrence is approximately 90% in equatorial and tropical regions [30]. The Red Sea and the Persian Gulf are permanent duct areas. The average duct heights are mostly between 10 m and 40 m [52]. Evaporation ducts can affect the stability of CBS-to-ship and ship-to-ship links. Due to their high availability, evaporative ducts are the most promising type for maritime communication applications.

For maritime environments, the modified refractivity of evaporation ducts can be expressed by a widely accepted logarithmic function [53]

$$M(z) = M_0 + 0.125z - 0.125\delta \ln\left(\frac{z + z_0}{z_0}\right), \tag{4}$$

where M_0 is the value of modified refractivity at the surface, δ is the evaporation duct height and z_0 is the aerodynamic roughness length, which can be assumed to be 1.5×10^{-4} m.

Elevated ducts are caused by the presence of marine layers and mainly occur along the trade winds and close to the coastline. Surface-based ducts are special cases of elevated ducts. The occurrence rate of elevated ducts can be up to 50%. The elevated duct height can reach hundreds of meters. Therefore, elevated ducts can affect the CBS-to-UAV/relay links and cause unexpected remote interference in cellular networks. The profile of elevated ducts can be modeled with a trilinear curve, and can be written as [54]

$$M(z) = M_0 + \begin{cases} g_1 z, & z \leq h_1, \\ g_1 h_1 + g_2 (z - h_1), & h_1 < z \leq h_2 \\ g_1 h_1 + g_2 (h_2 - h_1) + g_3 (z - h_2), & z > h_2 \end{cases} \tag{5}$$

where g_1, g_2, and g_3 are the slopes of the modified refractivity.

2.3. Radio Propagation Characteristics

Maritime radio propagation simulation requires modeling of atmospheric parameters and complex boundary problems. Based on the parabolic approximation of the Helmholtz wave equation, the parabolic equation methods could model the changes in refractivity in the atmosphere and simulate complex boundary conditions [35]. Moreover, the parabolic equation methods also have the potential to calculate radio propagation in complex and large-scale maritime environments.

With parabolic equation methods and atmospheric parameters, the path loss can be written as [53]

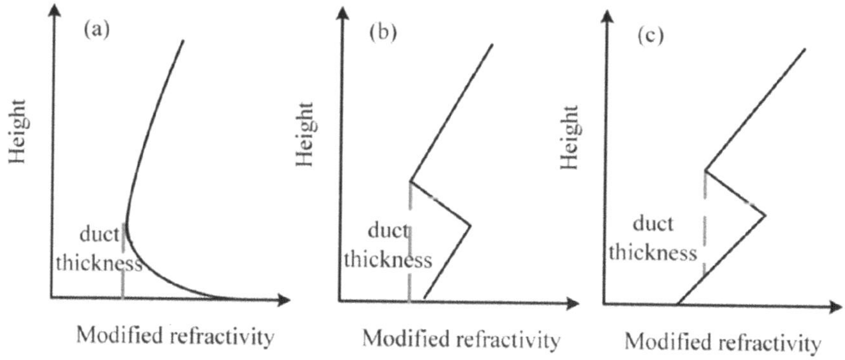

Figure 2. *Modified refractivity profiles of atmospheric ducts. (a) Evaporation duct. (b) Surface-based duct. (c) Elevated duct.*

$$PL = -20\log|u(x,z)| + 20\log(4\pi) + 10\log(x) - 30\log(\lambda), \tag{6}$$

where u denotes the field strength computed by parabolic equation methods, x is the distance, and λ denotes the wavelength.

The strong spatial and temporal variability of the lower atmosphere in coastal and near-shore areas can cause horizontal variability in atmospheric ducts. However, the prediction accuracy may be lost due to the assumption of horizontal homogeneity in the traditional parabolic equation methods. As shown in Figure 1, it is important to study a framework of meteorological information-aided radio propagation characterization, from maritime environment information awareness to the deployment decision. Therefore, an accurate and general algorithm with the help of meteorological reanalysis products is essential to deal with range-dependent atmospheric ducts.

3. SLICED PARABOLIC EQUATION ALGORITHM IN HORIZONTALLY INHOMOGENEOUS DUCT ENVIRONMENTS

In this section, we present a range-dependent duct model to make better use of meteorological reanalysis data. Then, we propose a sliced parabolic equation algorithm to simulate electromagnetic wave propagation in the maritime environment. The numerical solution and feasibility derivation are also provided. In the remainder of this section, we will present the workflow and implementation details.

3.1. Range-Dependent Duct Model

In this subsection, we will introduce the range-dependent refractivity and duct environments.

In the ideal case, assume that the meteorological data at different heights and distances can be measured by microwave refractometers and lidar. The refractivity $N(x,z)$ can be defined as

$$N(x,z) = \frac{77.6}{T(x,z)}\left(P(x,z) + \frac{4810e(x,z)}{T(x,z)}\right) \quad \text{N-units.} \tag{7}$$

However, these methods are high-cost. On the other hand, the fast developments of mesoscale numerical weather prediction can provide meteorological data with a high spatial and temporal resolution and make it possible to analyze the horizontal heterogeneity of ducts. It is possible to estimate the refractivity index and duct heights from standard meteorological data. Specifically, meteorological measurements can provide the temperature, humidity, wind speed, pressure, and sea surface temperature at certain heights. The environmental information is used as input parameters. Then, the profiles of temperature, humidity, and air pressure can be estimated with the Monin-Obukhov similarity theory [55], and the refractivity profiles can be obtained by (7). Finally, duct heights can be estimated.

For example, the fifth-generation European Center for Medium-Range Weather Forecasts (ECMWF) reanalysis data (ERA5) provide hourly products of meteorological variables. Its horizontal resolution is 0.25°. Therefore, we aim to use the up-to-date meteorological reanalysis product to study the spatial and temporal distributions of the duct environment.

Therefore, we adopt a range-dependent duct model, considering the actual horizontal resolution of meteorological data and computational accuracy. As shown in Figure 3, we assume that there are L slices of different atmospheric vertical structures in the horizontal direction. In each slice, $N(x, z)$ is almost constant in the x direction. The formulas of the range-dependent duct environments are as follows:

$$\begin{aligned} M_{slice-1}(x,z) &= N\left(\frac{x_1+x_2}{2}, z\right) + \left(\frac{z}{a}\right) \times 10^6, \quad (x_1 < x \le x_2), \\ M_{slice-2}(x,z) &= N\left(\frac{x_2+x_3}{2}, z\right) + \left(\frac{z}{a}\right) \times 10^6, \quad (x_2 < x \le x_3), \\ M_{slice-3}(x,z) &= N\left(\frac{x_3+x_4}{2}, z\right) + \left(\frac{z}{a}\right) \times 10^6, \quad (x_3 < x \le x_4). \end{aligned} \tag{8}$$

3.2. Sliced Parabolic Equation Algorithm

In the rectangular coordinate system, the wave equation satisfies the approximation of the Helmholtz equation,

$$\frac{\partial^2 \psi}{\partial x^2} + \frac{\partial^2 \psi}{\partial z^2} + k_0^2 n(x,z)^2 \psi = 0 \tag{9}$$

where $k_0 = 2\pi/\lambda$ denotes the wavenumber in a vacuum, n is the refractive index of the tropospheric medium, x is the propagation distance, and z is the height. The refractive index $n(x, z)$ varies with distance and height. pressed as

Substituting the simplified function $u(x,z) = e^{-ik_0x}\psi(x,z)$, the field $u(x,z)$ is exexpressed as

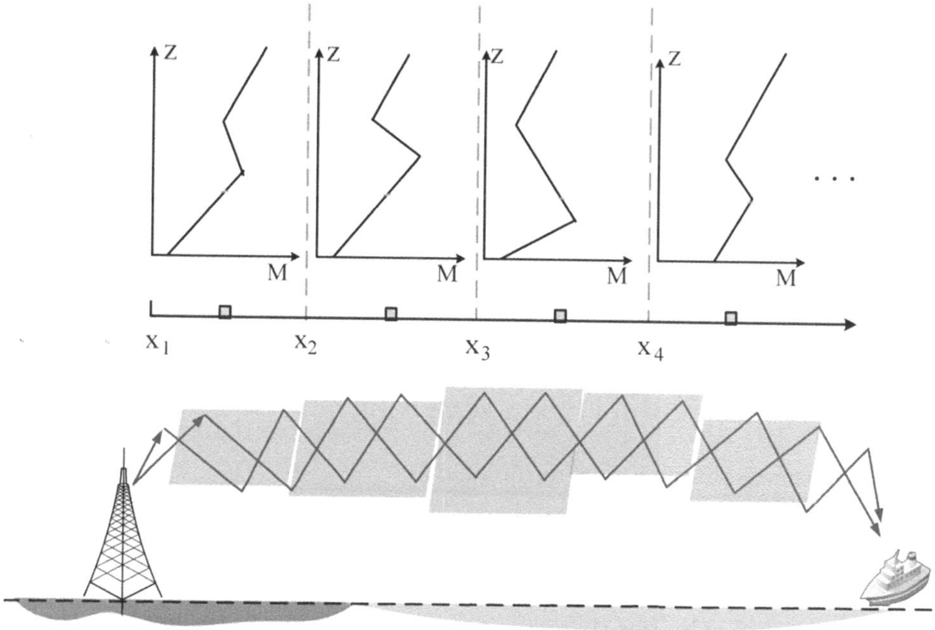

Figure 3. *The range-dependent duct model. The yellow squares represent the locations of meteorological measurements.*

$$\frac{\partial^2 u(x,z)}{\partial x^2} + 2ik_0 \frac{\partial u(x,z)}{\partial x} + \frac{\partial^2 u(x,z)}{\partial z^2} + k_0^2 \left(n(x,z)^2 - 1 \right) u(x,z) = 0. \tag{10}$$

The differential operator in (10) can be factored in terms of two pseudo-differential operators [35]. Then, the above equation can be rewritten as

$$\left[\frac{\partial}{\partial x} + ik_0(1-Q) \right] \left[\frac{\partial}{\partial x} + ik_0(1+Q) \right] u = 0 \tag{11}$$

where Q is expressed as

$$Q = \sqrt{\frac{1}{k_0^2}\frac{\partial^2}{\partial z^2} + n^2(x,z)} \tag{12}$$

The forward propagation electromagnetic waves are expressed as

$$\frac{\partial u}{\partial x} = -ik_0(1-Q)u \tag{13}$$

The approximate forms of the operator Q determine the scope of the elevation angle and the accuracy of calculation. Specifically, by defining $\varepsilon = n^2(x,z) - 1, \mu =$

$\frac{1}{k_0^2}\frac{\partial^2}{\partial z^2}$, Q is given by $Q = \sqrt{1+\varepsilon+\mu}$. The pseudo-differential operator of the Feit-Fleck type is given by

$$Q \approx \sqrt{1+\mu} + \sqrt{1+\varepsilon} - 1 \tag{14}$$

Then, (13) can be rewritten as

$$\frac{\partial u(x,z)}{\partial x} = ik_0\left[\sqrt{1+\frac{1}{k_0^2}\frac{\partial^2}{\partial z^2}}\right]u + ik_0[n(x,z)-2]u(x,z). \tag{15}$$

The wide-angle parabolic equation of the Feit-Fleck type can be expressed as

$$\frac{\partial u(x,z)}{\partial x} = \left[i\sqrt{k_0^2+\partial^2/\partial z^2} + ik_0(n(x,z)-2)\right]u(x,z) \tag{16}$$

The formal solution can be formulated as

$$u(x+\Delta x,z) = e^{\int_x^{x+\Delta x}\left[i\sqrt{k_0^2+\partial^2/\partial z^2}+ik_0(n(x,z)-2)\right]dx}u(x,z) \tag{17}$$

The split-step Fourier transform technique is widely used for parabolic equation's numerical solutions. It requires a large amount of computational power when $n(x,z)$ varies with distance.

In the L-th slice, we define $A = i\sqrt{k_0^2+\partial^2/\partial z^2}$, $B = ik_0(n(x,z)-2)$, and the temporary variable is expressed as

$$\Omega = B + A. \tag{18}$$

Next, we aim to solve the differentiation equation, which can be formulated as

$$u(x+\Delta x,z) = e^{\int_x^{x+\Delta x}\Omega(x,z)dx}u(x,z) \tag{19}$$

By applying the integral mean-value theorem in the i-th slice, it has an approximate formula as

$$\int_x^{x+\Delta x}\Omega dx = (B+A)\Delta x \tag{20}$$

Therefore, the general solution of the differentiation equation is given by

$$u(x+\Delta x,z) = e^{(B+A)\Delta x}u(x,z) \tag{21}$$

Then, the pseudo-differential operator is formulated as

$$e^{(B+A)\Delta x} = e^{A\Delta x}\cdot e^{B\Delta x} \tag{22}$$

Define $v(x,z) = e^{A\Delta x}u(x,z)$. The field $u(x,z)$ can be expressed as follows:

$$u(x+\Delta x,z) = v(x,z)e^{ik_0(n(x,z)-2)\Delta x}. \tag{23}$$

By using Taylor expansion, $v(x,z) = e^{A\Delta x}u(x,z)$ can be expressed as

$$\begin{aligned} v(x,z) \\ = \left[1 + i\Delta x\sqrt{k_0^2 + \frac{\partial^2}{\partial z^2}} - \frac{1}{2}(\Delta x)^2\left(k_0^2 + \frac{\partial^2}{\partial z^2}\right) + \ldots\right]u(x,z). \end{aligned} \tag{24}$$

The Fourier transform of u is defined as

$$U(x,p) = \mathcal{F}[u(x,z)] = \int_{-\infty}^{+\infty} u(x,z)e^{-ipz}dz \tag{25}$$

The inverse Fourier transform is given by

$$u(x,z) = \mathcal{F}^{-1}[U(x,p)] = \frac{1}{2\pi}\int_{-\infty}^{+\infty} U(x,p)e^{izp}dp \tag{26}$$

Then, we have $\mathcal{F}\left(\sqrt{k_0^2 + \frac{\partial^2}{\partial z^2}u}\right) = \sqrt{k_0^2 - p^2}U(x,p)$.

We apply the Fourier transform to both sides of (24) and have

$$\begin{aligned} V(x,p) \\ = \left[1 + i\Delta x\sqrt{k_0^2 - p^2} - \frac{1}{2}(\Delta x)^2\left(k_0^2 - p^2\right) + \ldots\right]U(x,p) \\ \cong e^{i\Delta\sqrt{k_0^2 - p^2}}U(x,p) \end{aligned} \tag{27}$$

The inverse Fourier transform of v is obtained as

$$v(x,z) = \mathcal{F}^{-1}\left[e^{i\Delta x\sqrt{k_0^2 - p^2}}U(x,p)\right] \tag{28}$$

By multiplying the refractive index term, the algorithm solution in the current slice is expressed as

$$u(x + \Delta x, z) = e^{ik_0(n(x,z)-2)\Delta x}\mathcal{F}^{-1}\left[e^{i\Delta x\sqrt{k_0^2 - p^2}}U(x,p)\right] \tag{29}$$

The initial field at the starting position ($x = 0$) is determined by the parameters of the antenna pattern. The initial field can be obtained by fast Fourier transform based on the aperture field and beam pattern. We use the Fourier shift theorems to embody the antenna height and elevation angle. Therefore, the key parameters of the antenna pattern are the height z_a, the 3 dB beamwidth θ_{BW}, and the elevation angle $\theta_{\text{tilt.}}$, respectively. The initial field in the p domain can be defined as [31]

$$U(0,p) = f(p)\exp\left(-ipz_a\right) - f^*(-p)\exp\left(ipz_a\right) \tag{30}$$

The initial field in the spatial z-domain is computed by taking the inverse fast Fourier transform. The Gaussian antenna pattern can represent various antenna types.

In this paper, the horizontally polarized Gaussian antenna pattern can be defined as

$$f(p) = \exp\left[-p^2 w^2/4\right] \tag{31}$$

where $w = \frac{\sqrt{2\ln 2}}{k\sin(\theta_{BW}/2)}$. The elevation angle is introduced by shifting the antenna pattern, i.e., $f(p) \to f(p - k\sin\theta_{\text{tilt}})$.

At each step, $e^{ik_0(n(x,z)-2)\Delta x}$ reflects the refraction effect, and $e^{i\Delta x\sqrt{k^2-p^2}}$ reflects the diffraction effect. Note that $\partial\left(n(x,z)^2\right)/\partial x \approx 0$ is assumed in the derivation of the numerical solution. It is essential to analyze the errors caused by the change in refractive index over distance in dynamic maritime environments.

3.3. Feasibility Analysis

In this subsection, we derive the approximate error expressions by the following theorems. With the Taylor expansion, we derive the actual field u, which can be expressed as

$$\begin{aligned} &u(x+\Delta x, z)_{\text{actual}} \\ &= u + \Delta x\frac{\partial u}{\partial x} + \frac{(\Delta x)^2}{2}\frac{\partial^2 u}{\partial x^2} + \frac{(\Delta x)^3}{6}\frac{\partial^3 u}{\partial x^3} + \dots \\ &\approx \left[1 + \Omega\Delta x + \frac{(\Delta x)^2}{2}\left(\frac{\partial\Omega}{\partial x} + \Omega^2\right)\right. \\ &\left. + \frac{(\Delta x)^3}{6}\left(\frac{\partial^2\Omega}{\partial x^2} + 2\frac{\partial\Omega}{\partial x}\Omega + \Omega\frac{\partial\Omega}{\partial x} + \Omega^3\right)\right] u(x,z), \end{aligned} \tag{32}$$

where $\frac{\partial u}{\partial x} = \Omega u$, and $\frac{\partial\Omega}{\partial x} = \frac{\partial A}{\partial x} + \frac{\partial B}{\partial x} = ik_0\frac{\partial n(x,z)}{\partial x}$.

According to the Monin-Obukhov similarity theory and meteorological accuracy, meteorological conditions in the horizontal direction are consistent in a certain distance. In each slice, we assume that the refractivity profile does not change with range, $\frac{\partial n(x,z)}{\partial x} \approx 0$. The field at $x + \Delta x$ can be expressed as

$$u(x+\Delta x, z)_{ideal} \approx \left[1 + \Omega\Delta x + (\Delta x)^2\Omega^2/2 + (\Delta x)^3\Omega^3/6\right] u(x,z) \tag{33}$$

By comparing u_{actual} with u_{ideal}, the relative error caused by the operator M can be written as

$$e_n = \frac{(\Delta x)^2}{2}\frac{\partial\Omega}{\partial x} + \frac{(\Delta x)^3}{6}\cdot\left(\frac{\partial^2\Omega}{\partial x^2} + 2\frac{\partial\Omega}{\partial x}\Omega + \Omega\frac{\partial\Omega}{\partial x}\right) + \dots \tag{34}$$

It can be seen that the prediction error of the wide-angle operator mainly depends on the step size Δx and $\frac{\partial n(x,z)}{\partial x}$. For horizontally inhomogeneous environments, $n(x,z)$ is almost constant in the x direction in each slice, i.e., $\frac{\partial\Omega}{\partial x} \approx 0$ and

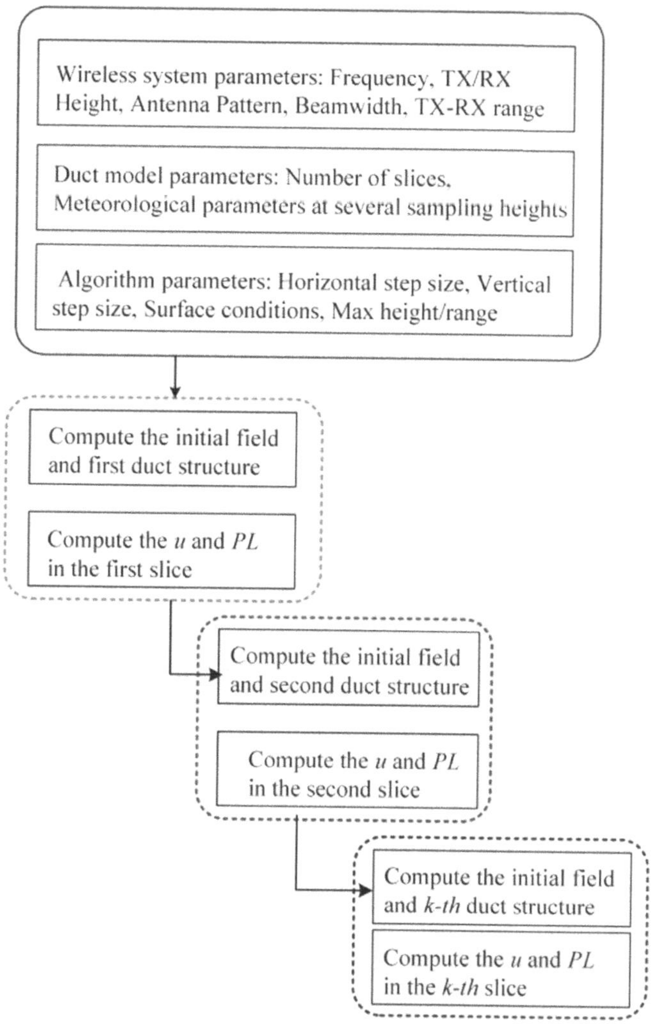

Figure 4. *The workflow of the sliced parabolic equation algorithm.*

$\frac{\partial^2\Omega}{\partial x^2} \approx 0$. Thus, $e_n \approx 0$. Therefore, the propagation simulation of the whole path can be completed through the calculation of each slice.

3.4. Workflow of the Proposed Algorithm

The workflow of the sliced parabolic equation algorithm is shown in Figure 4
.

In the first step, the input parameters consist of three parts. (1) We need to set the wireless system parameters, including the frequency, antenna height, antenna pattern, beamwidth, and link distance. (2) Algorithm parameters include the max height, max range, horizontal step size, vertical step size, and surface conditions.

(3) For the rangedependent duct model, the number of slices depends on the horizontal resolution of the meteorological data. The refractivity profiles of each slice can be directly provided from the reanalysis data, such as the duct height and duct strength. The refractivity profile can also be obtained by a bulk method based on the Monin-Obukhov similarity theory and flux-profile relationship [48]. The refractivity profile can be computed from the sea surface temperature, wind speed, air temperature, relative humidity, and atmospheric pressure at several sampling heights.

In the second step, we use the proposed algorithm to compute the field strength and path loss in each slice. In the first slice, the initial field strength is based on the antenna pattern. The initial field strength of the second slice is based on the end of the first slice. The full procedure is summarized in Algorithm 1.

Algorithm 1 . Sliced Parabolic Equation Algorithm.

Input:
Numerical solution parameters of the proposed algorithm,
Wireless system parameters,
Duct model parameters.
Output: Path loss distribution.
Initialize atmospheric vertical structures of L slices.
for each slice **do**
  Initialize $U(0, j\Delta p)$ and $u(0, m\Delta z)$ based on (30)–(31).
  for each step size **do**
    Compute $V(x + \Delta x, j\Delta p)$ by (27).
    Compute $v(x + \Delta x, m\Delta z)$ by (26).
    Compute $e^{ik_0(n(x,z)-2)\Delta x}$ in the L-th slice.
    Update $u(x + \Delta x, m\Delta z)$ by (29).
  end
end
Obtain the path loss distribution in the whole computational domain based on (6).

4. FIELD MEASUREMENT AND VERIFICATION

In this section, the accuracy of the proposed algorithm will be verified by 3.5GHz field measurement results. We first give a brief description, and then we discuss the distribution characteristics of the atmospheric ducts by analyzing the meteorological data. The comparison between the measured and simulated path loss is given in the remainder of this section.

4.1. Measurement Scenario

Two microwave communication links were set up in the near-shore region, and the radio signal strength was recorded from September 2013 to November 2016 . The radio measurement campaign was conducted by the Radio Communications Agency Netherlands. The 3.5GHz band is very important for the 5G wireless communication system. Therefore, the measurement frequency is set to the 3.5GHz band.

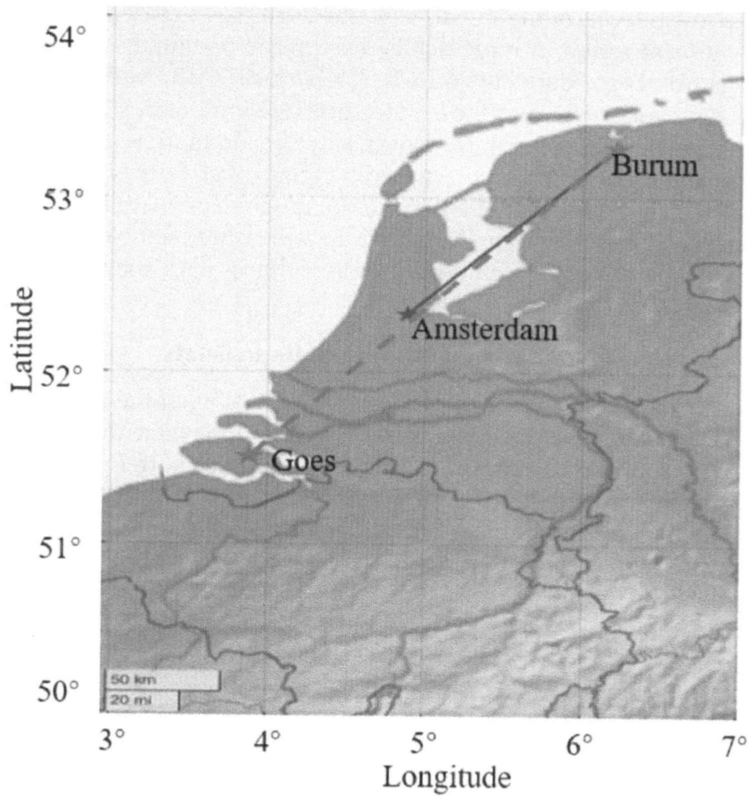

Figure 5. *The geographic information of the two wireless links.*

Table 1. *Details of the transmitter and receiver.*

Location	Antenna Height	Frequency	Distance
Burum	6 m	-	-
Amsterdam	107 m	3.449001 GHz	138 km
Goes	75 m	3.449005 GHz	253 km

Figure 5 illustrates the geographic information of the two wireless links. The receiver is located at 53.282°N, 6.214°E, in the Netherlands. The transmitter at Goes is located at 51.511°N, 3.884°E. The transmitter at Amsterdam is located at 52.336°N, 4.887°E. The two fixed wireless links are the 138 km Amsterdam-Burum path and the 253 km Goes-Burum path. These links are located in coastal areas and river deltas. The terrain is very flat in the experimental zone. The parameters of the antenna height and frequency are listed in Table 1.

The path loss of the 253 km link is approximately 220 dB under normal propagation conditions. On the other hand, the measurement system was able to cope with the path loss (approximately 150 dB) due to atmospheric ducts. Therefore, the upper and lower limits of the signal strength have been considered in the measurement system realization. A more detailed setup can be found in [16]. The RX/TX antenna gain is approximately 26.1dBi. The vertical 3 dB beamwidth is 8°. This measurement work improved the 3.5GHz propagation prediction models and can be applied for 5G deployment. Moreover, the measurement results indicated that short-term abnormal propagation caused by atmospheric ducts exists.

In this paper, we will focus on the effect of range-dependent atmospheric ducts on the radio wave propagation. Note that post-measurement corrections and calibrations were performed. To obtain accurate path loss data, the antenna gains and system losses were removed.

4.2. Distribution Characteristics of Atmospheric Ducts

First, we aimed to characterize the spatial variability of the atmospheric ducts in the experimental zone. We downloaded synchronous meteorological data from the ERA5 atmospheric reanalysis. It can provide the sea surface temperature, 2 m temperature, 10 m wind speed, 2 m relative humidity, and pressure. It provides the duct top height, the duct base height, and the minimum vertical gradient of refractivity inside the duct layer. The horizontal resolution of ERA5 is $0.25° \times 0.25°$ (approximately 30 km).

Figure 6 shows the distribution characteristics of the atmospheric ducts in the experimental zone for the period of 2013-2016. Figure 6a illustrates the atmospheric duct probability of the experimental zone. The color represents the probability of duct occurrence. For the wetland-to-sea path (the black line), the coastal zone has a higher probability of duct occurrence. This is because the eastern part of the ocean and the western part of the continent in the trade wind belt are high-probability zones of atmospheric ducts.

The above analysis can be used to identify signal enhancements caused by atmospheric ducts that last for more than three hours. In fact, atmospheric ducts occur more than 50% of the time in many areas, including the Red Sea, the Persian Gulf, and the South China Sea.

Figure 6b shows the mean top heights of atmospheric ducts from September 2013 to November 2016. The color represents the mean top height. The horizontal inhomogeneity of atmospheric ducts is significant in coastal regions. The top heights of the atmospheric ducts in the coastline area are distributed between 200 and 400 meters. In coastal areas, the mean top heights of the ducting layer vary significantly. This indicates that the strong spatial and temporal variability of the lower atmosphere in coastal and near-shore areas can cause horizontal variability in the ducting layer. Therefore, the effect of range-dependent atmospheric ducts should be considered carefully if the wireless link is along different latitudes.

4.3. Comparison with Measurements

To further verify the effectiveness of the sliced parabolic equation algorithm, the comparison between the measured and simulated path loss is given in this subsection.

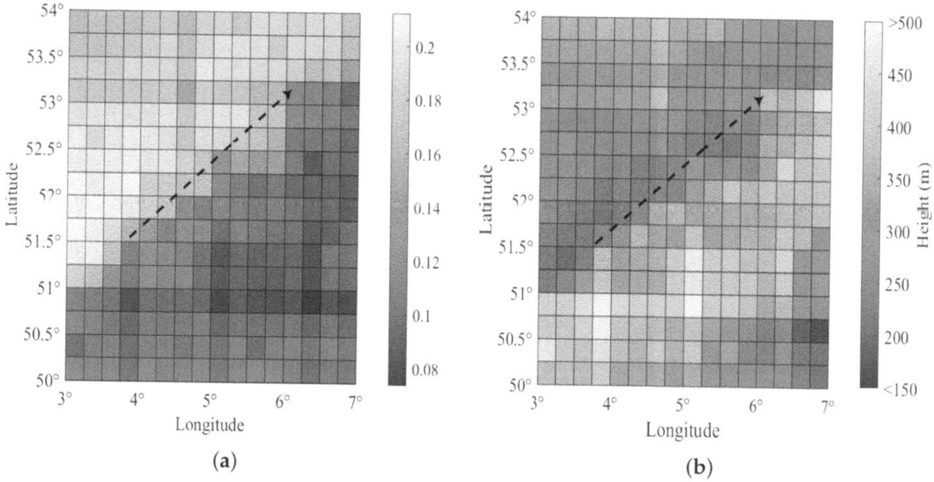

Figure 6. *Distribution characteristics of atmospheric ducts. (a) Atmospheric duct probability in the experimental zone for the period of 2013-2016. The color represents the probability of occurrence. (b) Mean top height of the ducting layer in the experimental zone for the period of 2013-2016. The color represents the mean top height of the ducting layer.*

In order to be consistent with the time resolutions of meteorological data, we select the hourly measurement data, such as the path loss at 02:00. The spatial resolutions of slices are based on the meteorological resolution. The refractivity conditions are based on the reanalysis meteorological data grid points along the propagation link. Therefore, the Goes-Burum path is divided into seven slices and the Amsterdam-Burum path is divided into four slices based on the meteorological resolution of the ERA5 data.

Figure 7a shows the Goes-Burum path loss data from August 26th to August 27th, 2016. The path loss of the 253 km link is approximately 220 dB under normal propagation conditions. Atmospheric ducts can occur for up to three hours and the signals were received with approximately 50 dB enhancement. Note that the signal enhancement lasted for approximately 3 hours. An example of the refractivity M-profiles of the Goes-Burum path is plotted in Figure 7b. The Goes-Burum path is divided into seven slices. Each distance is approximately 31 km. This indicates the appearance of a range-dependent elevated duct. It is obvious that the atmospheric ducts increase with range.

In this paper, we evaluate the proposed algorithm with the following two baselines: (a) the measured path loss and (b) the widely accepted parabolic equation method presented in references [32,34-36,48], which is based on an ideal homogeneous assumption (only using the refractivity profile near the TX/RX).

The dates of significant signal enhancements caused by atmospheric ducts are listed in Table 2. The input parameters of the simulation are listed in Table 3. These simulation parameters and hourly range-dependent refractivity profiles are provided to the algorithm to simulate path loss over the entire height and range. We mainly selected the time period of the occurrence of atmospheric ducts. The

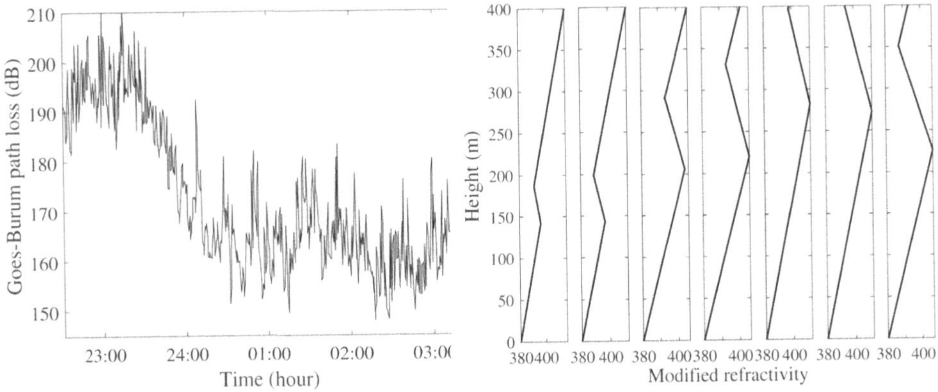

(a) (b)

Figure 7. *(a) Goes-Burum path loss data from 26 August to 27 August 2016. Atmospheric ducts can occur for up to three hours. (b) The refractivity conditions of the Goes-Burum path at 02:00.*

Table 2. *The occurrence time of significant signal enhancements caused by atmospheric ducts.*

Goes–Burum Path	Amsterdam–Burum Path
11 December 2013	11 December 2013
19 July 2014	30 April 2014
21 April 2015	7 June 2014
2 August 2015	1 August 2014
19 July 2016	10 August 2015
27 August 2016	7 September 2016

discriminant method indicates that the path loss is reduced by more than 40 dB compared to the average path loss. Atmospheric attenuation at 3.5GHz is small and can be neglected.

Figure 8 shows an hourly path loss comparison of the 253 km Goes-Burum link. The transmitting antenna height is 75 m and the receiver height is 6 m. When an atmospheric duct occurs, the path loss is approximately 170 dB at 253 km. Compared with the multiple measured path loss, the average prediction error of the traditional parabolic equation method is 9.4427 dB. However, the average prediction error of our proposed algorithm is 4.4286 dB. The proposed algorithm shows a good match with the measured path loss. The reasons are as follows. The duct condition on the propagation path varies widely. The sliced parabolic equation algorithm can reflect the inhomogeneity of the propagation path However, the homogeneous assumption exaggerates the duct effect. Therefore, the effect of range-dependent ducts on electromagnetic wave propagation is significant and should be considered in coastal areas.

Table 3. *Simulation parameters for the Goes-Burum link and Amsterdam-Burum link.*

Parameters	Values
Frequency	3.45 GHz
Antenna height	75 m (Goes) and 107 m (Amsterdam)
Polarization	vertical
3 dB Beamwidth	8°
Elevation angle	0°
Max range	253 km
Max height	400 m
Range step size	80 m
Height step	1 m

Based on the above simulation results, we compute the standard deviation of the duct heights of seven segments along the Goes-Burum path. Figure 9 shows the prediction error under different standard deviations. We can observe that the proposed algorithm achieves robust performance, while the performance of the traditional parabolic equation method varies widely. This is because the traditional method considering the homogeneous assumption is inaccurate and unstable in actual and dynamic atmospheric ducts.

Figure 10a compares the prediction performance on the 138 km Amsterdam-Burum path. The path loss under standard atmospheric conditions is approximately 202.31 dB. When the atmospheric duct occurs, the path loss is approximately 165 dB. We observe that the simulation results of the proposed algorithm follow the measured data closely in most cases. The proposed algorithm's average error is 3.1161 dB, while the average prediction error of the traditional method is 6.1810 dB. Since this path is on the sea surface or land, the horizontal inhomogeneity of the duct layers on the Amsterdam-Burum path is obvious. Most of the predicted values of the traditional method are unstable since it neglects the range-dependant atmospheric conditions in near-shore areas.

Figure 10b shows the relationship between the prediction performance in Figure 10a and the standard deviation of the duct heights. As one can see, the standard deviation of the duct height on the Amsterdam-Burum path is smaller than that on the Goes-Burum path. When the standard deviation is more than 15 m, the sliced parabolic equation algorithm can achieve better performance.

5. SIMULATION ANALYSIS

In this section, we further investigate the influence of different duct horizontal characteristics on the received signal strength.

5.1. Range-Dependent Evaporation Duct

In this subsection, the simulation results in horizontally inhomogeneous evaporation ducts are analyzed. The duct conditions are shown in Table 4. The propagation path is 210 km and the frequency is 3.5GHz. The transmitter height is 15 m. The modified refractivity profiles can be calculated by the bulk method after obtaining

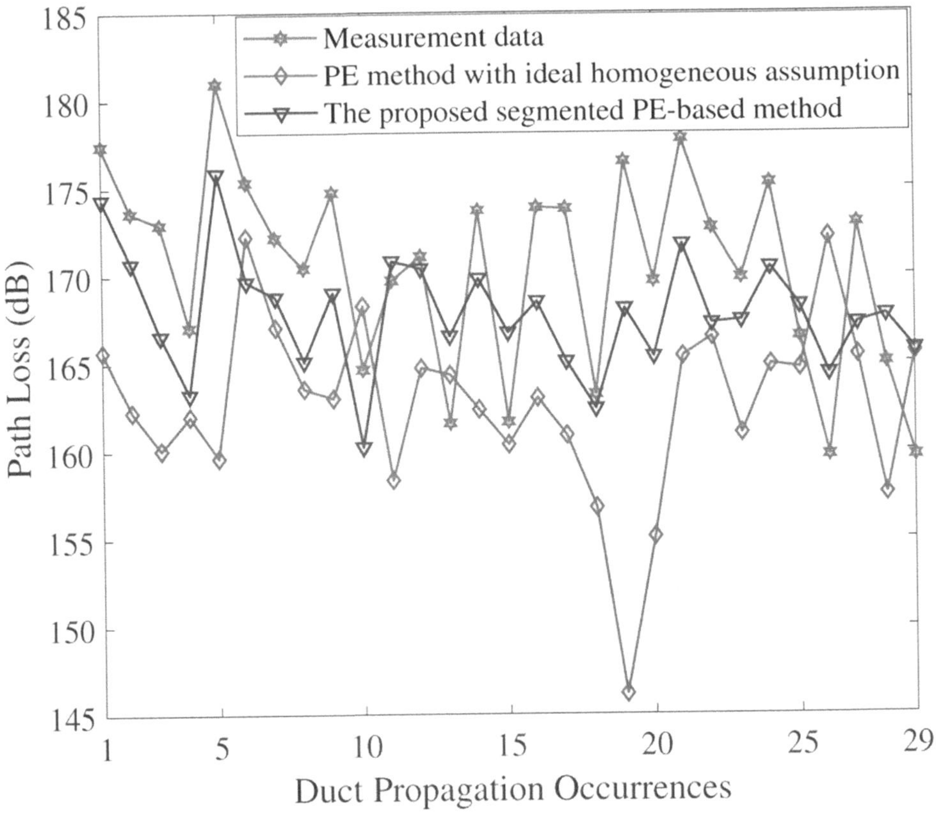

Figure 8. *Performance comparison on the Goes-Burum path. The proposed algorithm reduces the average prediction error by approximately* 5.01 dB.

the temperature, humidity, wind speed, and pressure. Under five different conditions, the evaporation duct heights (EDH) are 25 m, 32 m, 37 m, 19 m, and 16 m, respectively.

Assume that the 210 km path is divided into three slices. Case A is the horizontally homogeneous evaporation duct. The EDHs in the three slices are 25 m, 25 m, and 25 m. Case *B* and case *C* are simulated in horizontally inhomogeneous evaporation ducts. In case B, the EDH increases on the propagation path. The EDHs of the three slices are 25 m, 32 m, and 37 m. In case C, the EDH decreases on the propagation path. The EDHs of the three slices are 25 m, 19 m, and 16 m.

Figure 11 shows the electromagnetic wave propagation in a homogeneous evaporation duct. It is shown that the evaporation duct behaves similarly to a waveguide and can lead to decreased path loss. For example, the path loss is approximately 140 dB at 50 km when the receiving antenna height is 3 m. The path loss at 210 km is approximately 160 dB, which is only 20 dB higher than that at 50 km. If the receiving antenna is located outside the evaporation duct, then the path loss is approximately 20 dB higher than that inside the evaporation duct.

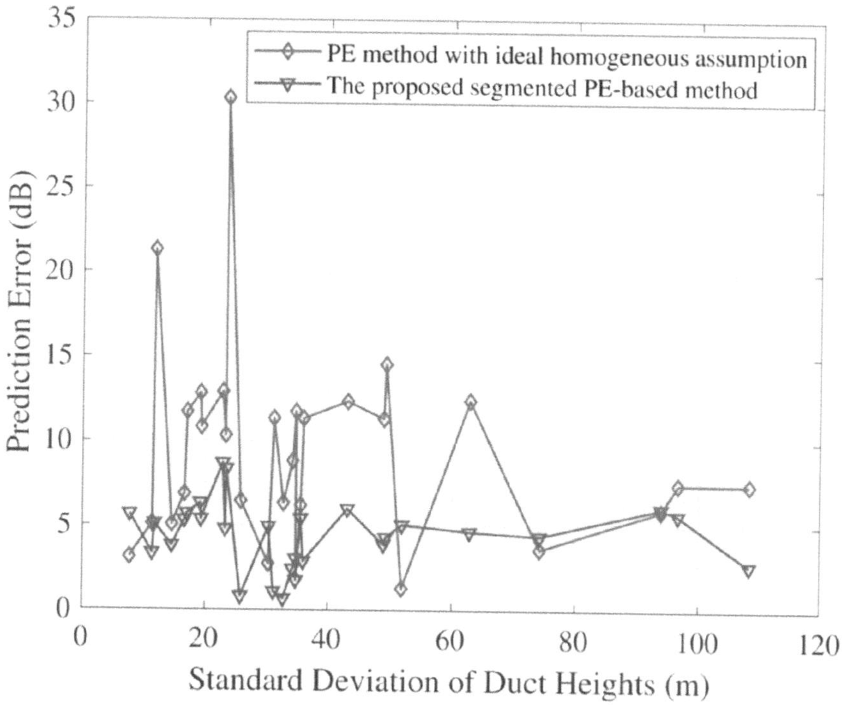

Figure 9. *Path loss prediction error vs. standard deviation of duct heights on the Goes–Burum path.*

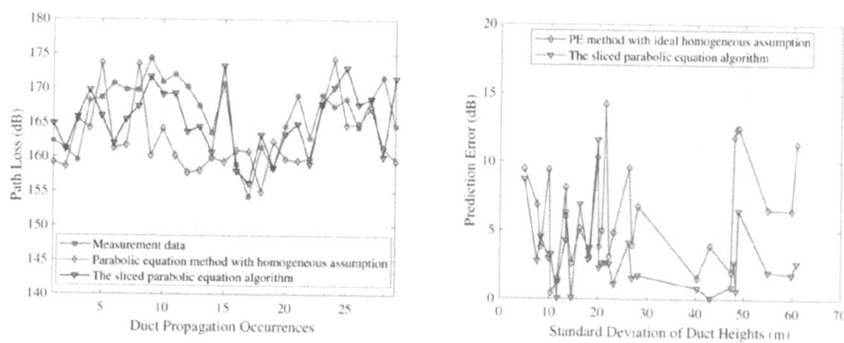

(a) (b)

Figure 10. *(a) Performance comparison on the Amsterdam-Burum path. The proposed algorithm reduces the average prediction error by approximately* 3.06 dB. *(b) Path loss prediction error vs. standard deviation of duct heights on the Amsterdam-Burum path.*

Figures 12 and 13 show the electromagnetic wave propagation in inhomogeneous evaporation ducts. The propagation path is 210 km. The slice range is 70 km. The electromagnetic wave can be trapped in the two cases. However, the duct effects are quite different. The reasons are as follows.

In Figure 12, the EDH increases along the propagation path ([25 m, 32 m, 37 m]) and the trapping feature becomes stronger. The electromagnetic wave that departs

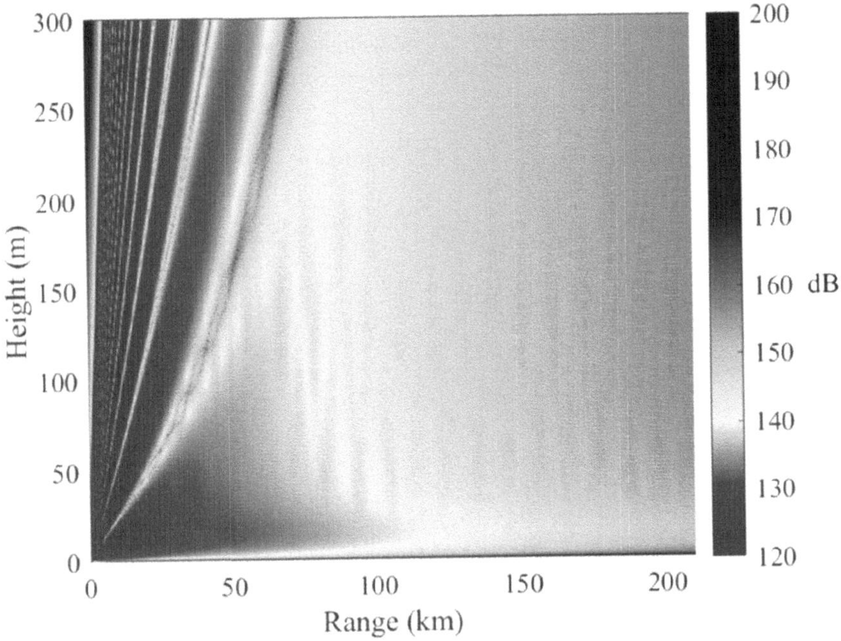

Figure 11. *Path loss in a homogeneous evaporation duct. In case A, the EDHs of the three slices are* 25 m, 25 m, *and* 25 m, *respectively.*

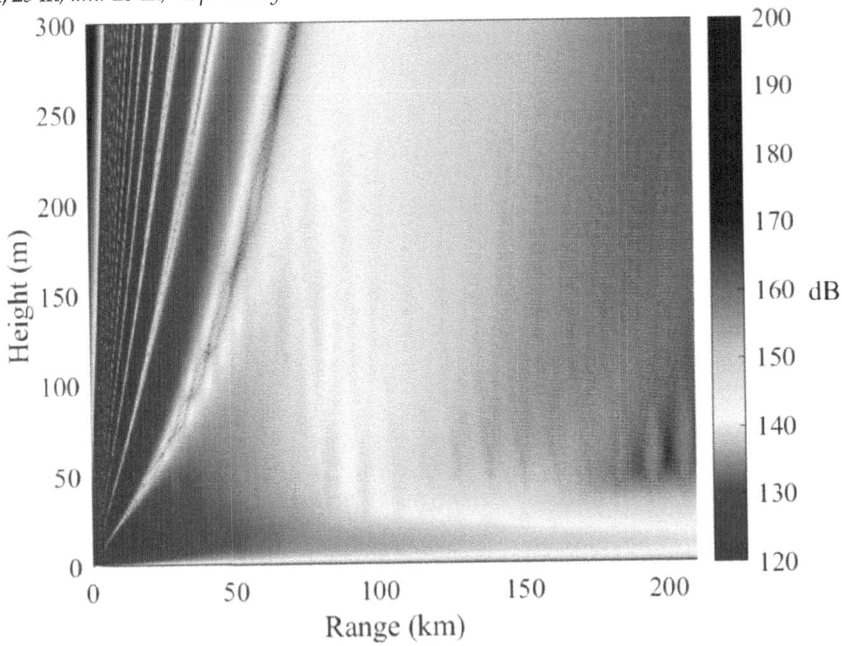

Figure 12. *Path loss in horizontally inhomogeneous evaporation ducts. In case B, the EDHs of the three slices are* 25 m, 32 m, *and* 37 m, *respectively.*

Table 4. *Simulation conditions for evaporation ducts.*

	Simulation Parameters
Frequency	3.5 GHz
Antenna height	15 m
Polarization	Horizontal
3 dB Beamwidth	5°
Elevation angle	0°
Max range	0–210 km
Max height	0–300 m
	Meteorological Information
Wind speed	8 m/s
Height of wind speed	10 m
Temperature of air (Ta)	21, 21.5, 22, 20.5, 20 °C
Height of temperature	2 m
Relative humidity	67.8%
Height of relative humidity	2 m
Sea level pressure	1013.25 hPa
Sea surface temperature	20 °C
	Corresponding evaporation duct height
Evaporation duct height	25, 32, 37, 19, 16 m

Table 5. *Simulation conditions for elevated ducts.*

	Simulation Parameters
Frequency	3.5 GHz
Antenna height	25 m
Polarization	Horizontal
3 dB Beamwidth	5°
Elevation angle	0°
Max range	0–210 km
Max height	0–300 m
	Elevated duct heights of three slices [Base height (m), Top height (m)]
Case D	[20, 55], [20, 55], [20, 55]
Case E	[20, 55], [23, 55], [27, 55]
Case F	[20, 55], [17, 55], [13, 55]

from the evaporation duct returns to the duct due to the refractivity. From the figure, we observe that the path loss fades with respect to the vertical height. In other words, the antenna heights and evaporation duct heights are key parameters when evaporation ducts are used for long-distance maritime communications.

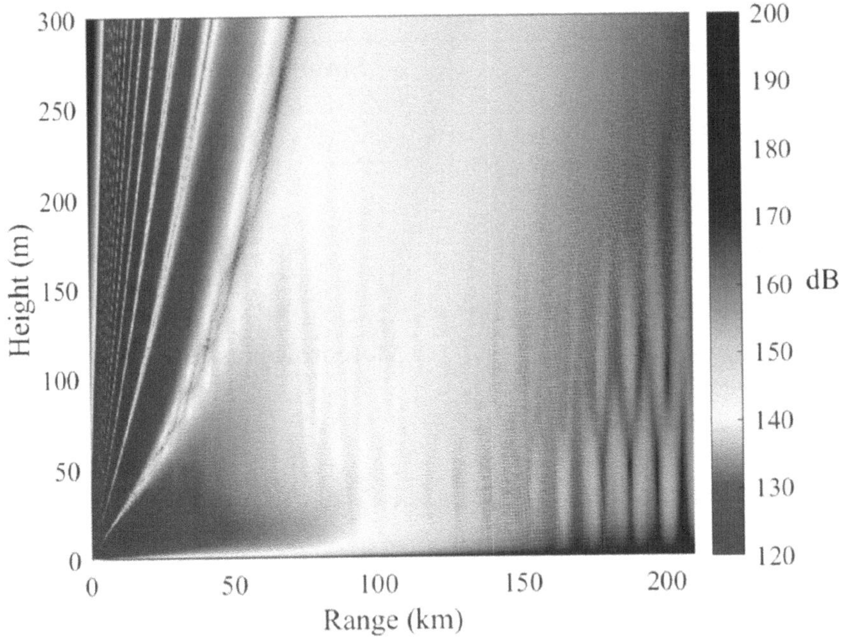

Figure 13. *Path loss in horizontally inhomogeneous evaporation ducts. In case C, the EDHs of the three slices are* 25 m, 19 m, *and* 16 m, *respectively.*

Table 6. *Advantages and disadvantages of the proposed algorithm.*

Advantages	It supports range-dependent evaporation ducts and elevated ducts. It supports standard meteorological reanalysis data. It can support boundary condition setting. It can support calm sea surface conditions. Detailed feasibility analysis. Long-distance measurement verification at 3.5 GHz.
Disadvantages	It does not support detailed sea surface roughness analysis. It does not support millimeter-wave band. It does not support complex inland terrain.

In Figure 13, the EDH decreases along the propagation path and the trapping feature becomes weaker. Therefore, some electromagnetic waves cannot be trapped and leave the duct layer. In the range of 140 km and 210 km, the duct effect is not obvious. The path loss is 177.80 dB at 210 km when the receiving antenna height is 3 m.

Figure 14 shows a comparison of the path loss in horizontally homogeneous and inhomogeneous evaporation ducts when the receiver height is 3 m. The numerical results indicate that the horizontal inhomogeneity of the evaporation duct exerts a larger influence on electromagnetic wave propagation. With increasing distance, the path loss difference gradually increases. There is a difference of approximately 30 dB in path loss between the two inhomogeneous cases. The

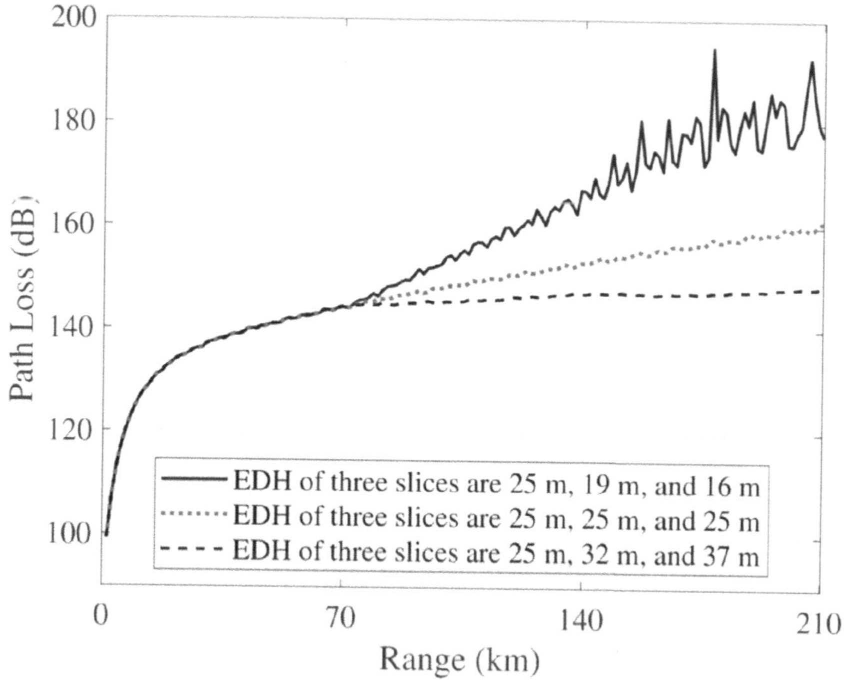

Figure 14. *Path loss comparison in horizontally homogeneous and inhomogeneous evaporation ducts, when the height of the receiver is* 3 m.

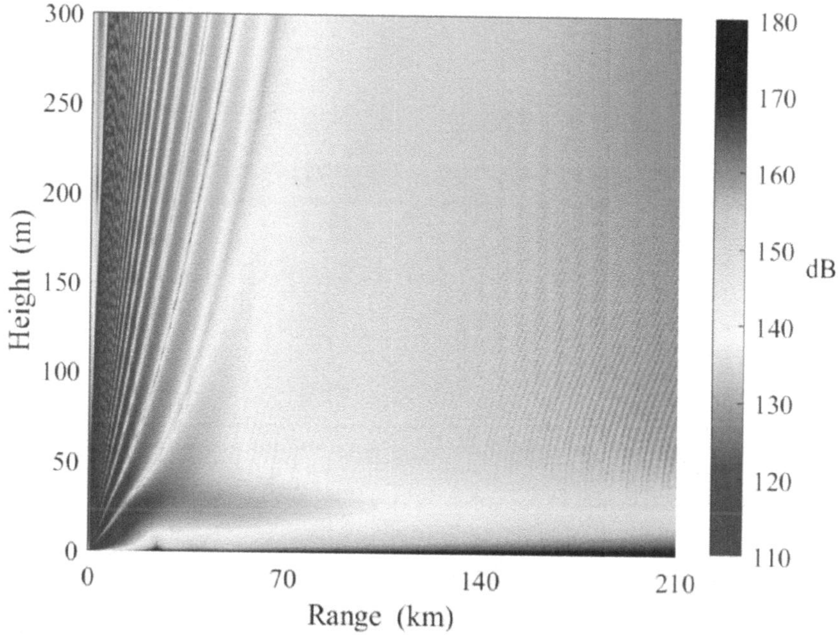

Figure 15. *Path loss in horizontally homogeneous elevated ducts (Case D in Table 5).*

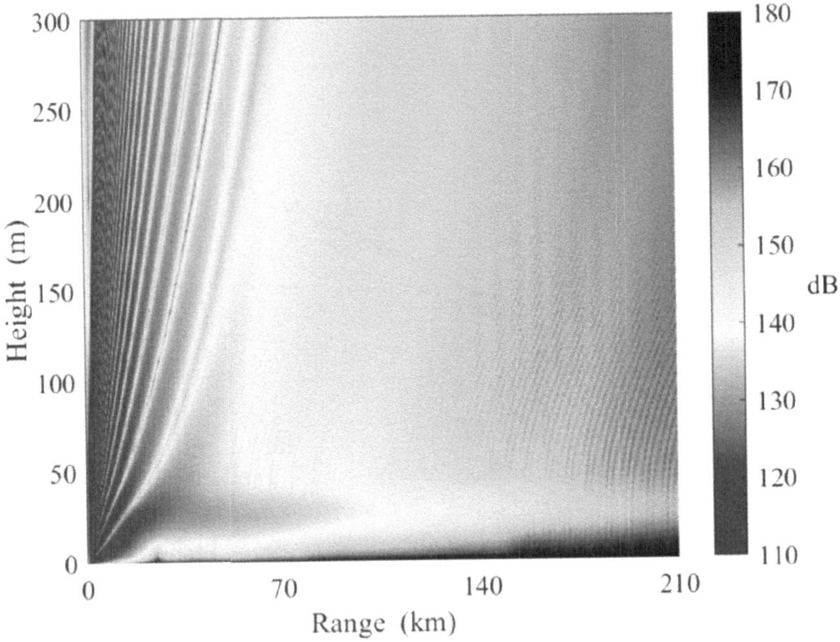

Figure 16. *Path loss in horizontally inhomogeneous elevated ducts (Case E in Table 5).*

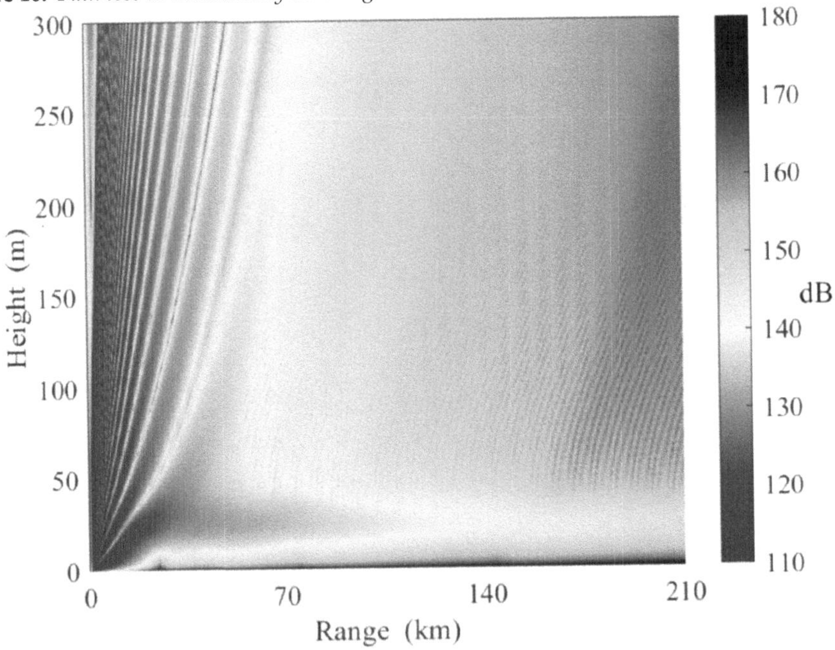

Figure 17. *Path loss in horizontally inhomogeneous elevated ducts (Case F in Table 5).*

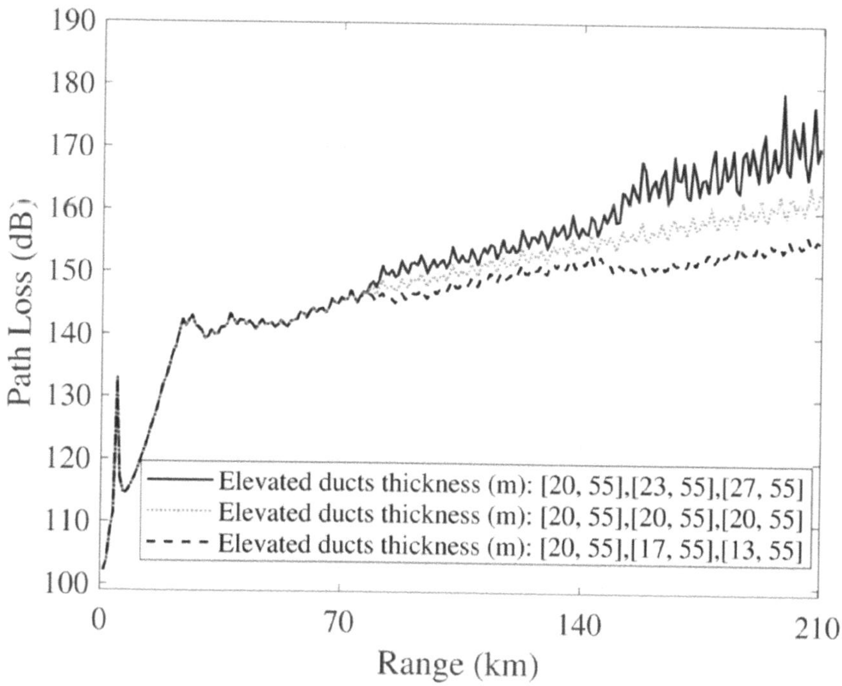

Figure 18. *Path loss comparison in horizontally homogeneous or inhomogeneous elevated ducts when the height of the receiver is* 9 m.

simulation results show that the duct horizontal characteristics have a great impact on the received signal strength.

The above simulation results show that the duct effect of propagation from the low EDH to the high EDH is better than that under the opposite conditions. The radio propagation in a low evaporation duct height environment has a more obvious horizontally inhomogeneous influence.

5.2. Range-Dependent Elevated Duct

In this subsection, we evaluate the performance of the proposed algorithm in horizontally inhomogeneous elevated ducts. The simulation parameters are shown in Table 5 . The propagation path is 210 km and the frequency is 3.5GHz. Note that the transmitter height is 25 m. Assume that the path is divided into three slices. According to the structure of elevated ducts in Section 2, we set the base heights and top heights of elevated ducts, which are listed in Table 5. Without loss of generality, we assume that $\frac{dM}{dz} = -43\text{M-unit}/\text{km}$.

Figure 15 shows the electromagnetic wave propagation in horizontally homogeneous elevated ducts. The base height and top height of the three slices are [20 m, 55 m], [20 m, 55 m], and [20 m, 55 m], respectively. From Figure 15, we can see the duct effects at heights between 20 and 55 m. When the receiving antenna height is less than 10 m, the path loss increases rapidly with distance. The reasons are as follows. Unlike evaporative ducts, there are standard refraction conditions between the elevated duct layer and the ground. Some electromagnetic waves cannot satisfy the angle of incidence conditions and leave the elevated duct.

In Figure 16, the strength of the elevated duct decreases along the path. The base height and top height of the three slices are [20 m, 55 m], [23 m, 55 m], and [27 m, 55 m], respectively. It can be observed that the signal can be trapped in elevated ducts in the first and second slices. When the receiving antenna height is 9 m, the path loss is approximately 156 dB at 140 km. However, the path loss increases rapidly over the third distance. The path loss is approximately 170 dB at 210 km. This is because the base height of the elevated duct in the third distance is higher than the transmitting antenna height.

Figure 17 shows the path loss distribution in horizontally inhomogeneous elevated ducts, where the conditions of elevated ducts improve along the path. The base height and top height of the three slices are [20 m, 55 m], [17 m, 55 m], and [13 m, 55 m], respectively. We can see that the electromagnetic wave can be trapped in elevated ducts along the entire path. Compared with Figures 15 and 16, the duct effect is much stronger in the range of 140 – 210 km. When the receiving antenna height is 9 m, the path loss at 160 km is approximately 151 dB.

Figure 18 shows a comparison of the path loss in horizontally homogeneous and inhomogeneous elevated ducts when the receiver height is 9 m. With increasing distance, the path loss difference gradually increases. There is a difference of approximately 15 dB in path loss between the two inhomogeneous cases. Moreover, it is noteworthy that the fluctuation in path loss is obvious in all cases. Compared with evaporation ducts, the number of reflection paths increases due to the signals propagated inside the elevated duct and in the region between the duct and the sea surface.

In order to apply the algorithm better, we list the advantages and disadvantages of the proposed algorithm in Table 6. It is generic with the help of meteorological measurements. It can predict the propagation loss under range-dependent refractivity conditions. Future work can focus on sea surface roughness analysis under atmospheric duct conditions. It will be an important direction to exploit atmospheric ducts to improve the cell coverage in maritime scenarios, such as BS-ship and UAV-ship communications.

6. CONCLUSIONS

In this paper, we focused on the effect of horizontally inhomogeneous ducts on maritime communications from both theoretical and measurement perspectives. Through a thorough investigation and analysis of radio propagation in atmospheric ducts, we proposed a sliced parabolic equation algorithm to support the actual duct environment. Then, we derived the corresponding numerical solution and the approximate error expression. The proposed algorithm was verified by the 3.5GHz long-distance radio measurement data. Results demonstrated that the proposed algorithm had better performance than the traditional parabolic equation method. We also provided the simulation results under duct conditions with different horizontal characteristics.

Author Contributions

Conceptualization, Y.W.; methodology, Y.W., T.Z. and T.X.; software, Y.W.; validation, Y.W. and H.H.; writing-original draft preparation, Y.W.; writing-review and editing, T.Z. and H.H.; supervision, T.Z.; All authors have read and agreed to the published version of the manuscript.

Acknowledgments

We are grateful to the authors from the Radio Communications Agency Netherlands [16], who shared the path loss results with us.

REFERENCES

1. Wang, Y.; Feng, W.; Wang, J.; Quek, T.Q. Hybrid Satellite-UAV-Terrestrial Networks for 6G Ubiquitous Coverage: A Maritime Communications Perspective. IEEE J. Sel. Areas Commun. 2021, 39, 3475-3490.
2. Li, X.; Feng, W.; Wang, J.; Chen, Y.; Ge, N.; Wang, C. Enabling 5 G on the Ocean: A Hybrid Satellite-UAV-Terrestrial Network Solution. IEEE Wirel. Commun. 2020, 27, 116-121.
3. Koo, H.; Chae, J.; Kim, W. Design and Experiment of Satellite-Terrestrial Integrated Gateway with Dynamic Traffic Steering Capabilities for Maritime Communication. Sensors 2023, 23, 1201.
4. Guan, S.; Wang, J.; Jiang, C.; Duan, R.; Ren, Y.; Quek, T.Q. MagicNet: The Maritime Giant Cellular Network. IEEE Commun. Mag. 2021, 59, 117-123.
5. Dong, X.; You, Z.; Shen, Y.; Lu, D.; Xu, Y.; Liu, J. A Two-Dimensional Sybil-Proof Mechanism for Dynamic Spectrum Access. IEEE Trans. Mob. Comput. 2023, 1, 1-15.
6. Alqurashi, F.S.; Trichili, A.; Saeed, N.; Ooi, B.S.; Alouini, M.-S. Maritime Communications: A Survey on Enabling Technologies, Opportunities, and Challenges. IEEE Internet Things J. 2023, 10, 3525-3547.
7. Yu, G.; Ding, X.; Liu, S. Joint Resource Management and Trajectory Optimization for UAV-Enabled Maritime Network. Sensors 2022, 22, 9763.
8. Jo, S.; Shim, W. LTE-Maritime: High-Speed Maritime Wireless Communication Based on LTE Technology. IEEE Access 2019, 7, 53172-53181.
9. Huo, Y.; Dong, X.; Beatty, S. Cellular Communications in Ocean Waves for Maritime Internet of Things. IEEE Internet Things J. 2020, 7, 9965-9979.
10. Yang, C.; Shi, Y.; Wang, J. The Preliminary Investigation of Communication Characteristics Using Evaporation Duct across the Taiwan Strait. J. Mar. Sci. Eng. 2022, 10, 1493.
11. Wang, J.; Qian, J.; Wang, J.; Gao, R.; Hu, Y.; Li, Y.; Feng, W. UAV Jitter May Have Merit: A Fading Analysis in Air-to-Sea Two Ray Channels. IEEE Wirel. Commun. Lett. 2022, 11, 2270-2274.
12. Dinc, E.; Alagoz, F.; Akan, O.B. Path-Loss and Correlation Analysis for Space and Polarization Diversity in Surface Ducts. IEEE Trans. Antennas Propag. 2016, 64, 4498-4503.

13. Chen, H.; Yin, F.; Huang, W.; Liu, M.; Li, D. Ocean Surface Drifting Buoy System Based on UAV-Enabled Wireless Powered Relay Network. Sensors 2020, 20, 2598.
14. Yu, J.; Chen, W.; Li, F.; Li, C.; Yang, K.; Liu, Y.; Chang, F. Channel Measurement and Modeling of the Small-Scale Fading Characteristics for Urban Inland River Environment. IEEE Trans. Wirel. Commun. 2020, 19, 3376-3389.
15. Ma, J.; Wang, J.; Yang, C. Long-Range Microwave Links Guided by Evaporation Ducts. IEEE Commun. Mag. 2022, 60, 68-72.
16. Colussi, L.C.; Schiphorst, R.; Teinsma, H.W.; Witvliet, B.A.; Fleurke, S.R.; Bentum, M.J.; Van Maanen, E.; Griffioen, J. Multiyear Trans-Horizon Radio Propagation Measurements at 3.5 GHz. IEEE Trans. Antennas Propag. 2018, 66, 884-896.
17. Rosson, P.; Estavoyer, J.; Lombard, L.; Miscopein, B.; Popon, X.; Doré, J.; Kténas, D.; Coquen, V.; Bris, R.J. Long-range Broadband wireless system for maritime communications in the 3.5GHz band. In Proceedings of the 2021 IEEE 94th Vehicular Technology Conference, Norman, OK, USA, 27-30 September 2021; pp. 1-6.
18. Zhang, H.; Zhou, T.; Xu, T.; Hu, H. Remote Interference Discrimination Testbed Employing AI Ensemble Algorithms for 6G TDD Networks. Sensors 2023, 23, 2264.
19. Dinc, E.; Akan, O.B. Beyond-line-of-sight Ducting Channels: Coherence Bandwidth, Coherence Time and Rain Attenuation. IEEE Commun. Lett. 2018, 19, 2274-2277.
20. Zhang, C.; Zhang, W.; Wang, W.; Yang, L.; Zhang, W. Research Challenges and Opportunities of UAV Millimeter-Wave Communications. IEEE Wirel. Commun. 2019, 26, 58-62.
21. Huang, L.-F.; Liu, C.-G.; Wang, H.-G.; Zhu, Q.-L.; Zhang, L.-J.; Han, J.; Zhang, Y.-S.; Wang, Q.-N. Experimental Analysis of Atmospheric Ducts and Navigation Radar Over-the-Horizon Detection. Remote Sens. 2022, 14, 2588.
22. Gunashekar, S.D.; Warrington, E.M.; Siddle, D.R. Long-term Statistics Related to Evaporation Duct Propagation of 2 GHz Radio Waves in the English Channel. Radio Sci. 2010, 45, 1-14.
23. Lee, Y.H.; Dong, F.; Meng, Y.S. Near Sea-Surface Mobile Radiowave Propagation at 5 GHz: Measurements and Modeling. Radioengineering 2014, 23, 824-830.
24. Dahman, G.S.; Couillard, D.; Grandmaison, M.; Poitau, G.; Gagnon, F. Improved 2-Ray Model for Overwater Propagation Channels: Modeling the Instantaneous Variations in the Received Signal Strength. IEEE Wirel. Commun. Lett. 2019, 8, 865-868.
25. Yang, C.; Wang, J.; Ma, J. Exploration of X-Band Communication for Maritime Applications in the South China Sea. IEEE Antennas Wirel. Propag. Lett. 2022, 21, 481-485.
26. Guo, X.; Zhao, D.; Zhang, L.; Wang, H.; Kang, S.; Lin, L. C Band Transhorizon Signal Characterisations in Evaporation Duct Propagation Environment over Bohai Sea of China. IET Microw. Antennas Propag. 2019, 13, 407-413.
27. Wang, J.; Zhou, H.F.; Li, Y.; Sun, Q.; Wu, Y.; Jin, S.; Quek, T.Q.; Xu, C. Wireless Channel Models for Maritime Communications . IEEE Access 2018, 6, 68070-68088.
28. Li, F.; Shui, Y.; Liang, J.; Yang, K.; Yu, J. Ship-to-ship Maritime Wireless Channel Modeling under Various Sea State Conditions Based on REL Model. Front. Mar. Sci. 2023, 10, 1134286.
29. Liu, Y.; Wang, C.; Chang, H.; He, Y.; Bian, J. A Novel Non-Stationary 6G UAV Channel Model for Maritime Communications. IEEE J. Sel. Areas Commun. 2021, 39, 2992-3005.

30. Dinc, E.; Akan, Ö.B. Beyond-line-of-sight communications with ducting layer. IEEE Commun. Mag. 2014, 52, 37-43.
31. Ozgun, O.; Apaydin, G.; Kuzuoglu, M.; Sevgi, L. PETOOL: MATLAB-based One-way and Two-way Split-step Parabolic Equation Tool for Radiowave Propagation over Variable Terrain. Comput. Phys. Commun. 2011, 182, 2638-2654.
32. Ozgun, O.; Sahin, V.; Erguden, M.E.; Apaydin, G.; Yilmaz, A.E.; Kuzuoglu, M.; Sevgi, L. PETOOL v2.0: Parabolic Equation Toolbox with evaporation duct models and real environment data. Comput. Phys. Commun. 2020, 256, 107454.
33. Abdelmoaty, A.; Dahman, G.; Bousselmi, A.A.; Poitau, G.; Gagnon, F. Using Vertically Separated MIMO in Ship-to-Ship Communications. IEEE Access 2020, 8, 103601-103609.
34. Zhang, H.; Zhou, T.; Xu, T.; Wang, Y.; Hu, H. Statistical Modeling of Evaporation Duct Channel for Maritime Broadband Communications. IEEE Trans. Veh. Technol. 2022, 71, 10228-10240.
35. Guo, Q.; Long, Y. Two-Way Parabolic Equation Method for Radio Propagation over Rough Sea Surface. IEEE Trans. Antennas Propag. 2020, 68, 4839-4847.
36. Zaidi, K.S.; Jeoti, V.; Drieberg, M.; Awang, A.; Iqbal, A. Fading Characteristics in Evaporation Duct: Fade Margin for a Wireless Link in the South China Sea. IEEE Access. 2018, 6, 11038-11045.
37. He, Y.; Wang, C.; Chang, H.; Huang, J.; Sun, J.; Zhang, W.; Aggoune, E.M. A Novel 3D Non-Stationary Maritime Wireless Channel Model. IEEE Trans. Commun. 2022, 70, 2102-2116.
38. Lim, T.H.; Wang, S.; Chong, Y.; Park, Y.B.; Ko, J.; Choo, H. High Altitude Ducts Causing Abnormal Wave Propagation in Coastal Area of Korea. Microw. Opt. Technol. Lett. 2020, 62, 643-650.
39. Zhang, Q.; Wang, S.; Shi, Y.; Yang, K. Measurements and Analysis of Maritime Wireless Channel at 8GHz in the South China Sea Region. IEEE Trans. Antennas Propag. 2023, 71, 2674-2681.
40. Hoeft, M.; Gierlowski, K.; Wozniak, J. Wireless Link Selection Methods for Maritime Communication Access Networks-A Deep Learning Approach. Sensors 2023, 23, 400.
41. Sit, H.; Earls, C.J. Deep Learning for Classifying and Characterizing Atmospheric Ducting within the Maritime Setting. Comput. Geosci. 2021, 157, 104919.
42. Zhang, H.; Zhou, T.; Xu, T.; Wang, Y.; Hu, H. FNN-Based Prediction of Wireless Channel with Atmospheric Duct. In Proceedings of the 2021 IEEE International Conference on Communications, Montreal, QC, Canada, 14-23 June 2021; pp. 1-6.
43. Yang, C.; Wang, J. The Investigation of Cooperation Diversity for Communication Exploiting Evaporation Ducts in the South China Sea. IEEE Trans. Antennas Propag. 2022, 70, 8337-8347.
44. Zhao, W.; Zhao, J.; Li, J.; Zhao, D.; Huang, L.; Zhu, J.; Lu, J.; Wang, X. An Evaporation Duct Height Prediction Model Based on a Long Short-Term Memory Neural Network. IEEE Trans. Antennas Propag. 2021, 69, 7795-7804.
45. Yang, C.; Shi, Y.; Wang, J.; Feng, F. Regional Spatiotemporal Statistical Database of Evaporation Ducts Over the South China Sea For Future Long-Range Radio Application. IEEE J. Sel. Top. Appl. Earth. Obs. Remote. Sens. 2022, 15, 6432-6444.

46. Sirkova, I. Brief Review on PE Method Application to Propagation Channel Modeling in Sea Environment. Cent. Eur. J. Eng. 2012, 2, 19-38.
47. Wei, T.; Feng, W.; Chen, Y.; Wang, C.-X.; Ge, N.; Lu, J. Hybrid Satellite-terrestrial Communication Networks for the Maritime Internet of Things: Key Technologies, Opportunities, and Challenges. IEEE Internet Things J. 2021, 8, 8910-8934.
48. Wang, Y.; Chen, Y.; Zhou, T.; Hu, H. A Traceable Approach to Remote Interference Management for New Radio. In Proceedings of the 2019 IEEE International Conference on Communications Workshops, Shanghai, China, 20-24 May 2019 ; pp. 1-6.
49. Wei, T.; Feng, W.; Wang, J.; Ge, N.; Lu, J. Exploiting the Shipping Lane Information for Energy-Efficient Maritime Communications. IEEE Trans. Veh. Technol. 2019, 68, 7204-7208.
50. Liu, X.; Xu, B.; Wang, X.; Zheng, K.; Chi, K.; Tian, X. Impacts of Sensing Energy and Data Availability on Throughput of Energy Harvesting Cognitive Radio Networks. IEEE Trans. Veh. Technol. 2023, 72, 747-759.
51. International Telecommunication Union. The Radio Refractive Index: Its Formula and Refractivity Data; Recommendation ITU-R P.453-13; ITU: Geneva, Switzerland, 2019.
52. International Telecommunication Union. Prediction Procedure for the Evaluation of Interference between Stations on the Surface of the Earth at Frequencies Above about 0.1GHz; Recommendation ITU-R P.452-16; ITU: Geneva, Switzerland, 2021.
53. Dinc, E.; Akan, O.B. Channel Model for the Surface Ducts: Large-Scale Path-Loss, Delay Spread, and AOA. IEEE Trans. Antennas Propag. 2015, 63, 2728-2738.
54. Gilles, M.A.; Earls, C.J.; Bindel, D.S. A Subspace Pursuit Method to Infer Refractivity in the Marine Atmospheric Boundary Layer. IEEE Trans. Geosci. Remote Sens. 2019, 57, 5606-5617.
55. Xu, L.; Yardim, C.; Mukherjee, S.; Burkholder, R.J.; Wang, Q.; Fernando, H.J.S. Frequency Diversity in Electromagnetic Remote Sensing of Lower Atmospheric Refractivity. IEEE Trans. Antennas Propag. 2022, 70, 547-558.

CHAPTER 9

A Novel Deep Learning Network with Deformable Convolution and Attention Mechanisms for Complex Scenes Ship Detection in SAR Images

Peng Chen[1], Hui Zhou[2], Ying Li[3], Peng Liu[1] and Bingxin Liu[1]

[1]*Navigation College, Dalian Maritime University, Dalian 116026, China*
[2]*School of Computer and Software, Dalian Neusoft Information University, Dalian 116023, China*
[3]*Environmental Information Institute, Dalian Maritime University, Dalian 116026, China*

ABSTRACT

Synthetic aperture radar (SAR) can detect objects in various climate and weather conditions. Therefore, SAR images are widely used for maritime object detection in applications such as maritime transportation safety and fishery law enforcement. However, nearshore ship targets in SAR images are often affected by background clutter, resulting in a low detection rate, high false alarm rate, and high missed detection rate, especially for small-scale ship targets. To address this problem, in this paper, we propose a novel deep learning network with deformable convolution and attention mechanisms to improve the Feature Pyramid Network (FPN) model for nearshore ship target detection in SAR images with complex backgrounds. The proposed model uses a deformable convolutional neural network in the feature extraction network to adapt the convolution position to the target sampling point, enhancing the feature extraction ability of the target, and improving the detection rate of the ship target against the complex background. Moreover, this model uses a channel attention mechanism to capture the feature dependencies between different channel graphs in the feature extraction network and reduce the false detection rate. The designed experiments on a public SAR image ship dataset show that our model achieves 87.9% detection accuracy for complex scenes and 95.1% detection accuracy for small-scale ship targets. A quantitative comparison of the proposed model with several classical and recently developed deep learning models on the same SAR images dataset demonstrated the superior performance of the proposed method over other models.

Keywords: *SAR image; ship target detection; deformable convolutional networks; channel attention*

1. INTRODUCTION

The continuous monitoring of ship targets in harbors and marine areas is an important practical task that is widely used in various maritime fields such as combating illegal fishing, oil spill monitoring, and traffic management [1,2]. Synthetic aperture radar (SAR) has become an important method for ship detection at sea because it is unaffected by weather, has a large imaging area, and has a constant resolution far from the observed object [3-5]. However, accurate ship detection remains challenging. Currently, several detection methods can easily miss ships that are in close proximity to each other.

Numerous manually extracted features have been used for ship detection [6,7]. Recently, deep learning models have been widely used for ship detection in SAR images [8-10]. Ship detection is a combination of target localization and classification. Target classification determines whether the input image contains the desired object, and target location identifies the location of the target [11-13]. For example, Wang et al. used SDD models to detect ships in complex backgrounds in SAR images and used alternate learning methods to improve the accuracy [14]. Kang used a fast region with a convolutional neural net- work (CNN) to obtain initial ship detection results and adjusted the final results using an adaptive threshold alarm rate [15].

In the SAR image ship target-detection task, target detection is better in a simple sea clutter background, because the grayscale characteristics of the target are significantly higher than those of the sea clutter [16-18]. However, in actual SAR image imaging processes, backgrounds such as ports, islands, and buildings appear in SAR images, which can cause confusion because of their high grayscale characteristics, resulting in low detection and high false alarm rates for deep learning target-detection algorithms [19]. The complex backgrounds mentioned in this study refer to SAR images with backgrounds, such as ports and islands, for ship targets in complex scenes. For the target-detection problem of complex scene interference, some scholars solve this problem from the perspective of improving the backbone network [20,21]. Wu et al. proposed a SAR image ship small target-detection algorithm to improve the network structure of the feature pyramid by redesigning the underlying residual units to solve the contradiction between the perceptual field and localization and introduced a balance factor to optimize the small target weights in the loss function [22]. Zhang et al. proposed a feature-fusion-based ship target-detection algorithm based on a multiscale single-shot detection framework with enhanced network feature extraction by adding deconvolution and pooling feature fusion modules [23]. He et al. proposed the Deformable Feature Fusion You Only Look Once (DFF-Yolov5) algorithm based on YOLOv5, which improved the YoloV5 model in two aspects: feature refinement and multifeature fusion in the feature extraction network [24]. Other scholars consider introducing attention mechanisms to solve the above problems [25, 26]. Liu et al. improved the target-detection method by integrating the detection frame length and width as parameters, performing curve optimization of the loss function, and combining it with the coordinate attention mechanism to detect ship targets [27]. Li et al. used multiple receptive field integration and channel domain attention to enhance the resistance of features to scale and environmental changes [28]. SAR ship detection can also occur in scenes with numerous ships, high density, and a small target size (less than 15 pixels $\times$15 pixels), which further increases the difficulty of target extraction and recognition. When using a deep learning model of ship targets for detection, the CNN can only extract relatively regular features in the target area. For SAR ship targets with complex backgrounds, standard CNNs are susceptible to interference from background coastal information when extracting features, which affects the expressiveness of the feature extraction network [14,29]. However, because SAR image ship targets have different scale information, the extraction of the semantic information of small targets is comparatively lower with an increase in the number of feature-mapping layers when feature extraction is performed by convolution. Therefore, the above method still has limitations for nearshore ship targets in complex backgrounds.

To counter the challenges of low detection rates of ship targets and small target detection in SAR images in complex scenes, classical Feature Pyramid Networks (FPNs) have been introduced by multiple authors to implement multiscale SAR ship detection [30]. For example, Lin et al. added a compressed incentive-based module to the top of an FPN to focus on more important channel features, thus enhancing the description of the top-level semantic information [31]. Li et al. embedded a convolutional block attention module into the feature fusion branch of an FPN, which enables a joint feature channel and spatial attention to facilitate the representation of multilevel ship features [32]. Zhao et al. proposed a field-of-view attentional FPN to enhance background discrimination by extending the field-of-view range of the contextual background [33]. Liu et al. designed a scalemigratable FPN to further enhance the fusion benefits of multiscale features and ultimately improve detection accuracy [34]. Recently, Li et al. proposed a feature self-attention-guided FPN to refine more representative multiscale features, thereby facilitating multiscale information flow [35]. However, the multiscale SAR ship detection performance of these existing detection models remains limited. On the one hand, most of them use classical regular convolution kernels to extract ship features, which cannot perform multiscale modeling of ship deformations owing to different incidence angles, resolutions, and other factors, resulting in limited multiscale feature expression capability. On the other hand, the multiscale feature fusion approach they adopt cannot provide global content, and the multiscale feature fusion methods they use cannot effectively perceive the global content, which leads to limited fusion benefits and is not conducive to a more comprehensive and adequate multiscale feature representation. This study made structural improvements to the FPN. First, a deformable CNN was used to replace the original CNN to improve the expression ability of the network features. Next, a target-detection network was used to extract features, and a channel attention mechanism was introduced to extract the weights of the feature channels to further enhance target feature extraction in complex backgrounds. Finally, through multiple sets of comparison experiments and an analysis of the detection results of high-density small target ship detection and SAR image ship target detection in complex environments, the practicality and effectiveness of the proposed algorithm for the detection of ship targets in complex backgrounds of SAR images were verified.

2. MATERIALS AND METHODS

2.1. Revisiting FPN

Because of the difference in imaging mechanisms between SAR and optical images, distinguishing ship targets from maritime false targets and background noise in SAR images is difficult and often confusing [36]. The use of an FPN can effectively eliminate the background environment and scattering noise, thereby improving the recognition accuracy of ship targets in SAR images. First, feature mapping was obtained by convolution using the backbone network, where the convolution layers formed feature-mapping layers from bottom to top in the following order: $\{C1, C2, C3, C4, C5\}$, and then by undergoing upsampling through the top-down pathway to obtain feature maps from higher pyramid layers. { C1, C2, C3, C4, C5} are laterally connected with the upsampling results through a 1×1 convolution kernel (256 channels) to form new feature maps $\{\mathrm{M2}, \mathrm{M3}, \mathrm{M4}, \mathrm{M5}\}$.

$$\begin{aligned}
\mathrm{M5} &= \mathrm{C5.Conv}\ (256, (1,1)), \\
\mathrm{M4} &= \mathrm{UpSampling}\ (\mathrm{M5}) + \mathrm{C4.Conv}\ (256, (1,1)), \\
\mathrm{M3} &= \mathrm{UpSampling}\ (\mathrm{M4}) + \mathrm{C3.Conv}\ (256, (1,1)), \\
\mathrm{M2} &= \mathrm{UpSampling}\ (\mathrm{M3}) + \mathrm{C2.Conv}\ (256, (1,1)).
\end{aligned}$$

Finally, to remove the confusion effect, feature maps from m2 to m5 were obtained using a 3×3 convolution, and a sequence of FPN feature maps { P2, P3, P4, P5 } was obtained. The FPN structure is illustrated in Figure 1 [37].

2.2. Improved Structure of FPN with Deformable CNNs

The backbone network of the original FPN target-detection algorithm used a CNN for feature extraction. Standard convolution is not sufficiently flexible for the shapereceptive field of the target, and the efficiency of convolution naturally decreases, whereas deformable convolution uses irregular shapes, which addresses this problem. Compared with the standard convolution kernel, the pixels used for convolution by the deformable convolution kernel are not shifted by a fixed step size in the x - and y-directions relative to the central pixel; however, a new convolution kernel is used to record x and the offset in the y-direction. Figure 2 shows the learning process for the deformable convolution. After the traditional convolution layer, a biased convolution layer was added. The convolution kernel

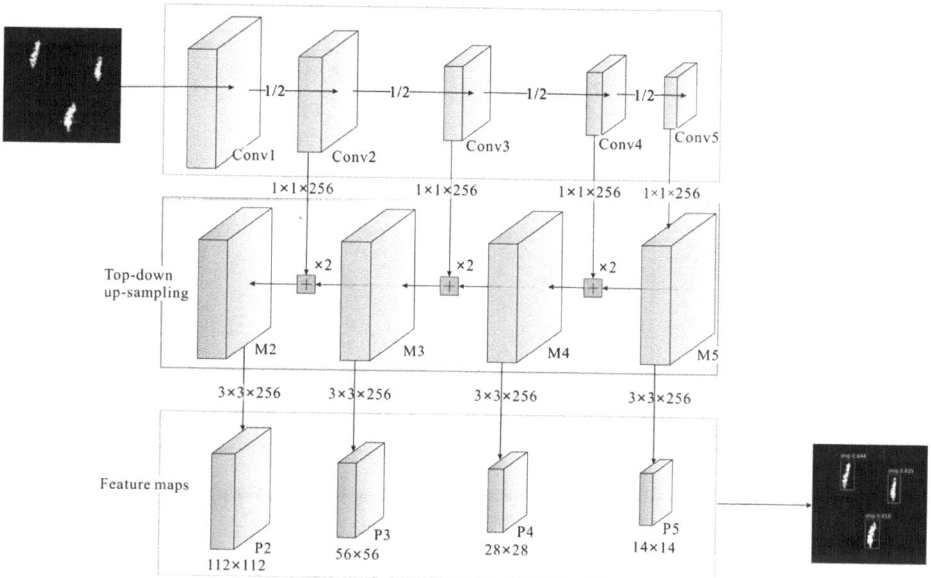

Figure 1. *Structure of the Feature Pyramid Network (FPN) network.*

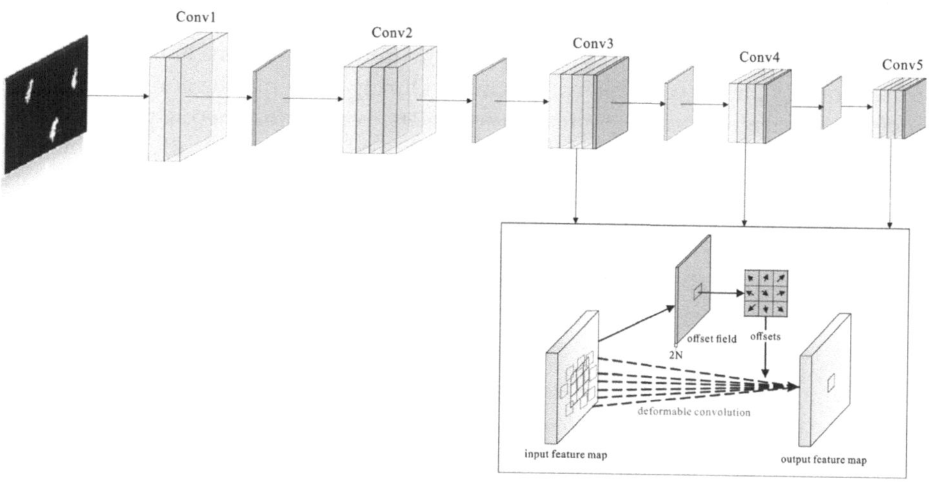

Figure 2. *Deformable convolutional neural network (CNN).*

of this layer was identical to that of an ordinary convolution kernel. The output deviation was the same as that of the input feature map. The generated channel dimension is 2 N, where N is the number of channels of the input feature map, which implies that the x-direction was recorded twice separately, as well as the offset features in the y-direction.

To improve the FPN target-detection model, $\{C1, C2, C3, C4, C5\}$ feature maps were first generated based on the input images using conventional convolution kernels. To avoid redundant computations, deformable convolution kernels were

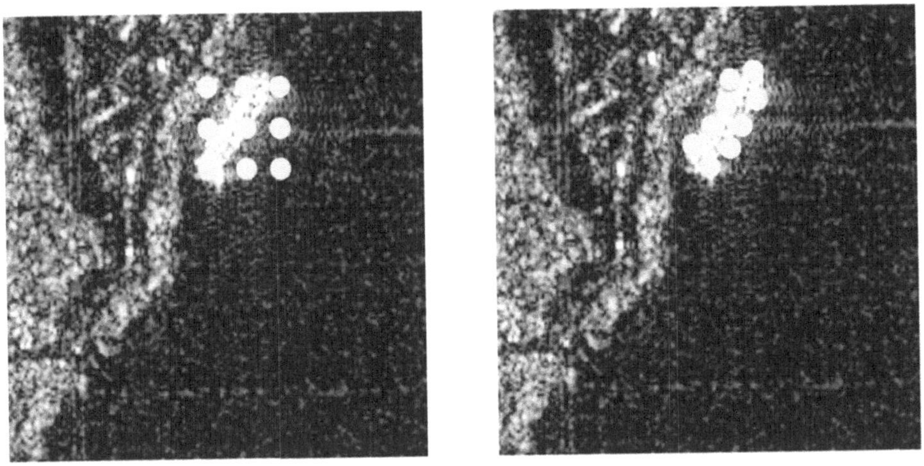

(a) CNN (b) deformable CNN

Figure 3. *Compared with the CNN (a), the deformable CNN (b) extracts the features of the input image by adjusting its shape according to the actual situation by shifting the convolutional kernel.*

used after the convolution layers conv3, conv4, and conv 5 of the FPN; that is, the $\{C3, C4, C5\}$ feature maps were used as the input, and another convolution layer was added to each feature layer to learn the deformation offsets in the x - and y-directions of the deformable convolution. During training, the convolutional kernels for generating output features and offsets were simultaneously learned using an interpolation algorithm and backpropagation [38]. Specifically, for the initial position p_0 of the input feature map, the output was obtained as y after an ordinary convolution operation.

$$y(p_0) = \sum_{p_n \in \Re} w(p_n) \cdot x(p_0 + p_n) \tag{1}$$

where w is the network weight parameter, $\Re$ is the specified convolution region, and p_n traverses this convolution region by adding a learning position offset Δp_n to the ordinary convolution y. The deformed convolution is formulated as follows:

$$y(p_0) = \sum_{p_n \in \Re} w(p_n) \cdot x(p_0 + p_n + \Delta p_n) \tag{2}$$

A deformable convolutional neural network can extract the features of an input image more effectively by offsetting the convolutional kernel and adjusting its shape according to the actual situation. As shown in Figure 3, the use of deformable CNNs can improve the feature representation capability of SAR image target-detection networks.

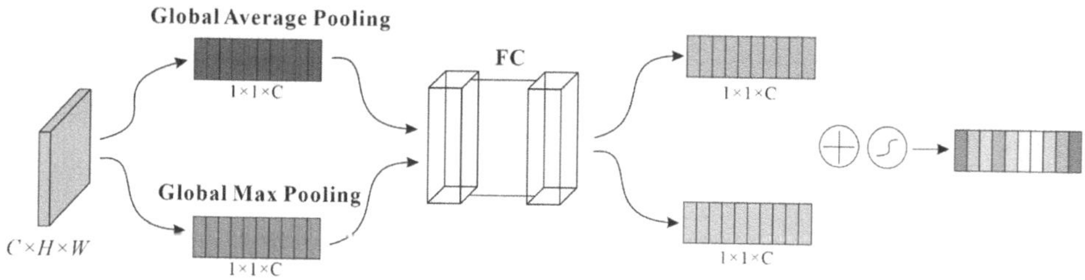

Figure 4. *Channel attention mechanism.*

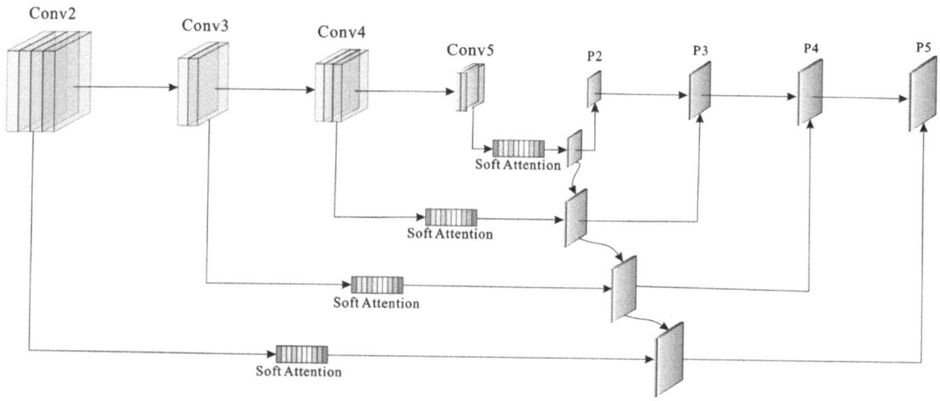

Figure 5. *Introducing the soft attention mechanism to FPN. The feature weight vector was used in each convolution layer, which obtained different levels of importance for different channels.*

2.3. Channel Attention Mechanism Introduced

In the complex backgrounds of SAR images, distinguishing nearshore ship targets is difficult because of increased interference. To improve the expression ability of features in images, in the target-detection model FPN, for feature-mapping layers of different scales, the channel attention mechanism is used to capture different channel maps, compute the feature dependencies between them, and calculate the weighted values of all the channel maps. The feature weight vector was learned to explicitly model the correlation between the feature channels. To compute the channel attention efficiently, we squeeze the spatial dimension of the input feature map. For aggregating spatial information, average pooling and max pooling have been commonly adopted to compute spatial statistics. as shown in Figure 4 . The dot product of the original feature layer (F) of any $H \times W \times C$ and the feature weight vector $\mathbf{w}$ were determined to obtain the feature layers with different levels of importance for different channels. The layers with channel weights were then merged at each layer in the FPN manner; that is, a new feature map layer is obtained, as shown in Figure 5.

First, using the feature layer (F) of any $H \times W \times C$ as the input, the global average pooling AvgPool and the maximum pooling MaxPool of the space were conducted, where the pool size was H × W, and the channel description row vectors $\mathbf{F}_{avg}$ and $\mathbf{F}_{max}$ of the two 1 × 1 × C were obtained. Two fully connected layers were shared, and the ReLU activation function was used to fit the complex correlations between the channels. Then, we added the description row vectors of the two channels and obtained the feature weight vector **w** of 1 × 1 × C using the sigmoid activation function. The original feature layer (H × W × C) and feature weight vector **w** were multiplied to obtain feature layers with different channel importance values, as shown in Figures 4 and 5, respectively. The region of interest was determined using a sliding window operation on the reconstructed feature map.

$$
\begin{aligned}
\mathbf{w} &= \text{sigmoid}\left(\text{TFC}\left(\mathbf{F_{avg}}\right) + \text{TFC}\left(\mathbf{F_{max}}\right)\right), \\
\mathbf{F_{avg}} &= \frac{1}{H \times W}\sum_{i=1}^{H}\sum_{j=1}^{W} \text{Avg Pool}(\mathbf{F}), \\
\mathbf{F_{max}} &= \frac{1}{H \times W}\sum_{i=1}^{H}\sum_{j=1}^{W} \text{MaxPool}(\mathbf{F}).
\end{aligned}
\tag{3}
$$

2.4. Loss Function

The overall loss of the model includes classification loss L_{class} and bounding box loss L_{box}, the classification loss adopts the cross-entropy loss function, and the bounding box loss includes the location loss L_1 the of ground truth b_i and predicted box $\hat{b}_{\sigma(i)}$ and Intersection over Union (IoU) loss $L_{\text{GIou.}}$.

$$
L_{\text{class}} = -\frac{1}{N}\sum_{i=1}^{N}\left(p_i \log \hat{p}_{\hat{\sigma} i} + (1 - p_i)\log\left(1 - \hat{p}_{\hat{\sigma} i}\right)\right), \tag{4}
$$

$$
L_{\text{box}}\left(y_i, \hat{y}_{\sigma(i)}\right) = \sum_{1}^{N}\left[\lambda_1 \left\|b_i - \hat{b}_{\sigma(i)}\right\| + \lambda_{GIoU} L_{GIoU}\left(b_i, \hat{b}_{\sigma(i)}\right)\right], \tag{5}
$$

where N is the number of prediction frames, p_i is the true category probability, and $\hat{p}_{\hat{o}i}$ represents the probability of predicting the $\hat{\sigma}_i$ th ship target. λ_1 and λ_{GIoU} are the corresponding loss function penalty factors.

IoU reflects the degree of coincidence between the prediction box and the ground truth, and the larger the coincidence, the greater the value; therefore, it works as an optimization function. GIoU introduces the minimum enclosure box and can avoid the problem that, when the prediction box and ground truth do not overlap, the gradient is 0 , and the model cannot be optimized. Given the coordinates of the ground truth are gt, and the calculated predicted box coordinates are **pb**, IoU can be obtained through calculations.

$$
\begin{aligned}
\mathbf{pb} &= \left(x^{p}_{\min}, x^{p}_{\max}, y^{p}_{\min}, y^{p}_{\max}\right), \\
\mathbf{gt} &= \left(x^{g}_{\min}, x^{g}_{\max}, y^{g}_{\min}, y^{g}_{\max}\right),
\end{aligned}
\tag{6}
$$

$$A_p = \left(x^p_{\max} - x^p_{\min}\right) \times \left(y^p_{\max} - y^p_{\min}\right),$$
$$A_g = \left(x^g_{\max} - x^g_{\min}\right) \times \left(y^g_{\max} - y^g_{\min}\right), \tag{7}$$

$$I_{pg} = \begin{cases} \left(x^I_2 - x^I_1\right) \times \left(y^I_2 - y^I_1\right), \text{ if } x^I_2 > x^I_1, y^I_2 > y^I_1 \\ 0, \text{ otherwise} \end{cases},$$
$$U_{pg} = A_p + A_g - I_{pg} \tag{8}$$

$$x^c_{\min} = \min\left(x^p_{\min}, x^g_{\min}\right), x^c_{\max} = \max\left(x^p_{\max}, x^g_{\max}\right),$$
$$y^c_{\min} = \min\left(y^p_{\min}, y^g_{\min}\right), y^c_{\max} = \max\left(y^p_{\max}, y^g_{\max}\right), \tag{9}$$

$$A_c = \left(x^c_{\max} - x^c_{\min}\right) \times \left(y^c_{\max} - y^c_{\min}\right), \tag{10}$$

$$IoU = \frac{I_{pg}}{U_{pg}}, GIoU = IoU - \frac{|A_c - U_{pg}|}{|A_c|} \tag{11}$$

$$L_{GIoU} = 1 - \text{GIoU}, \tag{12}$$

where $x^I_1 = \max\left(x^p_{\min}, x^g_{\min}\right)$, $x^I_2 = \min\left(x^p_{\min}, x^g_{\min}\right)$, $y^I_1 = \max\left(y^p_{\min}, y^g_{\min}\right)$, $y^I_2 = \min\left(y^p_{\min}, y^g_{\min}\right)$, I_{pg} is the intersection of the predicted box and ground truth, and U_{pg} is the union of the predicted box and ground truth.

3. RESULTS AND DISCUSSION

3.1. Implement Details

3.1.1. Dataset

The SAR marine dataset is currently the largest SAR dataset available for multiscale ship detection and was constructed by Wang et al. [39] and labeled by SAR experts. It entails 102 Chinese Gaofen-3 images and 108 Sentinel-1 images. The dataset comprises 43,819 ship chips with a resolution of 256 pixels in both range and azimuth. For Gaofen-3, the image modes included Ultrafine Strip-Map (UFS), Fine Strip-Map 1 (FSI), Full Polarization 1 (QPSI), Full Polarization 2 (QPSII), and Fine Strip-Map 2 (FSII). Furthermore, the resolution of the SAR images in the dataset ranged from 3 m to 10 m. For Sentinel-1, the imaging modes were S3 Strip-Map (SM), S6 SM, and IW-mode. The details of these images, including their resolutions, incidence angles, and polarizations, are summarized in Table 1. Ship objects have distinct scales and backgrounds. Furthermore, some ships were present in complex scenes, which were divided into three categories: offshore, island, and harbor, as shown in Figure 6. SAR ship detection can also occur in scenes with numerous ships, high density, and a small target size (less than 15 pixels $\times 15$ pixels; Figure 6). An overview of the dataset is presented in Table 2. The training, verification, and testing sets constitute 70%, 20%, and 10% of the dataset, respectively.

Table 1. *Detailed dataset information for original SAR imagery.*

Sensor	Polarization	Imaging Mode	Resolution Rg. × Az. (m)
GF-3	Single	UFS	3 × 3
	Dual	FSI	5 × 5
	Full	QPSI	8 × 8
	Dual	FSII	10 × 10
	Full	QPSII	25 × 25
Sentinel-1 SLC	Dual	SM	1.7 × 4.3~ 3.6 × 4.9

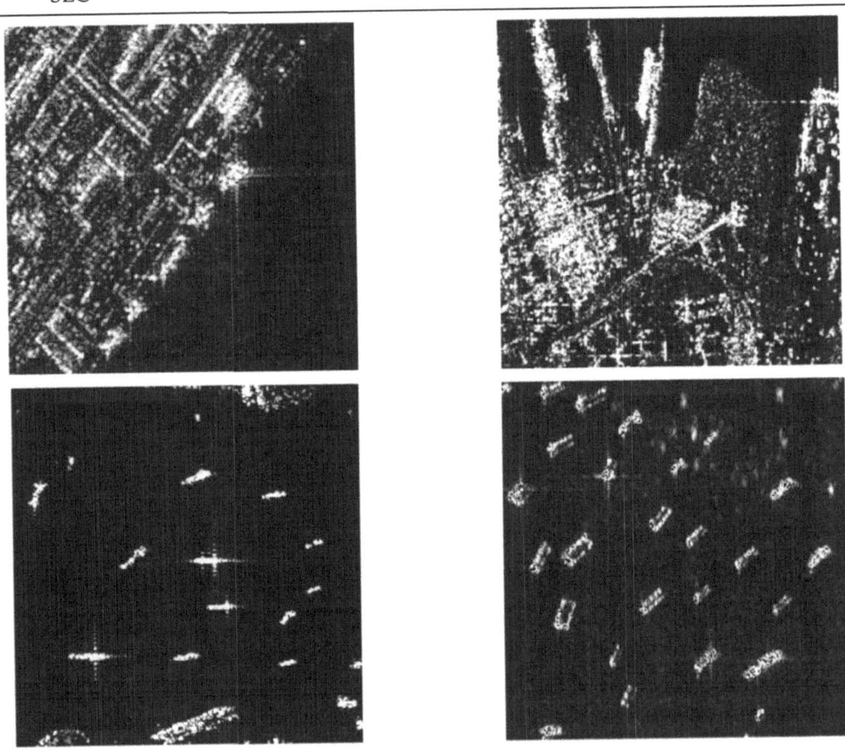

Figure 6. *Complex scenes and high-density small target ships in SAR images.*

Table 2. *Overview of the dataset.*

DataType	Image	Complex Scenes	High-Density Small Target Scenes
Training	21,420	12,840	8580
Verification	6120	3660	2460
Testing	3060	1836	1224

3.1.2. Evaluation Metrics

For detection evaluation, the evaluation indicators corresponding to each algorithm are calculated and categorized into the precision, recall rate, and F1 Score defined as follows:

$$\begin{aligned} \text{Precision} &= N_{TD}/N, \\ \text{Recall} &= N_{TD}/N_{GT}, \\ F1 - \text{score} &= \frac{2 \times \text{Precision} \times \text{Recall}}{\text{Precision} + \text{Recall}}, \end{aligned} \tag{13}$$

where N_{TD} is the number of ship targets detected correctly (true detection, TD), N_{GT} is the actual number of ship targets (ground truth, GT), and N is the total number of ship targets detected. The average precision (AP) with the adaptive IoU threshold was used as a metric based on the precision-recall (PR) curves, and AP is defined as

$$AP = \int_0^1 P(R)dR \tag{14}$$

Different IoU thresholds can be used to calculate different numbers of interest and detect different N_{TD}. Each IoU threshold corresponds to an AP value, and mAP denotes its mean, which assesses the detection effect of the model, where n denotes the number of scene categories. Overall, the mAP and precision-recall curves were employed to evaluate the proposed method.

$$mAP = \frac{\sum_{i=1}^{n} APi}{n}. \tag{15}$$

3.1.3. Implementation Details

The experiment platform was Ubuntu16.0, the GPU was NVIDIA Tesla V100, and the development platform was Paddle X. During the experiment, the empirical learning rate of the semantic model was 0.0001, the batch size was 24, and the dataset was randomly arranged in each iteration.

3.2. Experimental Process

Experiments verified the effects of the improvements and optimizations on various parts of the FPN structure, as shown in Tables 3 and 4. Since the channel attention mechanism enhances the model feature traction capability, the improved model with this mechanism introduced alone improves the mAP by 2.1% in complex scenes and 6.1% in high-density small target scenes over the original model. Similarly, after introducing the deformable convolutional network alone, the mAP improves by 7.7% in complex scenes and 9.5% in high-density small target scenes compared to the original model. This is because the complex background information blurs the ship target location information, and also the localization information of small targets is not obvious after multi-layer convolution, so it is important

Table 3. *Comparison of detection results in complex scenes.*

FPN	Channel Attention	Deformable CNN	Improved Loss Function	SAR Ships in Complex Scenes			
				Precision (%)	Recall (%)	F1 Score	mAP (%)
√				83.4	71.3	0768.	79.4
√	√			85.5	73.2	0.789	81.5
√		√		89.1	78.1	0.833	87.1
√	√	√		91.2	78.1	0.841	87.9
√	√	√	√	91.7	78.1	0.844	87.9

The check mark "√" indicates that the technique was used in training.

Table 4. *Comparison of detection results in high-density small target scenes.*

FPN	Channel Attention	Deformable CNN	Improved Loss Function	High-Density Small Target Scenes			
				Precision (%)	Recall (%)	F1 Score	mAP (%)
√				87.7	75.3	0.810	83.8
√	√			89.9	75.9	0.823	89.9
√		√		95.3	85.4	0.900	93.3
√	√	√		96.2	92.8	0.946	95.1
√	√	√	√	96.5	93.0	0.947	95.1

The check mark "√" indicates that the technique was used in training.

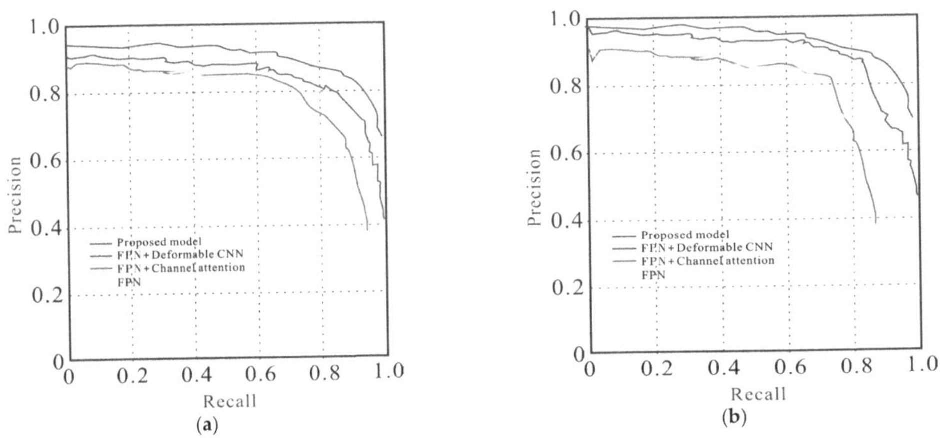

Figure 7. *Precision-recall (PR) curves for SAR ships in (a) complex and (b) high-density small target scenes.*

to enhance the location and feature information using variable convolution. In the final step, the two improvements were combined, and the regression loss function was replaced with a new GIoU loss function. As a result, the mAP increases to 87.9% in complex scenarios and 95.1% in high-density small target scenarios. The effectiveness of the proposed method even in complex scenarios is demonstrated.

The check mark " √ " indicates that the technique was used in training.

The check mark " √ " indicates that the technique was used in training.

The proposed model is compared with FPNs incorporating different mechanisms, following their training protocols, using the reported default parameters and iterations on each benchmark. Figure 7 shows the precision-recall curves of different models on the SAR dataset in different scenarios.

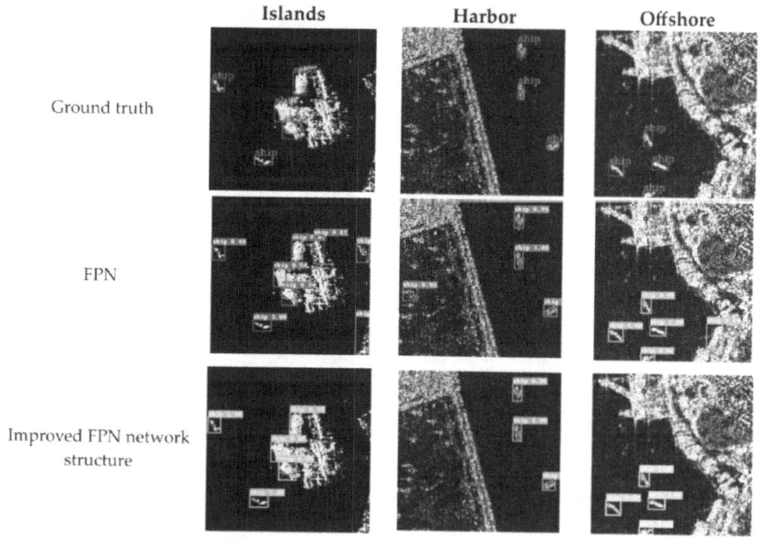

Figure 8. *Qualitative results of FPN and improved FPN on SAR Ships dataset in complex scenes. The first, second, and third rows represent the ground truth, FPN detection results, and improved FPN network structure detection results, respectively. Most of the prediction boxes are matched, and those error results from FPN were detected by our improved method.*

The results in Figures 8 and 9 show that the proposed model can detect multi-scale ship objects in various scenes. As shown in Figure 8, the proposed model can detect the ship when it is located in a complex scene. Moreover, as shown in Figure 9, the model can produce accurate detection results even when multiple dense ship targets are present.

3.3. Experiments on HRSID

Other validation experiments were performed on the public dataset HRSID. The original SAR image used to construct HRSID [40] includes 99 Sentinel-1B images, 36 TerraSAR- *X* images, and 1 TanDEM- *X* image. It is designed for ship detection based on CNN. It is also divided 70:30, respectively, into a training set and a testing set. According to statistics, the total number of ships marked in HRSID is 16,951. The number of small ships accounted for 54.8% of all ships.

FPNs were used to train these models to obtain baseline results. Then, channel attention and deformable CNNs were added to the models, and their original losses were replaced with GIoU in the final bounding box refinement stage. Similar to the above experiments, we gradually improved the models. The final results using the dataset HRSID are presented in Table 5.

The check mark " $\surd$ " indicates that the technique was used in training.

The F1 score and mAP of the proposed method herein are increased by 0.013 and 1.4% as compared with the original object detection model, respectively. The small improvement observed in the experiments demonstrates that the channel attention mechanism and deformable convolution are useful in target-detection

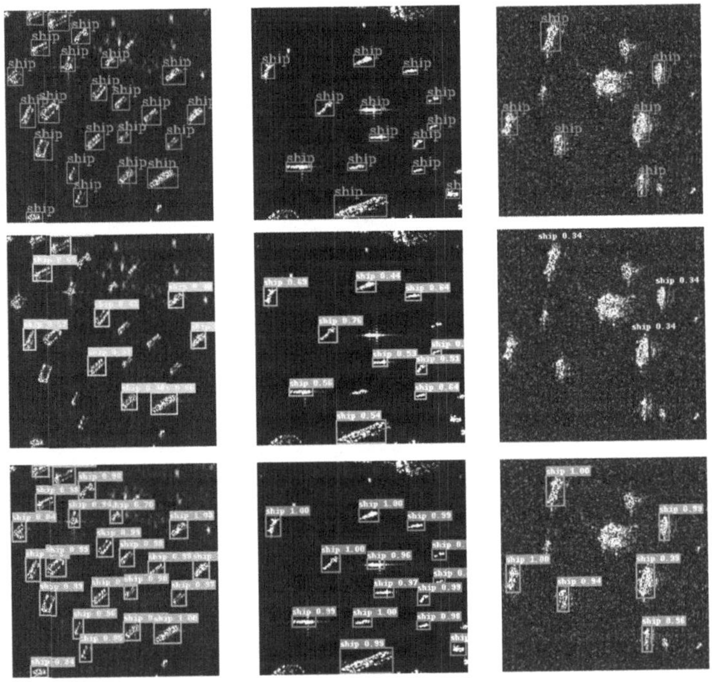

Figure 9. *Qualitative results of FPN and improved FPN on SAR ship datasets in the high-density small target dataset. The first, second, and third rows represent the ground truth, FPN detection results, and improved FPN network structure detection results, respectively. The improved FPN network structure has higher accuracy.*

Table 5. *Comparison among detection results in HRSID.*

FPN	Channel Attention	Deformable CNN	Improved Loss Function	Precision (%)	Recall (%)	F1 score	mAP (%)
√				88.2	92.1	0.901	88.2
√	√			88.7	92.6	0.906	88.7
√		√		89.2	93.0	0.911	89.3
√	√	√		89.3	93.2	0.912	89.3
√	√	√	√	89.6	93.3	0.914	89.6

The check mark "√" indicates that the technique was used in training.

tasks. On the other hand, it also shows that although deformable CNNs can adjust their shape by transforming the convolution kernel according to the actual situation, and channel attention can adjust weights to better extract features from the input image, they perform better in complex scenes or for small target detection.

3.4. Comparison with Other Models

Currently, the model proposed in this paper is compared with several classical and recently developed deep learning models, such as Yolov5 [27], CBAM Faster R-CNN [15], SSD [9], Mask R-CNN [36], and MS-FPN [30]. These models can achieve good detection results in general ship detection tasks. Figure 10 shows the ship recognition results of these models and results with the proposed model

Table 6. *Comparison of mAP from multiple models in complex scenes.*

Model	SAR Ships in Complex Scenes			
	Precision (%)	Recall (%)	F1 Score	mAP (%)
Yolo v5	78.2	76.1	0.771	76.3
CBAM Faster R-CNN	63.3	85.5	0.727	83.1
SSD	83.5	78.6	0.809	79.5
Mask R-CNN	91.7	77.9	0.842	85.8
MS-FPN	89.3	77.7	0.831	87.9
Proposed method	91.7	78.1	0.844	87.9

Table 7. *Comparison of mAP from multiple models in high-density small target scenes.*

Model	High-Density Small Target Scenes			
	Precision (%)	Recall (%)	F1 Score	mAP (%)
Yolo v5	92.8	87.1	0.898	84.4
CBAM Faster R-CNN	91.4	77.7	0.839	89.9
SSD	94.1	71.3	0.822	70.0
Mask R-CNN	95.9	93.8	0.948	91.4
MS-FPN	92.9	93.2	0.930	92.9
Proposed method	96.5	93.0	0.947	95.1

under different scenarios. The aforementioned models were executed on a Tesla V100, and the same training strategy was used to train the SAR ocean dataset with 105 iterations. The batch size was set to 32, and the initial learning rate was set to 0.00001 . Tables 6 and 7 show that the proposed model has the highest detection accuracy. The mAP reaches 87.9% in complex scenes, which is 2.1% higher than the next-best model, Mask R-CNN, and 4.8% higher than the CBAM Faster R-CNN, which also incorporates the attention mechanism This indirectly reflects that the proposed model was more effective in identifying ships under extremely complex scenes. Meanwhile, the mAP was 95.1% in high-density small target scenes, which was 3.7% higher than that of the suboptimal model Mask R-CNN and 5.2% higher than that of the attention mechanism model CBAM Faster R-CNN. This indirectly reflects the improvement of the proposed model in terms of multiscale ship detection performance.

4. CONCLUSIONS

For small target detection in complex scenes, which is prone to errors and missed detection problems, this study proposes a method based on an improved FPN targetdetection model that can significantly improve ship SAR detection performance in different complex scenes and at different scales, especially the small ones. The improved FPN model in this paper consists of the deformable CNN module and the channel attention mechanism module. The deformable CNN was used to replace the original CNN, and the channel attention mechanism was introduced to extract the weights of the feature channels to optimize the expression ability of the network features. The performance of the proposed model was verified on a

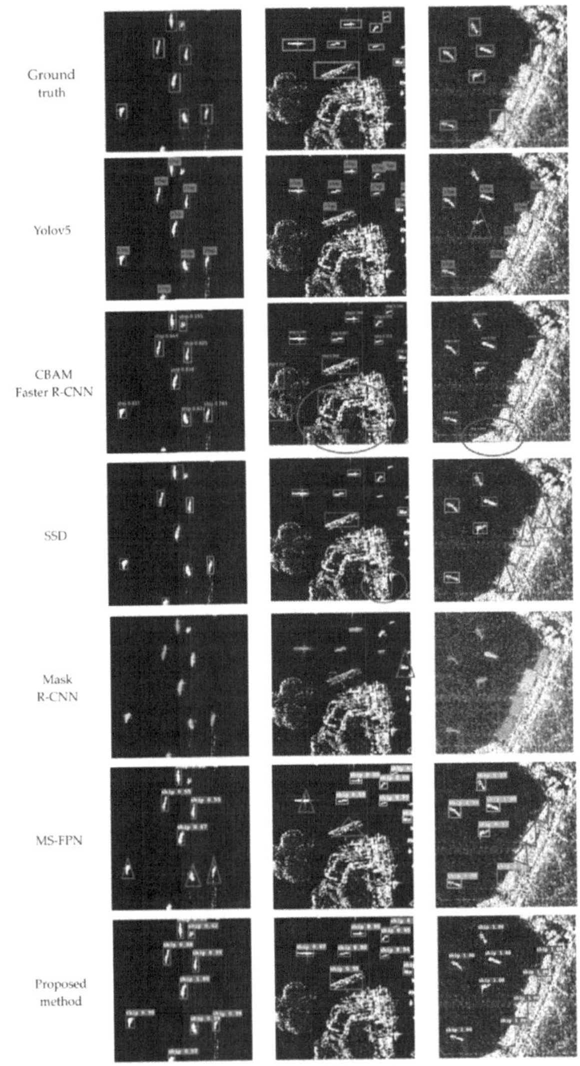

Figure 10. *Comparison of detection results for different models. The triangles indicate imprecise*

SAR ship dataset, showing that the mAP of ship detection via the proposed method reaches 94.7%, which can prove that the proposed method achieves a better performance than other state-of-the-art ship detection methods, such as CBAM faster R-CNN, SSD, YOLOV5, MS-FPN, and their variations.

To determine which module is most effective in improving the model, we compared the deformable CNN module, the channel attention mechanism module, their combination, and the replacement of the loss function. The results indicate that the deformable CNN module has the greatest impact, while combining the two modules further improves detection accuracy. However, we also observed that our proposed model fails to detect some very small ships against an interference background. The low recall rate in our ablation experiment suggests that there are missed detections. Additionally, introducing the GIoU loss function did not significantly improve detection accuracy.

Overall, the proposed model effectively detects ships in complex scenes and small ships in dense scenes. We have demonstrated that using deformable convolution instead of traditional convolution and introducing the channel attention mechanism is a more effective approach. However, it also needs to be noted that, although the proposed method has better detection performance and can effectively reduce false alarms. it cannot completely eliminate all false alarms for complex backgrounds. Further analysis and research are required.

Author Contributions

P.C. conceived and designed the algorithm and contributed to the manuscript and experiments; H.Z. was responsible for the construction of the ship detection dataset, constructed the outline of the manuscript, and made the first draft of the manuscript; Y.L. and B.L. supervised the experiments and were also responsible for the dataset; P.L. performed ship detection using machine learning methods. All authors have read and agreed to the published version of the manuscript.

REFERENCES

1. Xiao, X.; Zhou, Z.; Wang, B.; Li, L.; Miao, L. Ship Detection under Complex Backgrounds Based on Accurate Rotated Anchor Boxes from Paired Semantic Segmentation. Remote Sens. 2019, 11, 2506.
2. Yang, Z.; Yu, X.; Dedman, S.; Rosso, M.; Zhu, J.; Yang, J.; Xia, Y.; Tian, Y.; Zhang, G.; Wang, J. UAV Remote Sensing Applications in Marine Monitoring: Knowledge Visualization and Review. Sci. Total Environ. 2022, 838, 155939.
3. Li, J.; Qu, C.; Shao, J. Ship Detection in SAR Images Based on an Improved Faster R-CNN. In Proceedings of the 2017 SAR in Big Data Era: Models, Methods and Applications (BIGSARDATA), Beijing, China, 13-14 November 2017; pp. 1-6.
4. Fu, Q.; Luo, K.; Song, Y.; Zhang, M.; Zhang, S.; Zhan, J.; Duan, J.; Li, Y. Study of Sea Fog Environment Polarization Transmission Characteristics. Appl. Sci. 2022, 12, 8892.
5. Zhao, C.; Cheung, C.F.; Xu, P. High-Efficiency Sub-Microscale Uncertainty Measurement Method Using Pattern Recognition. ISA Trans. 2020, 101, 503-514.
6. An, Q.; Pan, Z.; You, H. Ship Detection in Gaofen-3 SAR Images Based on Sea Clutter Distribution Analysis and Deep Convolutional Neural Network. Sensors 2018, 18,
7. 7. Kang, M.; Ji, K.; Leng, X.; Lin, Z. Contextual Region-Based Convolutional Neural Network with Multilayer Fusion for SAR Ship Detection. Remote Sens. 2017, 9, 860.
8. Liu, W.; Wang, Z.; Liu, X.; Zeng, N.; Liu, Y.; Alsaadi, F.E. A Survey of Deep Neural Network Architectures and Their Applications. Neurocomputing 2017, 234, 11-26.
9. Zhang, J.; Lin, S.; Ding, L.; Bruzzone, L. Multi-Scale Context Aggregation for Semantic Segmentation of Remote Sensing Images Remote Sens. 2020, 12, 701.

10. Rostami, M.; Kolouri, S.; Eaton, E.; Kim, K. Deep Transfer Learning for Few-Shot SAR Image Classification. Remote Sens. 2019, 11, 1374.
11. Liu, L.; Chen, G.; Pan, Z.; Lei, B.; An, Q. Inshore Ship Detection in SAR Images Based on Deep Neural Networks. In Proceedings of the IGARSS 2018-2018 IEEE International Geoscience and Remote Sensing Symposium, Valencia, Spain, 22-27 July 2018; pp. 25 – 28.
12. Li, R.; Wang, X.; Wang, J.; Song, Y.; Lei, L. SAR Target Recognition Based on Efficient Fully Convolutional Attention Block CNN. IEEE Geosci. Remote Sens. Lett. 2020, 19, 21510590.
13. Zhou, G.; Yang, F.; Xiao, J. Study on Pixel Entanglement Theory for Imagery Classification. IEEE Trans. Geosci. Remote Sens. 2022, 60, 21706089.
14. Wang, Y.; Wang, C.; Zhang, H. Combining a Single Shot Multibox Detector with Transfer Learning for Ship Detection Using Sentinel-1 SAR Images. Remote Sens. Lett. 2018, 9, 780-788.
15. Kang, M.; Leng, X.; Lin, Z.; Ji, K. A Modified Faster R-CNN Based on CFAR Algorithm for SAR Ship Detection. In Proceedings of the 2017 International Workshop on Remote Sensing with Intelligent Processing (RSIP), Shanghai, China, 18-21 May 2017; pp. 1 – 4.
16. Jiao, J.; Zhang, Y.; Sun, H.; Yang, X.; Gao, X.; Hong, W.; Fu, K.; Sun, X. A Densely Connected End-to-End Neural Network for Multiscale and Multiscene SAR Ship Detection. IEEE Access 2018, 6, 20881-20892.
17. Cui, Z.; Wang, X.; Liu, N.; Cao, Z.; Yang, J. Ship Detection in Large-Scale SAR Images via Spatial Shuffle-Group Enhance Attention IEEE Trans. Geosci. Remote Sens. 2020, 59, 379-391.
18. Villano, M.; Krieger, G.; Steinbrecher, U.; Moreira, A. Simultaneous Single-/Dual-and Quad-Pol SAR Imaging over Swaths of Different Widths. IEEE Trans. Geosci. Remote Sens. 2019, 58, 2096-2103.
19. Wang, R.; Xu, F.; Pei, J.; Wang, C.; Huang, Y.; Yang, J.; Wu, J. An Improved Faster R-CNN Based on MSER Decision Criterion for SAR Image Ship Detection in Harbor. In Proceedings of the IGARSS 2019-2019 IEEE International Geoscience and Remote Sensing Symposium, Yokohama, Japan, 28 July-2 August 2019; pp. 1322-1325.
20. Hu, Q.; Hu, S.; Liu, S. BANet: A Balance Attention Network for Anchor-Free Ship Detection in SAR Images. IEEE Trans. Geosci. Remote Sens. 2022, 60, 21603357.
21. Hu, Q.; Hu, S.; Liu, S.; Xu, S.; Zhang, Y.-D. FINet: A Feature Interaction Network for SAR Ship Object-Level and Pixel-Level Detection. IEEE Trans. Geosci. Remote Sens. 2022, 60, 1-15.
22. Wu, T.; Li, B.; Luo, Y.; Wang, Y.; Xiao, C.; Liu, T.; Yang, J.; An, W.; Guo, Y. MTU-Net: Multilevel TransUNet for Space-Based Infrared Tiny Ship Detection. IEEE Trans. Geosci. Remote Sens. 2023, 61, 22538002.
23. Zhang, F.; Wang, X.; Zhou, S.; Wang, Y.; Hou, Y. Arbitrary-Oriented Ship Detection through Center-Head Point Extraction. IEEE Trans. Geosci. Remote Sens. 2021, 60, 21590405.
24. He, C.; Tu, M.; Xiong, D.; Tu, F.; Liao, M. Adaptive Component Selection-Based Discriminative Model for Object Detection in High-Resolution SAR Imagery. ISPRS Int. J. Geo-Inf. 2018, 7, 72.

25. Yang, M.; Wang, H.; Hu, K.; Yin, G.; Wei, Z. IA-Net: An Inception-Attention-Module-Based Network for Classifying Underwater Images From Others. IEEE J. Ocean. Eng. 2022, 47, 704-717.

26. Zhou, G.; Song, B.; Liang, P.; Xu, J.; Yue, T. Voids Filling of DEM with Multiattention Generative Adversarial Network Model Remote Sens. 2022, 14, 1206.

27. Ting, L.; Baijun, Z.; Yongsheng, Z.; Shun, Y. Ship Detection Algorithm Based on Improved YOLO V5. In Proceedings of the 2021 6th International Conference on Automation, Control and Robotics Engineering (CACRE), Dalian, China, 15-17 July 2021; pp. 483 – 487.

28. Li, L.; Zhou, Z.; Wang, B.; Miao, L.; An, Z.; Xiao, X. Domain Adaptive Ship Detection in Optical Remote Sensing Images. Remote Sens. 2021, 13, 3168.

29. Rezatofighi, H.; Tsoi, N.; Gwak, J.; Sadeghian, A.; Reid, I.; Savarese, S. Generalized Intersection over Union: A Metric and a Loss for Bounding Box Regression. In Proceedings of the IEEE/CVF Conference on Computer Vision and Pattern Recognition, Long Beach, CA, USA, 15-20 June 2019; pp. 658-666.

30. Sun, Z.; Meng, C.; Cheng, J.; Zhang, Z.; Chang, S. A Multi-Scale Feature Pyramid Network for Detection and Instance Segmentation of Marine Ships in SAR Images. Remote Sens. 2022, 14, 6312.

31. Lin, Z.; Ji, K.; Leng, X.; Kuang, G. Squeeze and Excitation Rank Faster R-CNN for Ship Detection in SAR Images. IEEE Geosci. Remote Sens. Lett. 2018, 16, 751-755.

32. Li, J.; Xu, C.; Su, H.; Gao, L.; Wang, T. Deep Learning for SAR Ship Detection: Past, Present and Future. Remote Sens. 2022, 14, 2712.

33. Zhao, Y.; Zhao, L.; Xiong, B.; Kuang, G. Attention Receptive Pyramid Network for Ship Detection in SAR Images. IEEE J. Sel. Top. Appl. Earth Obs. Remote Sens. 2020, 13, 2738-2756.

34. Liu, N.; Cui, Z.; Cao, Z.; Pi, Y.; Lan, H. Scale-Transferrable Pyramid Network for Multi-Scale Ship Detection in SAR Images. In Proceedings of the IGARSS 2019-2019 IEEE International Geoscience and Remote Sensing Symposium, Yokohama, Japan, 28 July-2 August 2019; pp. 1-4.

35. Li, X.; Li, D.; Liu, H.; Wan, J.; Chen, Z.; Liu, Q. A-BFPN: An Attention-Guided Balanced Feature Pyramid Network for SAR Ship Detection. Remote Sens. 2022, 14, 3829.

36. Nie, X.; Duan, M.; Ding, H.; Hu, B.; Wong, E.K. Attention Mask R-CNN for Ship Detection and Segmentation from Remote Sensing Images. IEEE Access 2020, 8, 9325-9334.

37. Lin, T.-Y.; Dollár, P.; Girshick, R.; He, K.; Hariharan, B.; Belongie, S. Feature Pyramid Networks for Object Detection. In Proceedings of the IEEE Conference on Computer Vision and Pattern Recognition, Honolulu, HI, USA, 21-26 July 2017; pp. 2117-2125.

38. Dai, J.; Qi, H.; Xiong, Y.; Li, Y.; Zhang, G.; Hu, H.; Wei, Y. Deformable Convolutional Networks. In Proceedings of the IEEE International Conference on Computer Vision, Venice, Italy, 22-29 October 2017; pp. 764-773.

39. Wang, Y.; Wang, C.; Zhang, H.; Dong, Y.; Wei, S. A SAR Dataset of Ship Detection for Deep Learning under Complex Backgrounds Remote Sens. 2019, 11, 765.

40. Wei, S.; Zeng, X.; Qu, Q.; Wang, M.; Su, H.; Shi, J. HRSID: A High-Resolution SAR Images Dataset for Ship Detection and Instance Segmentation. IEEE Access 2020, 8, 120234-120254.

CHAPTER 10

Building a Practical Multi-Sensor Platform for Monitoring Vessel Activity near Marine Protected Areas: Case Studies from Urban and Remote Locations

Samantha Cope[1], Brendan Tougher[1], Virgil Zetterlind[1], Lisa Gilfillan[2] and Andres Aldana[3]

[1] *ProtectedSeas, Anthropocene Institute, Palo Alto, CA 94306, USA*
[2] *WILDCOAST, Del Mar, CA 92014, USA*
[3] *Turneffe Atoll Sustainability Association, 62 Bella Vista, Belize City, Belize*

ABSTRACT

Monitoring vessel activity is an important part of managing marine protected areas (MPAs), but small-scale fishing and recreational vessels that do not participate in cooperative vessel traffic systems require additional monitoring strategies. Marine Monitor (M2) is a shore-based, multi-sensor platform that integrates commercially available hardware, primarily X-band marine radar and optical cameras, with custom software to autonomously track and report on vessel activity regardless of participation in other tracking systems. By utilizing established commercial hardware, the radar system is appropriate for supporting the management of coastal, small-scale MPAs. Data collected in the field are transferred to the cloud to provide a continuous record of activity and identify prohibited activities in real-time using behavior characteristics. To support the needs of MPA managers, both hardware and software improvements have been made over time, including ruggedizing equipment for the marine environment and powering systems in remote locations. Case studies are presented comparing data collection by both radar and the Automatic Identification System (AIS) in urban and remote locations. At the South La Jolla State Marine Reserve near San Diego, CA, USA, 93% of vessel activity (defined as the cumulative time vessels spent in the MPA) was identified exclusively by radar from November 2022 through January 2023. At the Caye Bokel Conservation Area, within the Turneffe Atoll Marine Reserve offshore of Belize, 98% was identified exclusively by radar from April through October 2022. Spatial and temporal patterns of radar-detected and AIS activity also differed at both sites. These case study site results together demonstrate the common and persistent presence of small-scale vessel activity near coastal MPAs that is not documented by cooperative systems. Therefore, an integrated radar system can be a useful tool for independent monitoring, supporting a comprehensive understanding of vessel activity in a variety of areas.

Keywords: *marine radar; Automatic Identification System (AIS); sensor integration; vessel monitoring; marine protected areas (MPAs); small-scale fishing*

1. INTRODUCTION

Recently, there has been a global focus on implementing marine protected areas (MPAs) to achieve large-scale conservation targets, such as protecting 30% of the world's marine area by 2030 [1]. These spatial goals require monitoring vast areas of the ocean [2,3], which can be supported by satellite-borne remote sensing (e.g., [4,5]). But small-scale, locally managed MPAs also require special attention due to their role in the economies of coastal communities [6]. These MPAs can enhance fish biomass [7] and support local food security [8], but limited staffing and budget capacity for monitoring are pervasive issues [9, 10]. Satellite-based monitoring solutions may not be appropriate for these areas due to data processing requirements and cost, and, at the time of writing, further development is needed to improve detection of small vessels amongst waves and other features [11-13]. Therefore, remote sensing solutions that are scaled for targeted use at the local level may be more practical.

The purpose of this paper is to describe and evaluate a coastal radar monitoring and reporting system that is appropriate for documenting small-scale vessel activity in nearshore areas with a specific focus on monitoring vessels that are found in the vicinity of coastal MPAs. The need for effective and appropriate technologies that support management of small-scale fisheries in developing countries and elsewhere is specifically addressed by international goals, including Sustainable Development Goals 14.a and 14.b [14]. The commercial availability of radar, along with its established and widespread use in the maritime domain, make it an appropriate tool for managers who may have limited access to resources. The system described in this paper is the Marine Monitor (M2) [15], a multisensor platform that uses X-band marine radar and optical imagery to detect and document vessel activity from shore. While these are common, well-established tools used in maritime surveillance, this paper presents a system specifically designed for conservation-based monitoring and reporting.

Cooperative vessel traffic systems, such as the Automatic Identification System (AIS), have been widely used to inform conservation-based assessments of human impacts on protected areas [16] , including fishing activities [5,17-20] and recreation and tourism [21-23]. The global availability of AIS, specifically, facilitates monitoring commercial fishing activity at a large scale (e.g., [24,25]), but participation is not universally mandated [16], not well enforced [26], and not commonly used in small-scale fishing [27]. Previous research in particular regions has shown that many vessels, even in highly populated areas, are not broadcasting AIS data [28-30] . Therefore, non-cooperative systems are likely better suited for monitoring MPAs where small-scale vessel activity is of interest. To evaluate the benefit of using coastal radar in these areas, data collected via radar are compared to data provided by AIS (integrated by the M2 system) at two case study sites: one in a highly developed urban area and one in a more remote location. These contrasting locations provide an opportunity to incorporate different types of small-scale vessel activity.

This paper is structured as follows: Section 1.1 summarizes related work using remote sensing systems to monitor non-cooperative vessel activity as it relates to MPAs. Section 1.2 includes an overview of the chosen system, including the technical capabilities of coastal radar that make it an appropriate choice for targeted monitoring of coastal MPAs Sections 2.1-2.6 provide details on the methods used to collect data. Section 2.7 describes the two case study sites and the methods used for analysis. Results from the case study sites are presented in Section 3, and Section 4 provides discussion of the case study results and overall challenges and the importance of monitoring small-scale vessel activity. A brief conclusion is provided in Section 5.

1.1. Related Work

Satellite-based remote sensing solutions are rapidly progressing to support maritime surveillance with a clear advantage being the ability to survey large swaths of the ocean [31]. Synthetic aperture radar (SAR) and optical imagery from satellites facilitate a multitude of image-processing techniques for detecting and classifying vessels in the images [32-34], including identification of fishing vessels [12] and fishing activity [35]. Lights used by some fisheries to attract target species can also be detected using the visible infrared imaging radiometer suite (VIIRS) and used to map the spatial and temporal extent of activity [35,36] and identify discrete vessel positions [37,38]. However, the temporal lag of vessel detections across subsequent images complicates route prediction and enforcement response on the water. For example, ref. [4] describes a patrol vessel in the Ascension Island MPA that was unable to intercept an SAR-detected vessel before it left the area. The issue of response time is especially relevant for small-scale MPAs when vessels can more quickly leave the boundaries. Higher resolution imagery may be expensive for MPA management applications [4], although development of more affordable systems, such as micro- and nano-satellites, is ongoing [39].

Using optical cameras and video feeds to monitor marine areas is a well-established practice for port surveillance and security [40]. The affordability of digital cameras and closed-circuit television systems has also made these tools useful for conservation-based monitoring in targeted, nearshore areas [41,42]. Stationary digital cameras that are programmed to capture images on a schedule have been used to document the presence of recreational fishing vessels in coastal areas of interest, such as MPAs and artificial reefs [43-46]. A common strategy is to deploy cameras near defined access points to a fishery, such as a boat ramp or areas where vessels must pass to reach their destination [42, 47, 48], and apply an automated vessel detection method on the images or video [49, 50]. These methods can facilitate real-time enforcement response when data are streaming and provide a robust record of vessel counts [42]. However, manual review of the data can be required to identify fishing effort or other specific activities [45,51], so cameras are often utilized as a secondary source of information [42, 49].

While the range of coastal radar is restricted compared to satellite systems, it is a traditional component in vessel traffic service systems for monitoring non-cooperating vessels in highly trafficked areas [52]. There are many examples of coastal radar as a component of a common operating picture that considers multiple inputs (e.g., [53-55]). Persistent target tracking, facilitated by the Automatic Radar Plotting Aid (ARPA), provides a continuous record of a unique vessel's trajectory, which is difficult to reconstruct from infrequent satellite images [56]. Some satellites can be tasked to a given area (albeit with limitations [57]) or revisit within minutes, but most have a longer revisit time [58,59]. Radar and AIS positions are both typically reported of the order of seconds, so association between the two is a common practice and helpful for providing identification information on radar-detected targets [60,61]. Conversely, radar provides situational awareness of targets that do not broadcast AIS data, such as smaller vessels [62]. Radar tracking is also unaffected by cloud cover and foggy conditions that can inhibit vessel detection in other types of imagery.

It is also important to mention remotely or autonomously controlled aerial, surface, or underwater vehicles. These systems typically patrol marine areas and capture images or other data that can be used to detect and identify vessels. At the time of writing, most of these systems are best suited for use in small, targeted areas [63] and provide complementary information to more traditional monitoring techniques [64]. However, with ongoing progress in edge computing and computer vision models, these vehicles can be effective tools for collecting data on vessel activity in MPAs [65].

1.2. System Overview

Coastal radar is well-suited for monitoring small-scale MPAs for three primary reasons: (1) radar systems can provide persistent monitoring that supports both real-time response and comprehensive analysis, (2) radar is a widely used and available technology, and (3) the range of vessel tracking is at a similar scale to smaller MPAs. The M2 system autonomously documents vessel activity by pairing vessel positions over time, provided by the radar system, with imagery of the vessel collected by a camera. In this way, spatio-temporal and identifying information on non-cooperative vessels is presented similarly to cooperative systems, such as AIS. The automated process for directing the camera to vessel locations expands on similar previous efforts (detailed in Section 2.4), thus generating high quality images so that MPA managers can effectively evaluate activity and, thus, compliance.

Figure 1. *Mobile Marine Monitor (M3) deployed in California, USA.*

The primary hardware components of M2 are the radar system, optical camera, and PC computer for processing, storing, and viewing data. Early deployments integrated these components with existing field stations that had a power source and internet connection.

Subsequent deployments required the development of an off-grid power system when MPAs were in remote areas or there was not an adaptable infrastructure in the area. The Mobile Marine Monitor (M3) integrates hardware components in a ruggedized, towable trailer containing a solar array and battery system (Figure 1). A microwave antenna also provides internet access onsite and the ability to stream remote data to the cloud. This provides a connected base of operations in the field and also the ability to relocate the system in response to changing monitoring needs. Finally, additional sensors, such as weather stations, provide in situ data, which can be helpful for interpreting vessel data and providing overall environmental monitoring at a remote location.

There have been software adaptations made to ensure users can effectively interact with the data collected. To maximize successful capture of tracking data and images, data are stored locally and on the cloud via independent streams when connected to the internet. This ensures that information is still captured should there be a transient local or cloud software or data failure or when an internet connection is intermittent. Additionally, a fully offline mode can be used for very

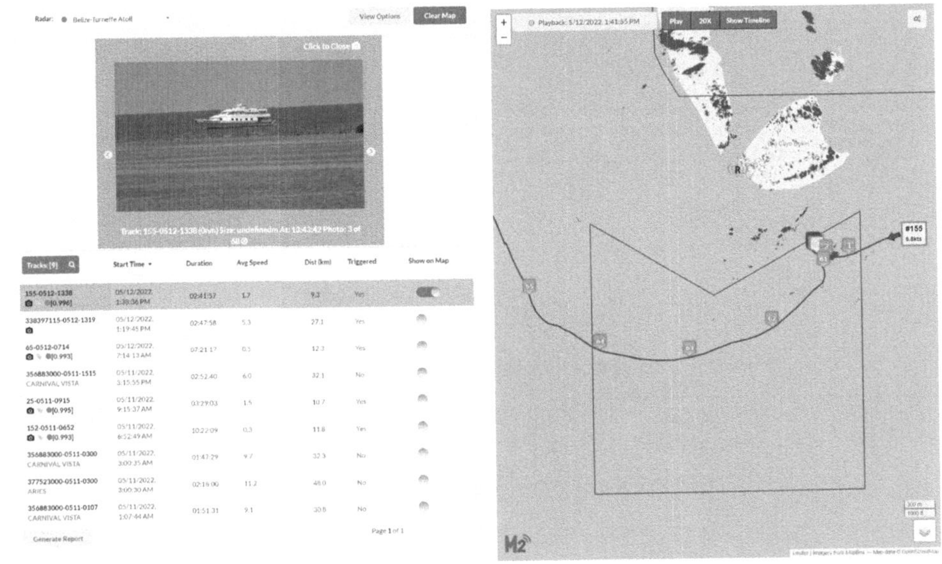

Figure 2. *Data viewer showing track list and associated attributes (left) and selected track displayed on map (right), along with radar imagery and area of interest in red. Green icons on the map indicate vessel locations when an image was captured. The yellow icon indicates where the vessel was when the image on the left was captured.*

remote areas where internet or other connections are not currently practical. In this mode, data are stored locally, including images, but can be exported to portable media (e.g., USB drive) and uploaded to the cloud from any internet-connected computer. In addition to accessing individual data streams from system components, users can dynamically engage all local and cloud-based data using an interactive browser-based interface (Figure 2).

2. MATERIALS AND METHODS

The primary system sensors and processes are shown in Figure 3. In addition to the radar and camera components, the system includes a very-high-frequency (VHF) antenna and AIS receiver to collect messages from AIS broadcasts in the area. Comparing known AIS records with potential duplicate records from another sensor can be helpful for differentiating between cooperating and non-cooperating vessels (e.g., [66,67]). Figure 3 highlights the similarity of radar and AIS data and processing.

2.1. Radar Tracking

The system utilizes a Furuno solid-state Doppler X-band radar paired with Nobeltec TimeZero Coastal Monitoring (TZCM) navigation software [68]. The availability of these components on the retail market helps minimize cost compared to custom-built systems and ensures access to adequate documentation and technical support. The detection range of the radar depends on the antenna height

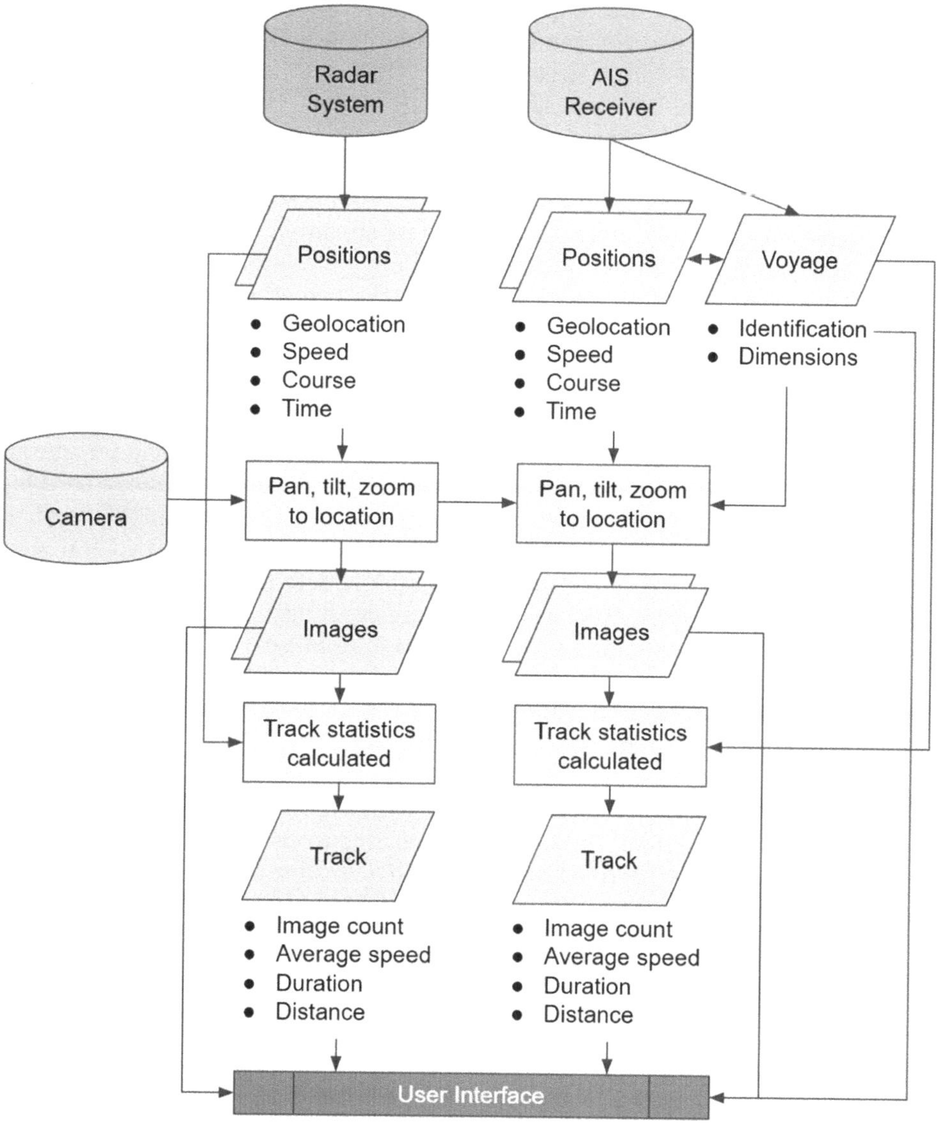

Figure 3. *Conceptual diagram of primary system sensors and processes, including the radar system, Automatic Identification System (AIS) receiver, and camera.*

above sea level according to

$$d_r = 1.23\sqrt{h} \quad (1)$$

where the antenna height (h) is in feet, and the distance to the radar horizon (d_r) is in nautical miles [69]. With system antennas typically installed at a minimum height of 16.4 feet (5 m), targets are generally detected within 5 nautical miles

(9.26 km) of the system, although the detection range extends further depending on vessel height, construction, and surface conditions [70].

Candidate vessels are detected and tracked autonomously by the ARPA-enabled radar system. With a radar sweep rate of 24-30 Hz, TZCM updates target positions from the radar imagery roughly once per second using a proprietary motion model based on the Kalman filter [71], a common algorithm for target tracking [61]. These estimated target positions, along with course and speed, for each vessel are provided over a network data stream and are hereafter referred to as detections. Global identification numbers are assigned to targets, thereby linking consecutive detections of unique targets, hereafter referred to as tracks. The Software calculates numerous track-level metrics across detections, such as average speed, and manages track life based on track persistence over time (dropping tracks if targets become latent or spatially move outside of the expected range of the system).

Additional data beyond detections and tracks are collected via an Internet Protocol (IP) data connection to the radar unit using Furuno's Software Development Kit (SDK) [72]. Radar images are captured every 6 s (as individual .png files) and stored for georeferencing and animation with track data. For each deployment location, the appropriate radar resolution is determined for optimal target detection and tracking. TZCM has dual-range capability, where users set a minimum and maximum resolution for short and long pulses, respectively [71]. This facilitates target tracking at both nearshore and more distant areas. The status of radar and network functioning are also logged over time to monitor overall system uptime and alert system managers to disruptions.

2.2. AIS Data and Track Association

Messages transmitted via AIS report instantaneous attributes, such as location, course, and speed, derived by onboard GPS sensors, providing similar data as those detections provided by the radar system. Static voyage data, such as vessel name, dimensions, and type, are also transmitted intermittently, although voyage data from smaller vessels using less powerful AIS transponders (Class B) may not be available or be received more infrequently [16,27]. Consecutive AIS detections from unique vessels are linked with voyage information (if available) to create AIS tracks. The similar structure of data collected by radar and provided by AIS facilitate integration of the two data streams (Figure 3), and the same track-level statistics are calculated for both.

Since the radar system tracks all vessels within its range, some radar tracks duplicate AIS tracks. Track association links coincident radar and AIS tracks according to criteria recommended by [61]: detections are associated when they have similar position, movement vector, and history. Using this guidance, radar and AIS detections are associated when they meet all of the following requirements (illustrated in Figure 4):

- detections are within 100 m in geolocation,
- difference in speed less than 1.5 knots (0.77 m/s),

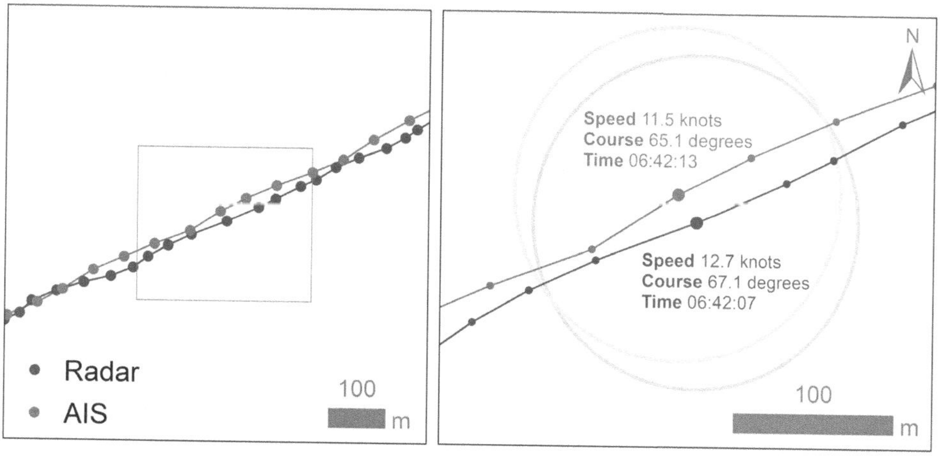

Figure 4. *Example of associated radar and AIS detections that meet the defined criteria based on similar position (shown with* 100 m *buffer around points), movement vector (speed and course), and history (time).*

- difference in course less than 10 degrees, and
- detected within 15 s.

2.3. Track Analysis

Using the descriptive track-level statistics, additional analysis helps highlight meaningful tracks. When systems are deployed, local site partners identify areas of interest where vessel activity is of concern, often MPA boundaries. Both radar and AIS detections are flagged if they are spatially within these areas, and M2 reports the total time a vessel spent within an area as a track-level statistic.

Site partners also identify vessel behaviors of interest that typically correspond to regulated activities within the area. For example, MPA managers may be interested in vessels traveling at very low speeds or idling in an area where fishing activities are prohibited as this behavior could be indicative of fishing [73,74]. Based on local knowledge, site partners identify speeds of interest and the time spent within the area that would suggest a prohibited activity may be occurring. Using these definitions, the system identifies radar and AIS tracks that meet the criteria, classifies them as alarm tracks, and sends notifications to subscribed users.

Occasionally the radar system can erroneously identify sea clutter and weather events as targets [70]. Within TZCM, exclusion areas are defined to minimize these false targets, but they can still occur in fluctuating environmental conditions. A model developed by machine learning, a technique helpful for analyzing vessel tracks [74,75], uses track statistics to calculate a confidence score between 0 and 1, with higher scores indicating those tracks more likely to be vessels. See [76] for a full description of the process. The confidence score is stored as a track statistic and can be used to classify and isolate only those radar tracks predicted to be true vessels. Tracks are typically classified with greater than 90% accuracy [76,77].

2.4. Camera Integration

Systems are configured with one or more pan, tilt, zoom (PTZ) IP cameras (Figure 5). While a stationary camera can be an affordable option for monitoring a fixed area (summarized in Section 1.1), dynamically directing a camera to vessel locations can provide a magnified image for identifying specific vessels and their activities. To capture the target vessel at the center of the frame, different techniques have been proposed, including directing the camera to radar- and AIS-reported locations [78,79] and object detection in live video [80,81]. The M2 system uses the former approach to simplify processing demands on the local PC and provides a number of configurable parameters so that the system can be easily adapted to different locations and vessel types (detailed in Section 2.5). By utilizing the Open Network Video Interface Forum (ONVIF) and Real Time Streaming Protocol (RTSP) for PTZ control and image capture, the system is compatible with most surveillance cameras [82]. This makes deployment especially cost effective when MPA managers have access to existing equipment.

Electro-optical (EO) cameras are commonly integrated with the system to capture images during the day using daily sunrise and sunset times at the system location to trigger camera operation. These cameras can also be configured to function at night if vessel lights are of interest as they can indicate certain fishing activities [67,83]. While infra-red (IR) cameras can capture higher quality images at night, they are currently costprohibitive for most locations when there is minimal non-AIS activity at night. When appropriate, dual cameras are supported, including those with a common pan/tilt head and co-boresight sensors, such as an EO/IR combination, and other sensor types, such as hyperspectral cameras.

2.5. Camera Calibration

Before the camera can autonomously pan, tilt, and zoom to vessel locations, sitespecific parameters must be set for each of these variables. The offset angle between true north and the camera's zero angle resting position must be obtained so that, when the camera is directed to target locations, it pans to the correct true north bearing. Reassigning the camera's zero angle resting position to true north is accomplished using built-in camera functionality. Then, an appropriate tilt angle can be calculated according to

$$\theta_{\text{tilt}} = (-1) \times \tan^{-1}\left(\frac{h}{d}\right) \tag{2}$$

where h is the camera's height above the water and d is the distance to the vessel (Figure 6). Additional tilt adjustments can be defined in bearing angle ranges (i.e., pie slices) to compensate for a camera mount that is not absolutely plum, common in rugged environments [80].

For determining appropriate zoom levels, calibration is performed for each unique camera model to estimate intrinsic lens parameters as a function of zoom percentage. First, images of a target board of known width are captured from a distance of 5 to 100 m at 5 -m increments and 5% zoom steps. For each distance

Figure 5. *Camera mounted with radar antenna on a telescoping mast.*

and zoom, the width of the target in pixels is determined from the captured image. Then, degrees per pixel $\left(\theta_{\text{pixel}}\right)$ can be calculated at each step according to

$$\theta_{\text{pixel}} = \tan^{-1}\left(w_{\text{measured}} \times \frac{\text{distance}}{2}\right) \times w_{\text{pixel}} \tag{3}$$

where w_{measured} is the measured width of the target board, and w_{pixel} is the width of the target in pixels. Next, the degrees of the field of view (θ_{FOV}) are calculated according to

$$\theta_{\text{FOV}} = \theta_{\text{pixel}} \times w_{\text{pixel}} \tag{4}$$

where w_{pixel} is the width of camera images in pixels. The field of view at each step is used to estimate the focal length (f) and ultimately an associated theoretical

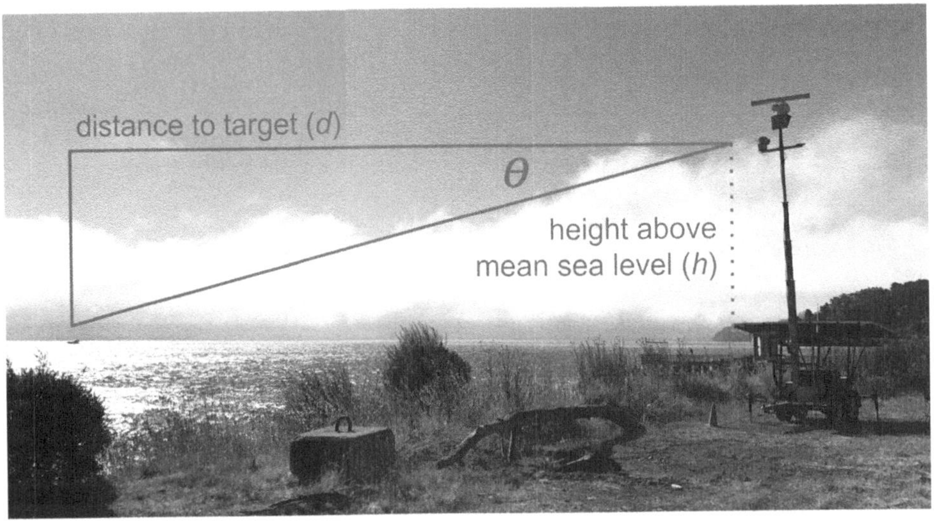

Figure 6. *Mathematical approach for calculating the appropriate camera tilt angle given the height of the camera above mean sea level at the site* (h) *and varying vessel distances* (d).

zoom level.

$$f = \text{ sensor size } \times \frac{\text{distance}}{\theta_{FOV}} \tag{5}$$

(6)

$$\text{zoom } (\%) = \frac{f_{\min} - f}{f_{\min} - f_{\max}} \times 100 \tag{7}$$

Finally, an adjustment factor is experimentally applied to (5) to fit the resulting theoretical zoom levels to the measured levels, if necessary, so that the ideal field of view is captured (Figure 7). Manufacturer specifications typically provide the minimum and maximum focal lengths, and these generally align with estimated values (Figure 7).

In addition to distance, the zoom calculation also considers available target information when determining image width and, thus, zoom level. For AIS targets, the base image width is set based on reported vessel length, typically included in voyage data [16]. For vessels without a reported length (and all radar targets), the software alternates between minimum and maximum vessel lengths that are set based on typical traffic. Site-specific AIS limits ensure that barges are adequately captured in images of smaller tow and push boats and that erroneous voyage data do not negatively impact vessel capture in photos. Image width is also adjusted based on the target's orientation relative to the camera,

$$\theta_A = \left| \theta_{\text{course}} - \theta_{\text{bearing}} \right|, \tag{8}$$

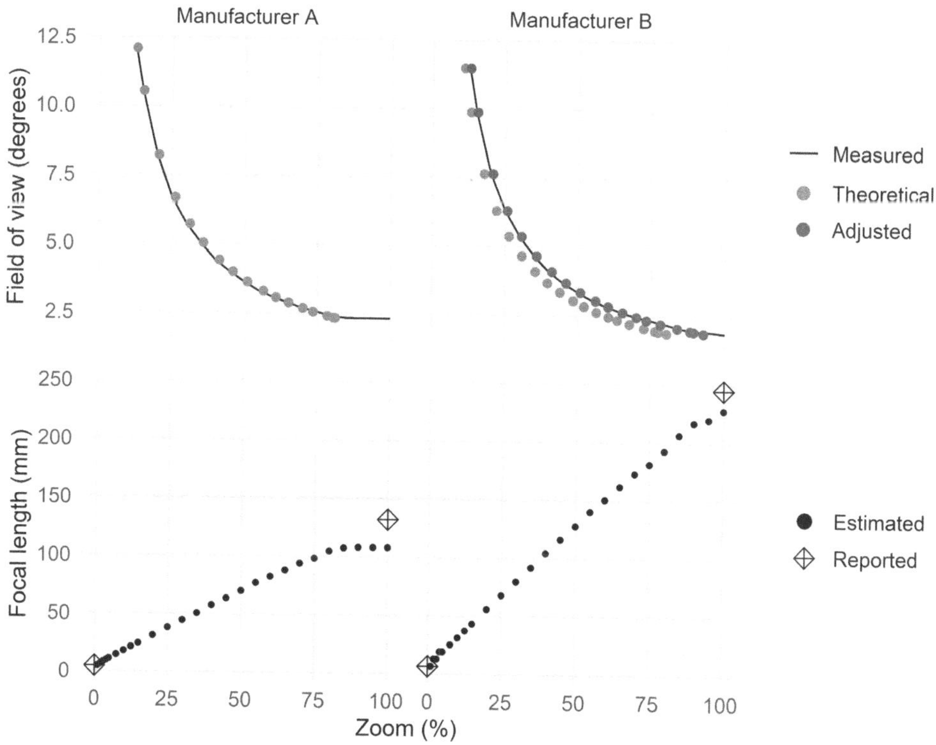

Figure 7. *Camera parameters at increasing zoom levels. Theoretical field of view values have been adjusted to best fit measured values, if necessary (top). Estimated focal lengths are compared to manufacturer-specified values (bottom).*

where θ_A is the aspect angle of the vessel based on its course and true bearing angle in relation to the camera (e.g., smaller widths are used when the vessel's bow or stern is pointed at the camera, subject to a minimum length for avoiding overzoom) (Figure 8). An adjusted vessel length $\left(l_{\text{adjusted}}\right)$ is calculated by

$$l_{\text{adjusted}} = \left|l_{\text{reported}} \times \sin\left(\theta_A\right)\right| \tag{9}$$

where l_{reported} is the vessel length reported by AIS or the set minimum and maximum lengths of radar targets. An additional width factor can be applied as a margin of error to best frame targets. See Figure 9 for image examples.

The final determined width is converted to degrees per pixel using target distance (3), which is then used with the lens calibration Equations (4) and (5) to calculate a zoom percentage (6). If the percentage is outside $0 - 100\%$, then the zoom is set to the minimum or maximum, respectively.

2.6. Target Selection and Cuing

At the time of deployment, geographic areas for desired image capture are defined. These can have arbitrary shapes and sizes to capture images where they

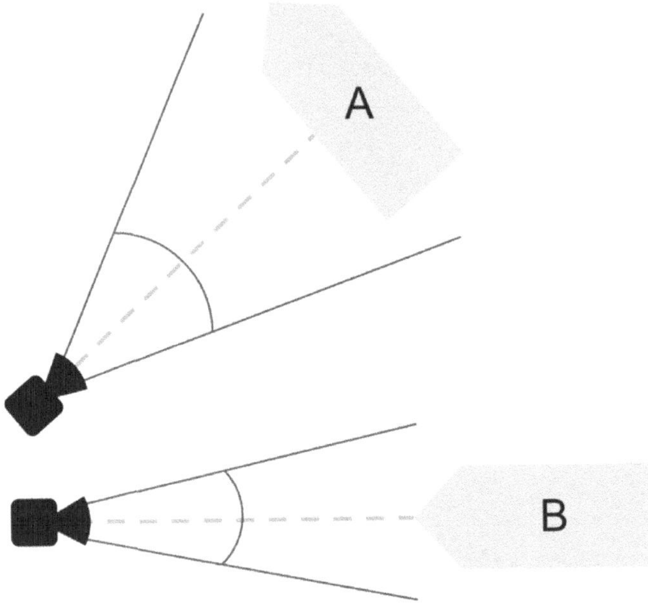

Figure 8. *Image width considers orientation of vessels with the camera. If vessels A and B were the same size and equidistant to the camera, vessel B would have a smaller image width and, thus, a higher zoom level than vessel A.*

will be useful, mask areas of high false targets, and protect the privacy of nearby private land or structures. A target selection algorithm executes every 15 s when one or more vessels are in the photocapture area. Targets with no prior images are given highest priority. When all candidate targets have been imaged, priority is given to the target with the longest time since the last image. All targets are subject to a capture delay of 120 s after the previous image and first detection (to reduce capture of transient false targets). Once a target has been photographed five times, the minimum delay for that target is increased to 5 min to reduce the overall count of images for tracks that persist over longer periods of time. If no targets meet the criteria, the camera rests and the nomination algorithm runs again in 15 s.

Considering camera pan and zoom speed is one obstacle for directing PTZ cameras to active targets [80]. To account for these variables, the target location, course, and speed are used to generate a predicted target location after a set delay (to account for camera pan speed, usually 2 s). For radar detections, the location is assumed to be the target center. Since AIS voyage data include vessel dimensions relative to its GPS antenna [16], the target location is additionally adjusted to the ship center for these vessels (only for camera cuing purposes) (Figure 10). The location is then transformed into a range and bearing angle from the camera, and it pans, tilts, and zooms according to the predetermined settings. The camera is re-cued with each new detection, which typically occurs every 2 – 6 s. Before the next target is nominated, an image is captured if there was at least one cue.

Figure 9. *Images captured autonomously of radar-detected vessels (left) and vessels broadcasting AIS (right).*

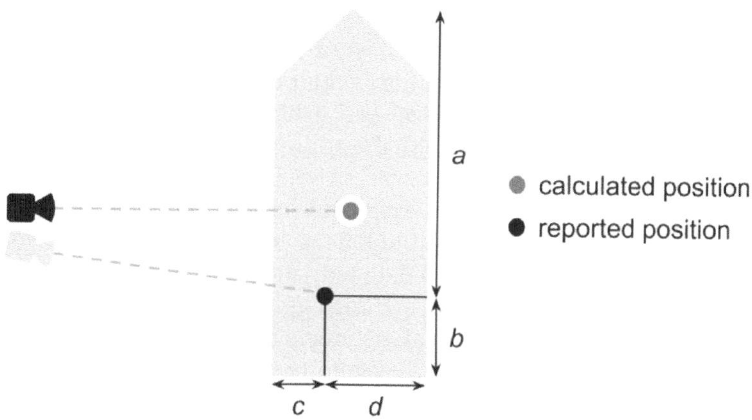

Figure 10. *Vessel dimensions (a, b, c, d) provided in AIS data with respect to the reported position [84]. Dimensions are used to calculate the position of the geometric center of the vessel for camera cuing purposes.*

To simplify access to image metadata, the system encodes the following information in the image filename at the time of capture: timestamp, system track identification number, most recent target location, speed, course, camera zoom percentage, vessel length (or assumed length), and vessel name (if known from AIS).

2.7. Case Study Analysis

The two case study sites, one in a highly developed coastal area and one offshore (Figure 11), provide an opportunity to evaluate and compare vessel traffic patterns from both radar and AIS at an urban versus remote location. Areas of high population density place a multitude of pressures on marine ecosystems [85], and poaching in MPAs is more likely to occur near urban areas [86]. The time and resources required to travel to more remote MPAs can deter human activity, and lower harvest rates have been recorded [87], but enforcement in these areas may be similarly limited [88]. The amount of vessels using or not using AIS is likely unique to a given location and a result of site-specific commercial and recreational activities. Therefore, these case studies are intended to assess the benefit of using radar at two distinct locations, which may not transfer to other areas (discussed further in Section 4).

The South La Jolla State Marine Reserve (SMR) is approximately 13 km^2 and directly adjacent to the city of San Diego, California, USA. As such, recreational activity is common, and there are more marine-related violations associated with the SMR than others in the area [89]. The SMR was designated in 2012 to preserve an area of high biodiversity, including a rocky reef ecosystem and kelp forest, and all harvesting of marine resources is prohibited, which includes all types of fishing [90]. As part of a larger network of MPAs in California, the SMR shares a boundary with the South La Jolla State Marine Conservation Area (SMCA) where some recreational fishing is allowed.

The Caye Bokel Conservation Area is approximately 55 km^2 and is part of the larger Turneffe Atoll Marine Reserve, designated in 2012 to improve management of local resources [91]. Located 50 km east of the central coastline of Belize in the Mesoamerican Reef region, there is little development on the atoll. Tourism directly benefits the larger economy of Belize, and fishing has significant social value to nearby communities [92]. Fishing with the appropriate license and permitted gears is allowed in the majority of the Reserve, but smaller zones prohibit extractive fishing. In the Caye Bokel Conservation Area, catch and release sport fishing is allowed with a valid license [93], and one traditional fisher is allowed to extract lobster, subject to catch limits and sharing catch data with management. Within the Conservation Area, there is a multi-species spawning aggregation site critical for sustaining commercial fish species [94] where all fishing is fully prohibited.

Based on deployment dates and system functioning, we collected data from November 2022 through January 2023 at the South La Jolla site and from April 2022 through October 2022 at the Caye Bokel site. Occasional gaps in monitoring occurred due to system maintenance, so we only considered data from days when both radar and AIS data were received. This resulted in 92 days of analysis at South La Jolla and 150 days at Caye Bokel. Radar tracks that duplicated AIS tracks, and those that were most likely false targets, were not included in analysis. Due to potential error in AIS track fusion and the machine learning model used to identify false targets, we consider radar data an estimate of activity sufficient to understand broad patterns in activity.

Tracks were classified by their source, radar or AIS, and further by whether or not they triggered a site-specific MPA alarm. Activity inside the MPA at each site was quantified in two ways: the number of unique track records detected ("tracks") and the duration of time they lasted ("hours"). Detections were identified as inside or outside the MPA, and the time between consecutive points along each unique

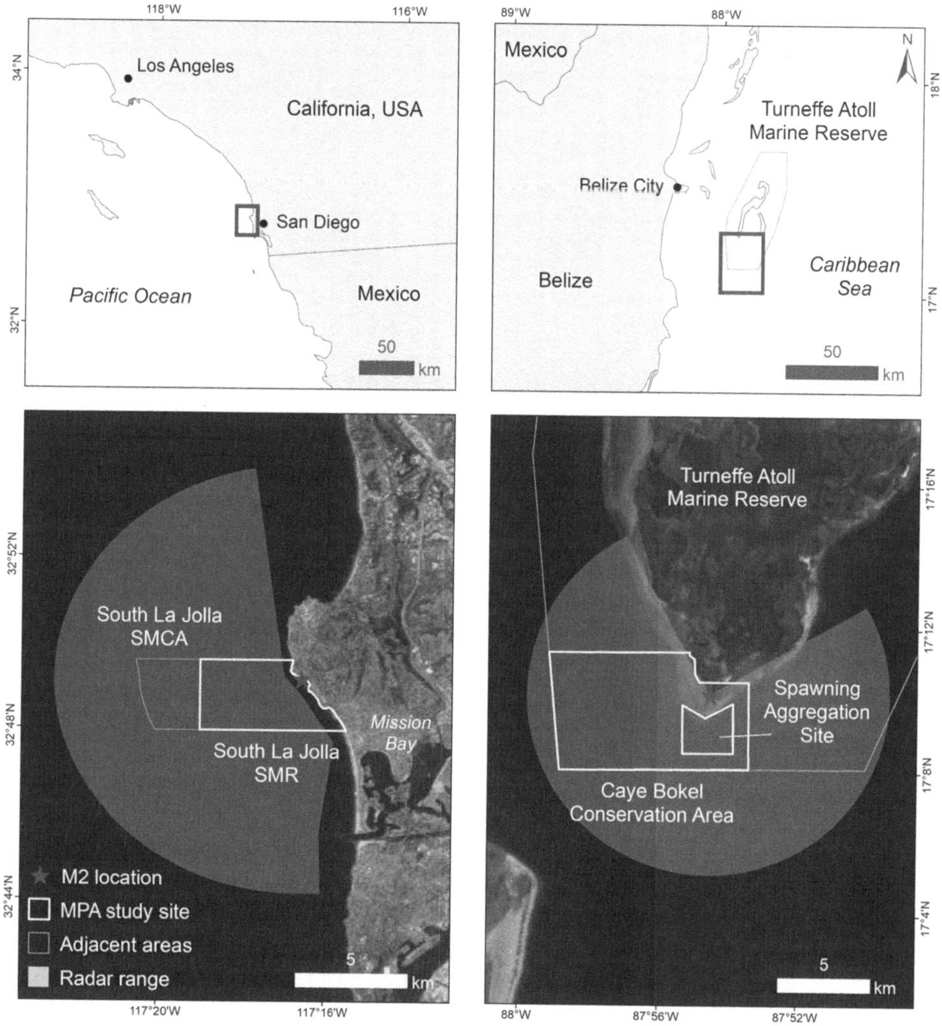

Figure 11. *Case study site locations and marine protected areas (MPAs). South La Jolla State Marine Reserve (SMR) (left) and the adjacent South La Jolla State Marine Conservation Area (SMCA). Caye Bokel Conservation Area (right), including a multi-species spawning aggregation site, within the Turneffe Atoll Marine Reserve. Basemap: ESRI World Imagery.*

track was summed for both areas. The cumulative time summed across all tracks, which did not consider temporal overlap, was intended to best reflect the pressure of an activity on an area (e.g., [77,95]).

Tracks are intended to estimate the number of vessels that entered the MPA. Therefore, when there were multiple AIS tracks with the same vessel identification number over a unique day, these were considered a single track record. The same method could not be performed with radar-detected tracks since the radar system does not include vessel identification information. Therefore, radar track counts

Figure 12. *Percentage of total tracks and track hours detected by radar or AIS outside and inside the MPAs at each case study site.*

may overestimate the number of unique vessels. Images captured by the camera help site partners identify vessels, and we report the number of images collected.

To assess activity over time, we calculated track counts and hours over each unique day of analysis. These daily values were summed and averaged over the full study period and per day of the week to evaluate weekly activity patterns. We also calculated track counts and hours within each hour of the day to evaluate daily patterns. Finally, we generated track heat maps using tracks both inside and outside MPA boundaries to show the spatial distribution of activity. We determined the resolution of the spatial grid (249 m) using the radar positional error model defined in [28]. All analyses were conducted in R [96].

3. RESULTS

Activity detected by AIS was common outside the MPAs, but, inside the MPAs, there was far more activity detected by radar compared to AIS at both case study sites (Figure 12). At South La Jolla, 94.8% of tracks observed in the MPA were detected by radar and accounted for 93.2% of track hours. The standard deviation values in Table 1 are meant to indicate the variability across days. At Caye Bokel, 94.5% of tracks in the MPA were detected by radar, accounting for 97.8% of track hours. There were over 200 images of radar-detected targets, on average, collected daily at South La Jolla and over 100 images at Caye Bokel.

Temporal patterns of daily activity detected by radar over the full study period were not especially similar to AIS patterns at either case study site (Figure 13). At South La Jolla, tracks and hours followed similar patterns over time, but activity detected by radar peaked in November 2022, while AIS activity peaked roughly one month later. At Caye Bokel, radar tracks and hours followed different temporal patterns: hours peaked near the beginning of May 2022, but tracks peaked roughly one month later. AIS tracks peaked in May 2022 (similar to radar hours), but AIS hours peaked much later in August 2022.

Weekly radar and AIS activity patterns at South La Jolla showed a general increase in both tracks and hours on weekends. The results were not similarly consistent at Caye Bokel (Figure 14): radar tracks in the MPA peaked on Wednesdays, but hours peaked on Sundays. The highest daily AIS track average occurred on Sundays, but hours peaked on Thursdays. At both sites, tracks that triggered the

Table 1. *Table 1. Total and daily average (with standard deviation) track counts, hours, and images collected of targets detected by radar or AIS inside the MPAs.*

		South La Jolla		Caye Bokel	
		Total	Daily	Total	Daily
Tracks					
	Radar	2284(94.8%)	25 ± 21	1404(94.5%)	9 ± 7
	AIS	125	1 ± 1	81	1 ± 1
Hours					
	Radar	321.0(93.2%)	3.5 ± 3.3	1248.8(97.8%)	8.3 ± 9.7
	AIS	23.6	0.3 ± 0.3	28.2	0.2 ± 0.4
Images					
	Radar	23,587	281 ± 183	18,997	142 ± 135
	AIS	7205	83 ± 50	1160	36 ± 52

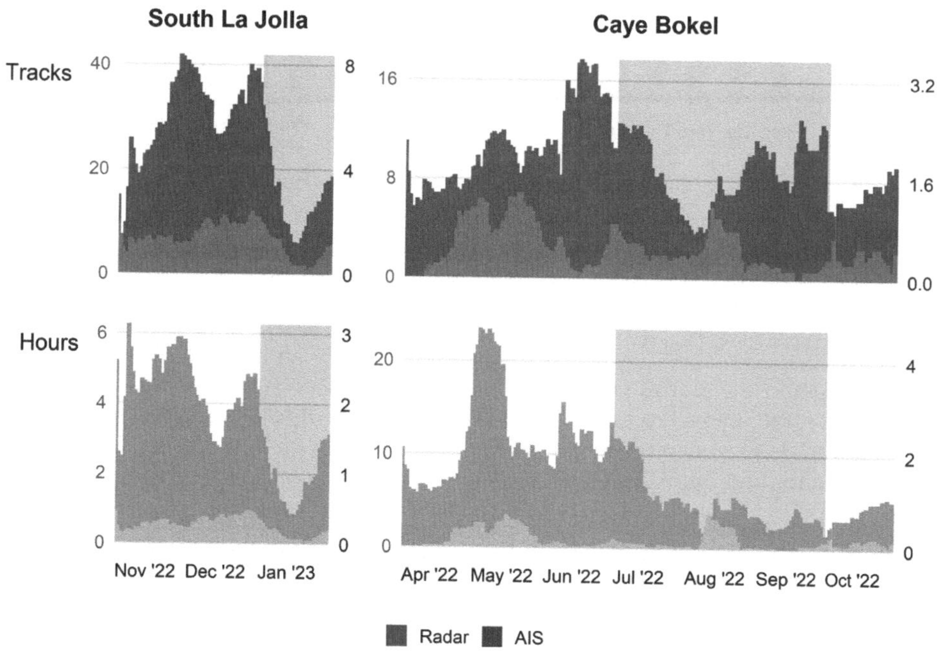

Figure 13. *Fourteen-day moving average of track counts and hours per day in the MPAs across the study period. Note that radar values are in red on the left side of each panel, and AIS values are in blue on the right side of each panel. Shaded areas indicate the closed season of local fisheries of interest: boat-based sport fishing for certain species of groundfish in southern California [97] and commercial, non-industrial fishing for Queen conch in Belize [91].*

site-specific alarm generally followed similar weekly patterns to the overall track counts.

Both the radar and AIS track counts followed similar patterns throughout the day at each site (Figure 15). At South La Jolla, tracks within the MPA began most often in the mid-day, although AIS tracks that triggered the alarm most often began early and late in the day. At Caye Bokel, the radar and AIS tracks, including those

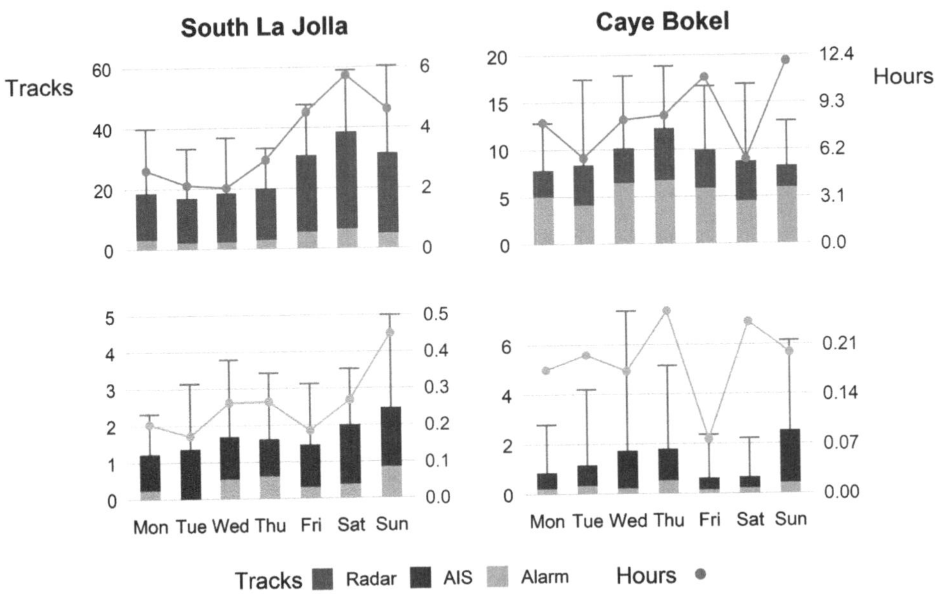

Figure 14. *Average daily tracks (with standard deviation) in the MPAs and average daily hours by day of the week. The percent of tracks that triggered the site-specific alarm are shown in yellow. At South La Jolla, the alarm is triggered when a track has a speed less than 3 knots* (1.5 m/s) *and remains in the MPA for more than* 5 min. *At Caye Bokel, the alarm is triggered when a track has a speed less than 5 knots* (2.6 m/s) *and remains in the MPA for more than* 10 min.

that triggered the alarm, peaked in the morning. At both sites, the hours spent in the MPA typically peaked along with the percentage of tracks triggering the alarm.

The highest concentration of radar tracks at South La Jolla was inside the MPA, while the highest concentration of AIS tracks was outside the southern boundary (Figure 16). At Caye Bokel, the radar tracks were concentrated in the MPA, including within the spawning aggregation site. There was a distinct pattern of AIS tracks just outside the eastern and southern boundaries.

4. DISCUSSION

The results from the case study sites clearly demonstrate that a single source of data on vessel activity from a cooperative system (in this case, AIS) does not capture the majority of activity in these urban and remote MPAs. While these results are specific to the case study locations, information on non-cooperative vessel activity is especially relevant when smallscale vessels (likely not to be using AIS) are of particular interest to MPA management. For example, recreational fishing is common in the vicinity of the South La Jolla SMR [43], and these private vessels are not required to participate in AIS. In the Turneffe Atoll region, both small private tour boats and traditional fishing boats are active [91] and also are not required to participate in AIS.

Daily track counts and hours were highly variable at both sites over their respective study periods (Figure 13), resulting in a high standard deviation of the

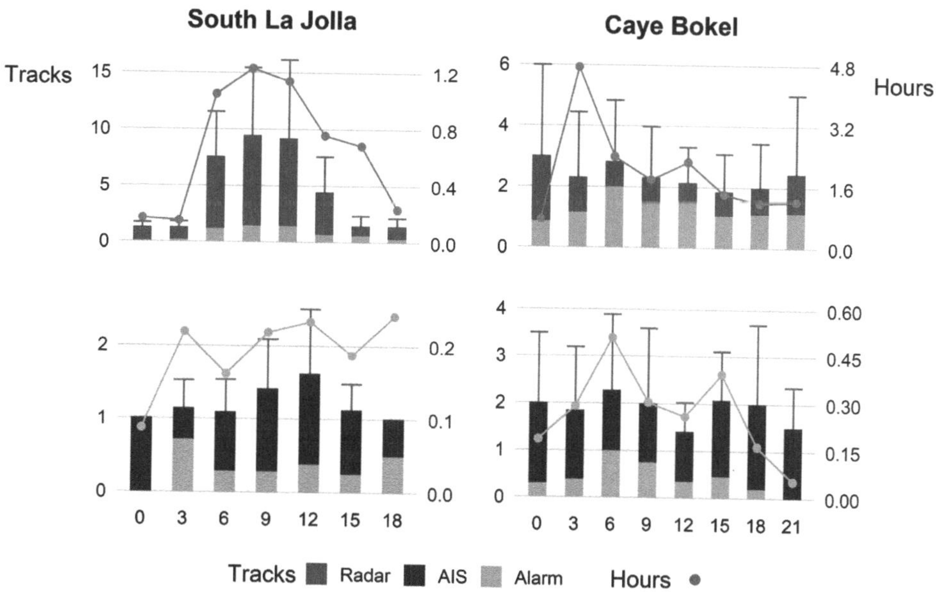

Figure 15. *Average daily tracks (with standard deviation) in the MPAs and average daily hours by the hour when first detected (grouped into 3-hour bins). Hours with zero tracks were ignored. The percent of tracks that triggered the site-specific alarm are shown in yellow. At South La Jolla, the alarm is triggered when a track has a speed less than 3 knots* (1.5 m/s) *and remains in the MPA for more than* 5 min. *At Caye Bokel, the alarm is triggered when a track has a speed less than 5 knots* (2.6 m/s) *and remains in the MPA for more than* 10 min.

daily values (Table 1). While the case studies were intended to evaluate broad patterns of activity, more detailed analysis of track and image data could provide further detail on specific vessel types and activities and related temporal patterns (similar to [28]).

Activity patterns at the South La Jolla SMR were similar to the expected patterns of recreational activity: most tracks occurred on weekend days and during daylight hours. In California overall, the vast majority of citations are issued to recreational fishers compared to commercial [98], and, in the SMR, recreational fishing occurs more often than commercial [99]. While the analysis for this case study did not explicitly identify fishing activity, the concentration of radar-detected tracks inside the MPA (many likely to be recreational vessels) suggests that the system is detecting the established high rates of poaching in the SMR [89, 100]. The study period was too short to capture changes in activity due to seasonality, although there was a steep decline in activity at the end of December 2022, which may have corresponded with the 1 January season closure of boat-based sport fishing for some local groundfish species [97]. Spiny lobster is an economically important fishery and a highly targeted species in the area [101], and both the commercial and recreational seasons were open for the full duration of the study period. It is important to note that all fishing is prohibited within the SMR even when fishing seasons are open.

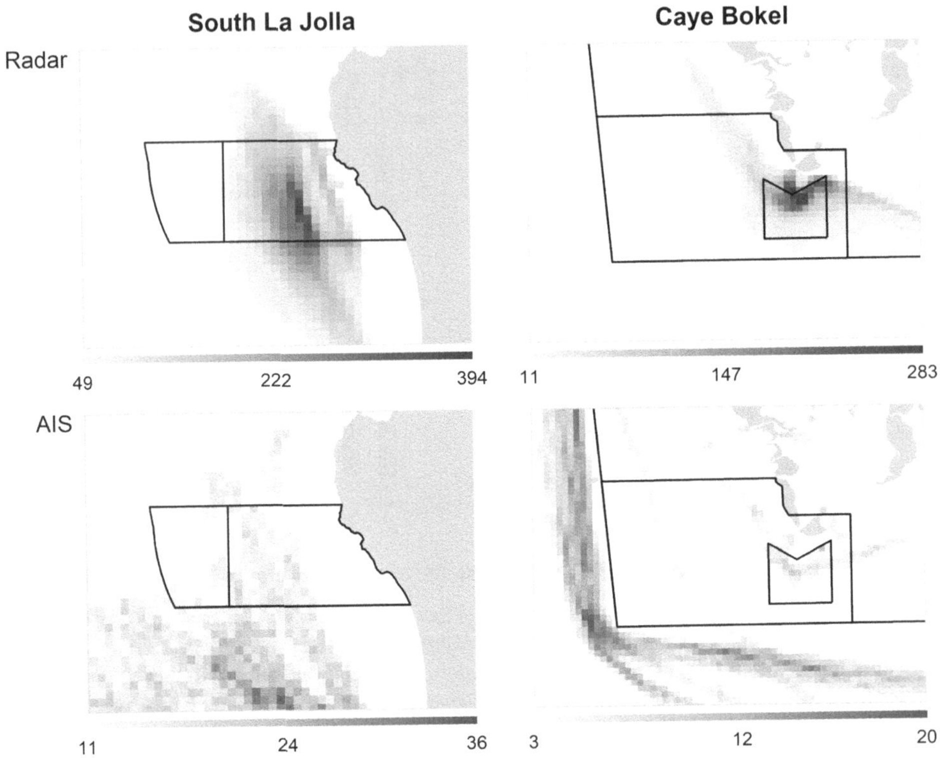

Figure 16. *Concentrated areas of all tracks across the study period. Only cells with track counts greater than the mean across all cells are shown. Black lines indicate boundaries of marine protected areas (MPAs).*

In the Caye Bokel Conservation Area, the temporal differences in track counts and hours may be due to the seasonality of tourism. Vessels spent the most time in the MPA in May 2022 before the typical increase in rainfall began in June [91]. Private liveaboard scuba-diving trips operate in the area and were occasionally observed overnighting near Caye Bokel. While the time vessels spent in the MPA was generally lower after May, there were still peaks of track counts in June and September. Despite the typical decrease in tourism during this time, commercial and traditional fishing still occur, including during a popular Queen conch season, which is open until July 1st or until the conch quota has been met, and which reopens on 1 October [91]. Since radar-detected tracks were concentrated in the spawning aggregation site, the camera is valuable for helping managers identify specific vessels. While all fishing is prohibited within the site, dive boats and vessels from nearby resorts are often observed near the area, only some of which broadcast AIS data. With different types of users accessing Caye Bokel, multiple monitoring strategies are beneficial.

At both sites, the spatial distribution of radar-detected tracks did not match the distribution of AIS tracks (Figure 16). At the South La Jolla site, many vessels likely originate from Mission Bay, south of the MPA (Figure 11). The concentration of AIS tracks suggests that these vessels travel farther from shore, while radar-detected vessels remain closer to the coast. At Caye Bokel, radar tracks were similarly concentrated nearshore. The concentration of AIS tracks just beyond the MPA boundaries suggests most vessels broadcasting AIS purposefully avoid the area. These differences highlight the importance of utilizing radar to monitor small-scale, non-AIS vessels; looking at spatial patterns of AIS activity alone would mask the majority of activity, including within coastal MPAs.

Incorporating radar greatly expanded the dataset on vessel activity beyond AIS in these MPAs, but there are limitations to its use. Unlike cooperative systems, data collected by radar do not provide vessel or voyage information. Previous research and ongoing work by Global Fishing Watch and others uses AIS data to identify track features that indicate certain activities, such as commercial fishing [24,74,102]. Radar-detected track lines of recreational and small-scale fishing likely have different characteristics than largerscale types of commercial fishing, but similar modeling approaches could be used to identify activities of interest (e.g., [103]). Image-processing techniques, such as object detection and classification, applied to images of radar-detected vessels, could also inform vessel identity. With the considerable number of images collected each day by the system (Table 1), integrating these strategies into the platform would help reduce the time required for manually reviewing track lines and images to identify radar-detected activity.

Both case studies showed that AIS and non-AIS vessels were active in the vicinity of MPAs. Therefore, radar can be complementary to cooperative vessel traffic systems in both urban and remote locations. Some vessels may intentionally disable their AIS transponder to evade detection when fishing illegally [25]. Vessels detected by non-cooperative systems, such as radar, could be compared with those broadcasting AIS, thus illuminating "dark" vessel activity (e.g., [66,104]). Future integration of vessel monitoring systems (VMS), commonly used to monitor commercial fisheries, into the M2 platform will support similar comparisons to data from that participatory system as well (e.g., [105, 106]). With ongoing development, it will be important to prioritize agnostic, generally applicable processes to maximize adoption by MPA managers and support adaptation into the future. Coastal areas will likely evolve with climate change, so systems and processes rigidly designed for unique locations and concerns may become ineffective [107]. The case studies presented here demonstrate how a coastal radar system can be used to document vessel activity in two different areas with different types of activity. This was accomplished by designing software to be used with different models of system hardware and defining simple parameters for identifying activities of interest.

The influence of human activity on coastal regions is substantial [108], and technologies for supporting management of small-scale fisheries in these areas have been called for in relation to many international goals [14]. The MPAs in the case studies presented here are relatively small compared to many that have been proposed and designated to meet global coverage targets [109] but are valuable for managing local fisheries. The localized remote sensing system described here is ideal for both targeted monitoring of small-scale vessel activity and integration with cooperative systems to ultimately provide a more comprehensive understanding of vessel traffic. Real-time access to data and the opportunity to summarize activity over time and space, demonstrated in the case studies, support active and adaptive MPA monitoring and enforcement in both urban and remote coastal areas.

5. CONCLUSIONS

The integrated radar system described in this paper is at the appropriate scale to meet the needs of local MPA managers. It provides persistent monitoring, uses common and accessible technologies, and tracks vessels within a range that matches the size of smaller MPAs. The case study results highlight how coastal radar systems can be sufficient for understanding both broad patterns in activities and fine-scale vessel behaviors in nearshore MPAs for which other non-cooperative systems may not be suitable at this time. The autonomous data collection provides a continuous record of vessel activity over space and time, and the purposeful similarity in data structure to AIS, an established source of vessel data for conservation-based assessments, also helps facilitate adoption and understanding. With approachable tools, technology can be effectively utilized to support management of MPAs and small-scale fisheries.

Author Contributions

Conceptualization, S.C., V.Z. and B.T.; methodology, V.Z. and S.C.; software, V.Z.; validation, L.G. and A.A.; formal analysis, S.C.; investigation, B.T. and V.Z.; resources, B.T., L.G. and A.A.; data curation, S.C. and V.Z.; writing-original draft preparation, S.C. and V.Z.; writingreview and editing, S.C., B.T., V.Z., L.G. and A.A.; visualization, S.C.; supervision, B.T. and V.Z.; project administration, B.T., L.G. and A.A. All authors have read and agreed to the published version of the manuscript.

Acknowledgments

The authors wish to acknowledge the staff at the Anthropocene Institute, WILDCOAST, and the Turneffe Atoll Sustainability Association (TASA) that have assisted with M2 overall and at the South La Jolla and Turneffe Atoll sites specifically. This work would not be possible without those that first identified and prepared the site locations, helped install hardware in the field, and maintain system functioning. We wish to specifically recognize Joe Cooper of WILDCOAST for overall project support and his wealth of local knowledge in southern California and the ChattenBrown family. At Turneffe Atoll, thank you to all the TASA Conservation Officers for their daily sacrifice and commitment to protecting Belize's natural wealth. The authors also wish to thank Global Conservation for their enduring support at these sites and others around the world. Conflicts of Interest: The authors declare no conflict of interest.

REFERENCES

1. Convention on Biological Diversity. First Draft of the Post-2020 Global Biodiversity Framework CBD/WG2020/3/3. 2021 Available online: https://www.cbd.int/doc/c/abb5/591f/2e46096d3f0330b08ce87a45/wg2020-03-03-en.pdf (accessed on 13 April 2023).
2. Maxwell, S.M.; Ban, N.C.; Morgan, L.E. Pragmatic approaches for effective management of pelagic marine protected areas. Endanger. Species Res. 2014, 26, 59-74.
3. Wilhelm, T.; Sheppard, C.R.C.; Sheppard, A.L.S.; Gaymer, C.F.; Parks, J.; Wagner, D.; Lewis, N. Large marine protected areasadvantages and challenges of going big. Aquat. Conserv. 2014, 24, 24-30.
4. Rowlands, G.; Brown, J.; Soule, B.; Boluda, P.T.; Rogers, A.D. Satellite surveillance of fishing vessel activity in the Ascension Island Exclusive Economic Zone and Marine Protected Area. Mar. Policy 2019, 101, 39-50.
5. White, T.D.; Ong, T.; Ferretti, F.; Block, B.A.; McCauley, D.J.; Micheli, F.; de Leo, G.A. Tracking the response of industrial fishing fleets to large marine protected areas in the Pacific Ocean. Conserv. Biol. 2020, 34, 1571-1578.
6. Albers, H.J.; Ashworth, M.F. Economics of marine protected areas: Assessing the literature for marine protected area network expansions. Annu. Rev. Resour. Econ. 2022, 14, 533-554.
7. Bayley, D.T.I.; Mogg, A.O.M.; Purvis, A.; Koldewey, H.J. Evaluating the efficacy of small-scale marine protected areas for preserving reef health: A case study applying emerging monitoring technology. Aquat. Conserv. 2019, 29, 2026-2044.
8. Mascia, M.B.; Claus, C.A.; Naidoo, R. Impacts of marine protected areas on fishing communities. Conserv. Biol. 2010, 24, 1424 – 1429.
9. Gill, D.A.; Mascia, M.B.; Ahmadia, G.N.; Glew, L.; Lester, S.E.; Barnes, M.; Craigie, I.; Darling, E.S.; Free, C.M.; Geldmann, J.; et al. Capacity shortfalls hinder the performance of marine protected areas globally. Nature 2017, 543, 665-669.
10. Appleton, M.R.; Courtiol, A.; Emerton, L.; Slade, J.L.; Tilker, A.; Warr, L.C.; Malvido, M.A.; Barborak, J.R.; de Bruin, L.; Chapple, R.; et al. Protected area personnel and ranger numbers are insufficient to deliver global expectations. Nat. Sustain. 2022, 5, 1100-1110.
11. Kanjir, U.; Greidanus, H.; Oštir, K. Vessel detection and classification from spaceborne optical images: A literature survey. Remote Sens. Environ. 2018, 207, 1-26.
12. Snapir, B.; Waine, T.W.; Biermann, L. Maritime vessel classification to monitor fisheries with SAR: Demonstration in the North Sea. Remote Sens. 2019, 11, 353.
13. Song, J.; Kim, D.; Kang, K. Automated procurement of training data for machine learning algorithm on ship detection using AIS information. Remote Sens. 2020, 12, 1443.
14. Morgera, E.; Ntona, M. Linking small-scale fisheries to international obligations on marine technology transfer. Mar. Policy 2018, 93, 295-306.
15. Marine Monitor: Solution for Situational Awareness of the Nearshore Marine Environment. Available online: https://m2 marinemonitor.com/ (accessed on 18 April 2023).
16. Robards, M.D.; Silber, G.K.; Adams, J.D.; Arroyo, J.; Lorenzini, D.; Schwehr, K.; Amos, J. Conservation science and policy applications of the marine vessel Automatic Identification System (AIS)_A review. Bull. Mar. Sci. 2016, 92, 75-103.

17. McCauley, D.J.; Woods, P; Sullivan, B.; Bergman, B.; Jablonicky, C.; Roan, A.; Hirshfield, M.; Boerder, K.; Worm, B. Ending hide and seek at sea. Science 2016, 351, 1148-1150.

18. Dureuil, M.; Boerder, K.; Burnett, K.A.; Froese, R.; Worm, B. Elevated trawling inside protected areas undermines conservation outcomes in a global fishing hot spot. Science 2018, 362, 1403-1407.

19. Tassetti, A.N.; Ferrà, C.; Fabi, G. Rating the effectiveness of fishery-regulated areas with AIS data. Ocean Coast. Manag. 2019, 175, 90-97.

20. Perry, A.L.; Blanco, J.; García, S.; Fournier, N. Extensive use of habitat-damaging fishing gears inside habitat-protecting marine protected areas. Front. Mar. Sci. 2022, 9, 811926.

21. Iacarella, J.C.; Burke, L.; Davidson, I.C.; DiBacco, C.; Therriault, T.W.; Dunham, A. Unwanted networks: Vessel traffic heightens the risk of invasions in marine protected areas. Biol. Conserv. 2020, 245, 108553.

22. Boljat, H.U.; Grubišić, N.; Slišković, M. The impact of nautical activities on the environment-A systematic review of research. Sustainability 2021, 13, 10552

23. Meijles, E.W.; Daams, M. N.; Ens, B. J.; Heslinga, J.H.; Sijtsma, F.J. Tracked to protect-Spatiotemporal dynamics of recreational boating in sensitive marine natural areas. Appl. Geogr. 2021, 130,102441

24. Kroodsma, D.A.; Mayorga, J.; Hochberg, T.; Miller, N.A.; Boerder, K.; Ferretti, F.; Wilson, A.; Bergman, B.; White, T.D.; Block, B.A.; et al. Tracking the global footprint of fisheries. Science 2018, 359, 904-908.

25. Welch, H.; Clavelle, T.; White, T.D.; Cimino, M.A.; Van Osdel, J.; Hochberg, T.; Kroodsma, D.; Hazen, E.L. Hot spots of unseen fishing vessels. Sci. Adv. 2022, 8, eabq2109.

26. Bunwaree, P. The illegality of fishing vessels 'going dark' and methods of deterrence. Int. Comp. Law Q. 2023, 72, 179 – 211.

27. Taconet, M.; Kroodsma, D.; Fernandes, J.A. Global Atlas of AIS-Based Fishing Activity-Challenges and Opportunities; Food and Agriculture Organization (FAO): Rome, Italy, 2019. Available online: https://agris.fao.org/agris-search/search.do?recordID=XF2020000847 (accessed on 14 April 2023).

28. Cope, S.; Hines, E.; Bland, R.; Davis, J.D.; Tougher, B.; Zetterlind, V. Application of a new shore-based vessel traffic monitoring system within San Francisco Bay. Front. Mar. Sci. 2020, 7, 86.

29. Iacarella, J.C.; Burke, L.; Clyde, G.; Wicks, A.; Clavelle, T.; Dunham, A.; Rubidge, E.; Woods, P. Application of AIS- and flyoverbased methods to monitor illegal and legal fishing in Canada's Pacific marine conservation areas. Conserv. Sci. Pract. 2023 , e12926.

30. O'Hara, P.D.; Serra-Sogas, N.; McWhinnie, L.; Pearce, K.; Le Baron, N.; O'Hagan, G.; Nesdoly, A.; Marques, T.; Canessa, R. Automated identification system for ships data as a proxy for marine vessel related stressors. Sci. Total Environ. 2023, 865, 160987.

31. Soldi, G.; Gaglione, D.; Forti, N.; Di Simone, A.; Daffin, F.C.; Bottini, G.; Quattrociocchi, D.; Millefiori, L.M.; Braca, P.; Carniel, S.; et al. Space-based global maritime surveillance. Part I: Satellite technologies. IEEE Aerosp. Electron. Syst. Mag. 2021, 36, 8-28.

32. Marino, A.; Sanjuan-Ferrer, M.J.; Hajnsek, I.; Ouchi, K. Ship detection with spectral analysis of synthetic aperture radar: A comparison of new and well-known algorithms. Remote Sens. 2015, 7, 5416-5439.

33. Stasolla, M.; Mallorqui, J.J.; Margarit, G.; Santamaria, C.; Walker, N. A comparative study of operational vessel detectors for maritime surveillance using satellite-borne synthetic aperture radar. IEEE J. Sel. Top. Appl. Earth Obs. Remote Sens. 2016, 9, 2687-2701.

34. Reggiannini, M.; Righi, M.; Tampucci, M.; Duca, A.L.; Bacciu, C.; Bedini, L.; D'Errico, A.; Di Paola, C.; Marchetti, A.; Martinelli, M.; et al. Remote sensing for maritime prompt monitoring. J. Mar. Sci. Eng. 2019, 7, 202.

35. Ruiz, J.; Caballero, I.; Navarro, G. Sensing the same fishing fleet with AIS and VIIRS: A seven-year assessment of squid jiggers in FAO major fishing area 41. Remote Sens. 2019, 12, 32.

36. Li, Y.; Song, L.; Zhao, S.; Zhao, D.; Wu, Y.; You, G.; Kong, A.; Xi, X.; Yu, Z. Nighttime fishing vessel observation in Bohai Sea based on VIIRS fishing vessel detection product (VBD). Fish. Res. 2023, 258, 106539.

37. Elvidge, C.D.; Zhizhin, M.; Baugh, K.; Hsu, F. Automatic boat identification system for VIIRS low light imaging data. Remote Sens 2015, 7, 3020-3036.

38. Tsuda, M.E.; Miller, N.A.; Saito, R.; Park, J.; Oozeki, Y. Automated VIIRS boat detection based on machine learning and its application to monitoring fisheries in the East China Sea. Remote Sens. 2023, 15, 2911.

39. Yao, Y.; Jiang, Z.; Zhang, H.; Zhou, Y. On-board ship detection in micro-nano satellite based on deep learning and COTS component. Remote Sens. 2019, 11, 762.

40. Christopher, K. Port Security Management, 2nd ed.; CRC Press: Boca Raton, FL, USA, 2014.

41. Watson, G.J.; Murray, J. M.; Schaefer, M.; Bonner, A. Successful local marine conservation requires appropriate educational methods and adequate enforcement. Mar. Policy 2015, 52, 59-67.

42. Hartill, B.W.; Taylor, S.M.; Keller, K.; Weltersbach, M.S. Digital camera monitoring of recreational fishing effort: Applications and challenges. Fish Fish. 2020, 21, 204-215.

43. Parnell, P.E.; Dayton, P.K.; Fisher, R.A.; Loarie, C.C.; Darrow, R.D. Spatial patterns of fishing effort off San Diego: Implications for zonal management and ecosystem function. Ecol. Appl. 2010, 20, 2203-2222.

44. Keller, K.; Steffe, A.S.; Lowry, M.; Murphy, J.J.; Suthers, I.M. Monitoring boat-based recreational fishing effort at a nearshore artificial reef with a shore-based camera. Fish. Res. 2016, 181, 84-92.

45. Wood, G.; Lynch, T.P.; Devine, C.; Keller, K.; Figueira, W. High-resolution photo-mosaic time-series imagery for monitoring human use of an artificial reef. Ecol. Evol. 2016, 6, 6963-6968.

46. Lancaster, D.; Dearden, P.; Haggarty, D.R.; Volpe, J.P.; Ban, N.C. Effectiveness of shore-based remote camera monitoring for quantifying recreational fisher compliance in marine conservation areas. Aquat. Conserv. 2017, 27 , 804-813.

47. Powers, S.P.; Anson, K. Estimating recreational effort in the Gulf of Mexico red snapper fishery using boat ramp Cameras: Reduction in federal season length does not proportionally reduce catch. N. Am. J. Fish. Manag. 2016, 36, 1156-1166.

48. Hartill, B.W.; Payne, G.W.; Rush, N.; Bian, R. Bridging the temporal gap: Continuous and cost-effective monitoring of dynamic recreational fisheries by web cameras and creel surveys. Fish. Res. 2016, 183, 488-497.

49. Wawrzyniak, N.; Hyla, T.; Popik, A. Vessel detection and tracking method based on video surveillance. Sensors 2019, 19, 5230.

50. Carić, H.; Cukrov, N.; Omanović, D. Nautical tourism in marine protected areas (MPAs): Evaluating an impact of copper emission from antifouling coating. Sustainability 2021, 13, 11897.

51. Harasti, D.; Davis, T.R.; Jordan, A.; Erskine, L.; Moltschaniwskyj, N. Illegal recreational fishing causes a decline in a fishery targeted species (Snapper: Chrysophrys auratus) within a remote no-take marine protected area. PLoS ONE 2019, 14, e0209926.

52. International Association of Marine Aids to Navigation and Lighthouse Authorities . IALA VTS Manual Ed. 8.2. International Association of Marine Aids to Navigation and Lighthouse Authorities. 2022. Available online: https: / / www.iala-aism.org / product / m0002 / (accessed 10 June 2023).

53. Xiao, Z.; Fu, X.; Zhang, L.; Goh, R.S.M. Traffic pattern mining and forecasting technologies in maritime traffic service networks: A comprehensive survey. IEEE Trans. Intell. Transp. Syst. 2019, 21, 1796-1825.

54. Xiao, Z.; Fu, X.; Zhao, L.; Zhang, L.; Teo, TK.; Li, N.; Zhang, W.; Qin, Z. Next-generation vessel traffic services systems-From "passive" to "proactive". IEEE Intell. Transp. Syst. Mag. 2022, 15, 363-377.

55. Li, J.; Yang, Y.; Li, X.; Sun, J.; Li, R. Knowledge-transfer-based bidirectional vessel monitoring system for remote and nearshore images. J. Mar. Sci. Eng. 2023, 11, 1068.

56. Dekker, R.; Bouma, H.; den Breejen, E.; van den Broek, B.; Hanckmann, P.; Hogervorst, M.; Mohamoud, A.; Schoemaker, R.; Sijs, J.; Tan, R.; et al. Maritime situation awareness capabilities from satellite and terrestrial sensor systems. In Proceedings of the Maritime Systems and Technology Conference and Exhibition, (MAST Europe), Gdansk, Poland, 4-6 June 2013.

57. Khan, C.B.; Goetz, K.T.; Cubaynes, H.C.; Robinson, C.; Murnane, E.; Aldrich, T.; Sackett, M.; Clarke, P.J.; LaRue, M.A.; White, T.; et al. A biologist's guide to the galaxy: Leveraging artificial intelligence and very high-resolution satellite imagery to monitor marine mammals from space. J. Mar. Sci. Eng. 2023, 11, 595.

58. Martellato, E.; Piccirillo, A.M.; Ferraioli, G.; Rotundi, A.; Della Corte, V.; Palumbo, P.; Alcaras, E.; Appolloni, L.; Aulicino, G.; Bertini, I.; et al. A new orbiting deployable system for small satellite observations for ecology and Earth observation. Remote Sens. 2022, 14, 2066.

59. Lin, Y.; Li, J.; Zhou, Z.; Luo, Q.; Wang, C.; Yan, X. Simulation analysis of the satellite coverage performance for identifying illegal fishing in target sea area. In Proceedings of the 2022 Global Reliability and Prognostics and Health Management (PHM-Yantai), Yantai, China, 13-16 October 2022.

60. Lin, B.; Huang, C. Comparison between ARPA radar and AIS characteristics for vessel traffic services. J. Mar. Sci. Technol. 2006, 14, 7.

61. Kazimierski, W.; Stateczny, A. Radar and Automatic Identification System track fusion in an electronic chart display and information system. J. Navig. 2015, 68, 1141-1154.

62. Zainuddin, S.; Pasya, I.; Rashid, N.E.; Zaiaami, N.; Maryam, A.; Abdullah, R.; Ali, M.S. Maritime radar: A review on techniques for small Vessel detection. Int. J. Electr. Electron. Syst. Res. 2019, 14, 30-45.

63. López, J.J.; Mulero-Pázmány, M. Drones for conservation in protected areas: Present and future. Drones 2018, 3, 10.
64. Provost, E.J.; Butcher, P.A.; Coleman, M.A.; Kelaher, B.P. Assessing the viability of small aerial drones to quantify recreational fishers. Fish. Manag. Ecol. 2020, 27, 615-621.
65. Molina-Molina, J.C.; Salhaoui, M.; Guerrero-González, A.; Arioua, M. Autonomous marine robot based on AI recognition for permanent surveillance in marine protected areas. Sensors 2021, 21, 2664.
66. Kurekin, A.A.; Loveday, B.R.; Clements, O.; Quartly, G.D.; Miller, P.I.; Wiafe, G.; Agyekum, K.A. Operational monitoring of illegal fishing in Ghana through exploitation of satellite Earth observation and AIS data. Remote Sens. 2019, 11, 293.
67. Park, J.; Lee, J.; Seto, K.; Hochberg, T.; Wong, B.A.; Miller, N.A.; Takasaki, K.; Kubota, H.; Oozeki, Y.; Doshi, S.; et al. Illuminating dark fishing fleets in North Korea. Sci. Adv. 2020, 6, eabb1197.
68. TimeZero Coastal Monitoring. Available online: https://www.coastalmonitoring.com/ (accessed on 18 April 2023).
69. French, A.P. How far away is the horizon? Am. J. Phys. 1982, 50, 795-799.
70. Bole, A.G.; Wall, A.D.; Norris, A. Radar and ARPA Manual: Radar, AIS, and Target Tracking for Marine Radar Users, 3rd ed.; Butterworth-Heinemann: Oxford, UK, 2014; ISBN 978-0080977522.
71. Nobeltec TimeZero User Guide; TIMEZERO, 2023. Available online: https://userguide.mytimezero.com/tz-coastal-monitoring (accessed on 18 April 2023).
72. Furuno Software Development Kit. Available online: https://www.furunousa.com/en/sdk (accessed on 18 April 2023).
73. Lee, J.; South, A.B.; Jennings, S. Developing reliable, repeatable, and accessible methods to provide high resolution estimates of fishing-effort distributions from vessel monitoring system (VMS) data. ICES J. Mar. Sci. 2010, 67, 1260-1271.
74. de Souza, E.N.; Boerder, K.; Matwin, S.; Worm, B. Improving fishing pattern detection from satellite AIS using data mining and machine learning. PLoS ONE 2016, 11, e0158248.
75. Huang, H.; Hong, F.; Liu, J.; Liu, C.; Feng, Y.; Guo, Z. FVID: Fishing vessel type identification based on VMS trajectories. J. Ocean Univ. China 2018, 18, 403-412.
76. Cope, S.; Zetterlind, V.; Tougher, B. Integrating marine radar in a multi-sensor platform for remote, unsupervised vessel tracking in the nearshore environment. In Proceedings of the Global Oceans 2020: Singapore-U.S. Gulf Coast, Biloxi, MS, USA, 5-20 October 2020; pp. 1-5.
77. Cope, S.; Tougher, B.; Morten, J.; Pukini, C.; Zetterlind, V. Coastal radar as a tool for continuous and fine-scale monitoring of vessel activities of interest in the vicinity of marine protected areas. PLoS ONE 2022, 17, e0269490.
78. Jianmin, Z.; Jie, W. Intelligent vessel dynamics video monitoring system based on AIS data. In Proceedings of the 2009 IEEE International Conference on Systems, Man and Cybernetics, San Antonio, TX, USA, 11-14 October 2009.
79. Yoshihara, K.; Yamamoto, S.; Hashimoto, T. Real-time tracking of offshore vessel using pan-tilt camera for keeping a lookout. Mar. Eng. 2018, 53, 429-434.

80. Loomans, M.J.H.; de With, P.H.N.; Wijnhoven, R.G.J. Robust automatic ship tracking in harbours using active cameras. In Proceedings of the 2013 IEEE International Conference on Image Processing, Melbourne, Australia, 15-18 September 2013.

81. Helgesen, H.H.; Leira, F.S.; Fossen, T.I.; Johansen, T.A. Tracking of ocean surface objects from unmanned aerial vehicles with a pan/tilt unit using a thermal camera. J. Intell. Robot. Syst. 2018, 91, 775-793.

82. Sabirin, H.; Bäse, G. Video surveillance of today: Compressed domain object detection, ONVIF web services based system component communication and standardized data storage and export using VSAF-A walkthrough. In Video Surveillance; Lin, W., Ed.; Intech: London, UK, 2011; pp. 37-54, ISBN 978-953-307-436-8.

83. Nguyen, K.Q.; Winger, P.D. Artificial light in commercial industrialized fishing applications: A review. Rev. Fish. Sci. Aquac. 2018, 27, 106-126.

84. International Telecommunication Union. Recommendation M.1371-5 Technical Characteristics for an Automatic Identification System Using Time Division Multiple Access in the VHF Maritime Mobile Frequency Band; International Telecommunication Union: Geneva, Switzerland, 2014. Available online: https://www.itu.int/rec/R-REC-M.1371-5-201402-I/en (accessed on 10 June 2023).

85. Todd, P.A.; Heery, E.C.; Loke, L.H.L.; Thurstan, R.H.; Kotze, D.J.; Swan, C. Towards an urban marine ecology: Characterizing the drivers, patterns and processes of marine ecosystems in coastal cities. Oikos 2019, 128, 1215-1242.

86. Weekers, D.P.; Zahnow, R.; Mazerolle, L. Conservation criminology: Modelling offender target selection for illegal fishing in marine protected areas. Br. J. Criminol. 2019, 59, 1455-1477.

87. Nickols, K.J.; White, J.W.; Malone, D.; Carr, M.H.; Starr, R.M.; Baskett, M.L.; Hastings, A.; Botsford, L.W. Setting ecological expectations for adaptive management of marine protected areas. J. Appl. Ecol. 2019, 56, 2376-2385.

88. Albers, H.J.; Preonas, L.; Capitán, T.; Robinson, E.J.Z.; Madrigal-Ballestero, R. Optimal siting, sizing, and enforcement of marine protected areas. Environ. Resour. Econ. 2020, 77, 229-269.

89. California Department of Fish and Wildlife. California's Marine Protected Area Network Decadal Management Review; California Department of Fish and Wildlife: Sacramento, CA, USA, 2022. Available online: https://wildlife.ca.gov/Conservation/Marine/MPAs/Management/Decadal-Review (accessed on 16 April 2023).

90. California Department of Fish and Wildlife. MPA Outreach Materials; California Department of Fish and Wildlife: Sacramento, CA, USA, 2016. Available online: https://wildlife.ca.gov/Conservation/Marine/MPAs/Outreach-Materials (accessed on 16 April 2023).

91. Belize Fisheries Department. Turneffe Atoll Marine Reserve Management Plan; Belize Fisheries Department: Belize City, Belize, 2011. Available online: http://www.turneffeatollmarinereserve.org/reserve-management (accessed on 16 April 2023).

92. Rader, D.; Requena, N.; Wade, B.; Burden, M. Combining rights-based management with science-based marine protected area networks to sustain fisheries against climate impacts in Belize. In Adaptive Management of Fisheries in Response to Climate

Change; FAO Fisheries and Aquaculture Technical Paper No. 667; Bahri, T., Vasconcelos, M., Welch, D.J., Johnson, J., Perry, R.I., Ma, X., Sharma, R., Eds.; Food and Agriculture Organization (FAO): Rome, Italy, 2021; pp. 115-126, ISBN 978-92-5-133890-2.

93. Belize Fisheries Department.Statutory Instrument No. 111 of 2012 Fisheries (Turneffe Atoll Marine Reserve) Regulations; Belize Fisheries Department: Belize City, Belize, 2012. Available online: http://www.turneffeatollmarinereserve.org/reserve-management (accessed on 16 April 2023).

94. Belize Fisheries Department. Statutory Instrument No. 161 of 2003 Fisheries (Spawning Aggregation Site Reserves) Order; Belize Fisheries Department: Belize City, Belize, 2003. Available online: http://www.spagbelize.org/Legislation.aspx (accessed on 16 April 2023).

95. Stelzenmüller, V.; Rogers, S.I.; Mills, C.M. Spatio-temporal patterns of fishing pressure on UK marine landscapes, and their implications for spatial planning and management. ICES J. Mar. Sci. 2008, 65, 1081-1091.

96. R: A Language and Environment for Statistical Computing; R Core Team: Vienna, Austria, 2022. Available online: https://www.Rproject.org/ (accessed on 17 April 2023).

97. California Department of Fish and Wildlife. Ocean Sport Fishing Regulations; California Department of Fish and Wildlife: Sacramento, CA, USA , 2022. Available online: https://wildlife.ca.gov/Fishing/Ocean/Regulations/Sport-Fishing (accessed on 17 April 2023).

98. California Department of Fish and Wildlife. Law Enforcement Division Citation Summary 2011; California Department of Fish and Wildlife: Sacramento, CA, USA, 2011. Available online: https://nrm.dfg.ca.gov/FileHandler.ashx?DocumentID=86746\&inline (accessed on 17 April 2023).

99. Ford, T.; Burdick, H.; Barilotti, A.; House, P. Aerial Monitoring of Ocean Vessels in Southern California, March 2017-Annual Project Report; The Bay Foundation, LightHawk, Vantuna Research Group, 2017. Available online: http://www.santamonicabay.org/wp-content/uploads/2014/04/The-Bay-Foundation-Aerial-Monitoring-March-2017-Final.pdf (accessed on 17 April 2023).

100. Kemsley, A.; Pukini, C. Marine Protected Area Watch and Marine Monitor (M2) RADAR technology: Case studies in anthropogenic use monitoring in California's marine protected areas. In Proceedings of the OCEANS 2021: San Diego-Porto, San Diego, CA, USA, 20-23 September 2021; pp. 1-5.

101. Hovel, K.A.; Neilson, D.J.; Parnell, E. California Spiny Lobster (Panulirus interruptus), California South Coast MPA Baseline Study, 2012 to 2013; South Coast Lobster Research Group. 2015. Available online: https://data.ca.gov/dataset/california-spiny-lobsterpanulirus-interruptus-california-south-coast-mpa-baseline-study-2-2013 (accessed on 17 April 2023).

102. Guan, Y.; Zhang, J.; Zhang, X.; Li, Z.; Meng, J.; Liu, G.; Bao, M.; Cao, C. Identification of fishing vessel types and analysis of seasonal activities in the northern South China Sea based on AIS data: A case study of 2018. Remote Sens. 2021, 13 , 1952.

103. Behivoke, F.; Etienne, M.; Guitton, J.; Randriatsara, R.M.; Ranaivoson, E.; Léopold, M. Estimating fishing effort in small-scale fisheries using GPS tracking data and random forests. Ecol. Indic. 2021, 123, 107321.

104. Roger, M.; Guida, R. Classification-aided SAR and AIS data fusion for space-based maritime surveillance. Remote Sens. 2020, 13, 104.

105. Hsu, F.; Elvidge, C.D.; Baugh, K.; Zhizhin, M.; Ghosh, T.; Kroodsma, D.; Susanto, A.; Budy, W.; Riyanto, M.; Nurzeha, R.; et al. Cross-matching VIIRS boat detections with vessel monitoring system tracks in Indonesia. Remote Sens. 2019, 11, 995.

106. Li, J.; Cai, Y.; Zhang, P.; Zhang, Q.; Jing, Z.; Wu, Q.; Qiu, Y.; Ma, S.; Chen, Z. Satellite observation of a newly developed light-fishing "hotspot" in the open South China Sea. Remote Sens. Environ. 2021, 256, 112312.

107. Toimil, A.; Losada, I.J.; Nicholls, R.J.; Dalrymple, R.A.; Stive, M.J.F. Addressing the challenges of climate change risks and adaptation in coastal areas: A review. Coast. Eng. 2020, 156, 103611.

108. Allan, H.; Levin, N.; Kark, S. Quantifying and mapping the human footprint across Earth's coastal areas. Ocean Coast. Manag. 2023, 236, 106476.

109. Toonen, R.J.; Wilhelm, T.; Maxwell, S.M.; Wagner, D.; Bowen, B.W.; Sheppard, C.R.C.; Taei, S.M.; Teroroko, T.; Moffitt, R.; Gaymer, C.F.; et al. One size does not fit all: The emerging frontier in large-scale marine conservation. Mar. Pollut. Bull. 2013, 77, 7-10.

INDEX

P

Q

R

S

T

U

V

W

Z

#1